BIOTECHNOLOGY, PATENTS AND MORALITY

To Etienne Vermeersch

quia fecit mihi magna

Biotechnology, Patents and Morality

Edited by
SIGRID STERCKX
Department of Philosophy and Moral Science
University of Ghent

Ashgate

Aldershot • Brookfield USA • Singapore • Sydney

Published by
Ashgate Publishing Ltd
Gower House
Croft Road
Aldershot
Hants GU11 3HR
England

Ashgate Publishing Company
Old Post Road
Brookfield
Vermont 05036
USA

British Library Cataloguing in Publication Data

Biotechnology, patents and morality
    1. Biotechnology - Patents - Congresses 2. Biotechnology -
    Moral and ethical aspects - Congresses
    I. Sterckx, Sigrid
    174.2'5

Library of Congress Catalog Card Number: 97-73211

ISBN 1 84014 158 1

Reprinted 1999

Printed in Great Britain by Biddles Limited, Guildford and King's Lynn

Contents

Part Six: The debate 255

Part Seven: Comments and conclusions 281

List of contributors

Daniel Alexander obtained a BA in Physics & Philosophy at Oxford University College in 1985, followed by a Diploma in Law at Central London Polytechnic and a LLM at Harvard Law School in 1987. He was called to the bar in 1988. He has a practice, and specialises principally in issues such as intellectual property, European Community, scientific commercial law, environmental and administrative law. Mr Alexander is joint editor of books, writer of a number of articles in journals, and frequently gives lectures and talks on legal and scientific topics as mentioned above. Since 1989 he has been involved in a number of high-profile cases.

Tom Claes obtained a degree in moral science at the University of Ghent (Belgium) in 1986, followed by a doctorate in moral science at the same university. From 1990 to 1994 he was employed as Junior Research Assistant, and consequently became project researcher for the Flemish Council of Science Policy at the University of Ghent. In 1994 Dr Claes became Senior Research Assistant. His main fields of scientific interests are: ethical theory, the relation between culture and morality, inter-cultural ethics, comparative ethics, bio-ethics and cognitive science. He is a member of the editorial committee of *'Cultural Dynamics'* (SAGE-London).

Désiré Collen obtained a degree in medicine at the Catholic University of Leuven (Belgium). He became a Doctor of Science in 1973 at the same university. He received the title 'Doctor honoris causa' at the Erasmus University of Rotterdam, The Netherlands (1988), at the Free University of Brussels, Belgium (1994), and at the University Notre Dame, USA (1995). Currently Désiré Collen is Professor at the faculty of medicine at the Catholic University of Leuven, and visiting Professor at Harvard Medical School, USA. In addition, he holds the position of Deputy Head of the University Clinic UZ in Leuven. Professor Collen is Director of the Centre of Molecular and Vascular Biology, as well as Director of the Centre of Transgene Technology and Gene Therapy. He is also President of the Department of Molecular and Cardiovascular Research, and of the D. Collen Research Foundation. For his contributions to research, he received several prizes and awards.

R. Stephen Crespi entered the patent profession after an initial training in chemistry and some years' laboratory experience in the pharmaceutical and food industries. He has worked in private practice, in industrial patent departments and in the public sector, where he was Patents Controller of the British Technology Group (NDRC) from 1976 to 1988. He has maintained a specialist interest in patenting in the life sciences, particularly in biotechnology, and has been active in publication in this field. He is co-author of 'Biotechnology and Patent Protection' (OECD Report 1985), and author of 'Patenting in the Biological Sciences' (John Wiley 1982) and 'Patents - a Basic Guide to Patenting in Biotechnology' (Cambridge University Press 1988). Mr Crespi practises as a European Patent Attorney and as a consultant to patent firms and to industry. He has assisted the European Commission as an expert on the proposed 'Directive on the legal protection of biotechnological inventions'.

Johan De Tavernier is Professor at the Department Moral Philosophy of the Catholic University of Leuven, Belgium.

Dani De Waele obtained a PhD in Genetics in 1976. Afterwards she deliberately moved into research areas that offered more space for reflecting upon science. At the faculty of Philosophy she worked in temporary multidisciplinary research projects on knowledge acquisition processes in the laboratory of genetics where she previously worked. Research like the one published under the title: 'Tiny little gods in science: hidden norms in daily scientific activity' led to a comparative analysis of explicit and implicit research guidelines in a laboratory of genetics with e.g. the 'Règles pour la direction de l'esprit' and the 'Discours de la méthode' of Descartes. Following this she worked several years in consecutive temporary multidisciplinary projects concerning the socio-economic and environmental-ethical aspects of biotechnology, this on behalf of the so-called 'social partners' and of the Flemish government. At present Dani De Waele is involved in a research project entitled 'Biotechnology as a challenge for an inquiry into values. Towards a new paradigm for approaching the problematic interaction between nature and culture', a project financed by the Flemish government. She has authored and co-authored many scientific and lay publications, and has given a number of lectures on afore-mentioned issues, more recently particularly in the field of biotechnology.

Steve Emmott is qualified as a barrister. He started as a British civil servant and following this ran a business of his own. He consequently started working for 'Friends of the Earth' in Brussels. At present Mr Emmott is the legal adviser to 'The Genetics Forum', the only NGO in the UK dealing with the whole range of genetic engineering issues. He also temporarily replaces the Biotechnology Co-ordinator of the Green Group in the European Parliament.

Gerry Evers-Kiebooms is Professor in the Department of Psychology of the Catholic University of Leuven (Belgium). She also holds the position of Head of the Unit Psychosocial Genetics of the Centre of Human Genetics in Leuven. Her main research interests are: risk perception; decision making; psychological aspects of genetic disease; genetic risk and genetic testing and public education about human genetics.

Leo Fretz studied philosophy. He taught social ethics and political philosophy at the Delft University of Technology from 1972 to 1994. He has published on Sartre's philosophy in France, Germany, Italy, The Netherlands and in the United States. Dr Fretz also publishes on ethics, bio-ethics and political philosophy. In 1993 he became the chairman of the first public debate in the Netherlands on the social and moral dilemma's regarding transgenic animals. In 1995 he was the chairman of the second public debate, on the theme 'Predictive genetic research, where are we going?'.

Larissa Gruszow obtained a Diploma of Electrotechnical Engineer at Warsaw Polytechnicum (Poland) in 1956. She carried on to become a Patent Attorney at Gevers Patent Attorney Bureau in Brussels, Belgium. From 1977 onwards Mrs Gruszow has been employed at the European Patent Office, at the Directorate International Legal Affairs. Her field of activity covers the Patent Cooperation Treaty (PCT), and the legal aspects of biotechnological inventions.

Lars Klüver obtained a MSc in Environmental Biology and Ecology at the University of Copenhagen, Denmark. In 1985 he started the Consultancy Klüver & Prestegaard. He made 8 audio-visual productions for the Danish Ministry of Environment, and worked out a communication strategy for the National Museum of Denmark. In 1986 Mr Klüver became involved with the Danish Board of Technology in Copenhagen; from 1987 until 1992 he was responsible for the biotechnology program, a program on technology assessment and public debate. He was Leader of the Project Unit in the secretariat of the Board, responsible for the yearly working plan, the methodology and the performance of the various projects. Since December 1994 Mr Klüver holds the position of Director of the Secretariat of the Danish Board of Technology.

Michiel Linskens is studying for a Doctorate in Biology at the University of Utrecht (The Netherlands). During 1988 and 1989 he carried out an inventory study on genetic engineering of animals at the Agricultural University of Wageningen, followed by a study on genetic engineering of animals and ethics at the Leiden State University. Since 1991 Mr Linskens is employed as Policy Adviser Biotechnology to the Dutch Society for the Protection of Animals.He has published many articles in a variety of newspapers and magazines on genetic engineering of animals, and is co-author of a book, and author of a brochure on that issue.

Isabelle Meister studied biology at the Swiss Federal Institute of Technology in Zürich and specialised in biotechnology. As she did not want to go into research work she started to work as a consultant in biotechnology and pesticide issues for Greenpeace Switzerland. From 1991 until the end of 1994, Mrs Meister was in charge of the Greenpeace Switzerland genetic engineering campaign. Since January

1995 she has worked as team leader of the Greenpeace International genetic engineering campaign.

Jan Mertens is currently working as a policy adviser of MEP Mrs Magda Aelvoet (Green Group at the European Parliament). He also works for IPE, the research department of the Flemish Green Party AGALEV. Within IPE, his main focus is on international politics. Previously Mr Mertens coordinated AGALEV's working party on animal welfare issues. He was very closely involved in the determination of the Green Party's biotechnology policy as well.

Rio D. Praaning started law studies at Leiden University in the Netherlands. His professional background covers positions such as Researcher and Assistant Director of the East-West Institute (The Hague, The Netherlands); Director of the Netherlands Atlantic Commission; Senior Consultant and Director Government Relations, Dewe Rogerson Europe (London, Brussels). Presently Mr Praaning holds the position of Chief Executive Officer of the Praaning Meines Consultancy Group s.a. in Brussels, Belgium. He is adviser to several governmental, non-governmental and industrial organisations on environmental and industrial matters, as well as adviser to patients organisations on European affairs. He also advises national and local authorities on international political matters.

Ronald Schapira obtained a Law degree (J.D.) at Columbia University in New York, and followed post-graduate courses in Trademark and Copyright Law. He is also a Bachelor of Engineering (Chemical Engineering). He is admitted to the Bar in New York State, the US Patent and Trademark Office and the French Patent Office. Mr Schapira has twenty-five years of experience in private practice, in New York City as well as in Paris, and is in-house counsel for major European and American corporations.

Ulrich Schatz studied Law and obtained the degree of Dr Jur at the University of Munich (Germany) in 1964. After being employed at the Max-Planck-Institut für ausländisches und internationales Patent-, Urheber- und Wettbewerbsrecht, he became Principal Administrator at the EEC-Commission in Brussels. From 1970 to 1977 Dr Schatz was Member of the Board of Directors of the International Patent Institute in The Hague. In 1977 he was appointed Principal Director, International Affairs, at the European Patent Office in Munich, a position which he still holds at the moment. As reflected in his work experience, his specialty is European and international patent law. He has published a series of articles in the field of patent legislation, in particular on the subject of exhaustion of rights.

Jozef Schell has obtained a degree in Zoology and a PhD in Microbiology (Comparative Biochemistry) at the University of Ghent (Belgium) and Utrecht (The Netherlands). In 1970 he was appointed full professor and Director of the Laboratory of Genetics at the University of Ghent. His main research interest is Molecular

Biology and Genetic Engineering of Plants (Plant Bacterial Interactions). In 1982 he founded, with Professor Marc Van Montagu, the company Plant Genetic Systems Inc., which has won high acclaim in the international scientific community. Presently Professor Schell is Director of the Department of Genetic Principles of Plant Breeding at the 'Max-Planck-Institut für Züchtungsforschung' in Cologne. He is Elected member of Academies in Belgium, Germany, UK, Sweden, Hungaria, USA and India. His memberships of Scientific Advisory Boards, Councils and Boards of Institutions are too numerous to mention. Furthermore Professor Schell has been awarded many different national and international prizes and distinctions. In addition to all this it should be mentioned that he is an Editorial Board Member of several scientific journals and publications.

Sigrid Sterckx obtained a MSc in Moral Science in 1994, at the University of Ghent (Belgium). A few months later she became Research Assistant at the Department Philosophy and Moral Science of this university. From March 1995 until April 1996, she carried out a research project for the Commission of the European Communities (DG XII), entitled 'Biotechnology, patents and morality: Towards a consensus'. She has recently been appointed Research Assistant of the (Flemish) Fund for Scientific Research, at the Department of Philosophy and Moral Science at the University of Ghent. At present Ms Sterckx is preparing a PhD. The title of her dissertation will be 'Biotechnology, Patent law and Ethics. Ethical aspects of the patenting of biotechnological inventions: a comparative inquiry into the justifiability of a technical concept and a moral concept of patentability.' She is also a member of several committees regarding these issues and serves as an external expert to the Advisory Committee for Bioethics of the Federal Government.

Harriet M. Strimpel is senior scientist and patent agent at Bromberg & Sunstein, a law firm located in Boston Massachusetts that specialises in business litigation and intellectual property matters. Since joining the firm in 1991, she has prosecuted numerous patent applications before the United States Patent and Trademark Office in the Biotechnology, Medical and Chemical Arts, and has obtained patents for US and overseas clients. She has also served as the expert witness for the plaintiff in a federal trial involving the civil rights of a biotechnologist whose home and research results were destroyed by fearful local officials, obtaining for the plaintiff the highest settlement recorded for a civil rights case in Massachusetts. Dr Strimpel obtained her doctorate from Oxford University (molecular virology) and performed postdoctoral research at the Hebrew University, Israel (virology), at the National Institute for Medical Research, United Kingdom (gene structure and expression), and at Biogen Corporation, USA (expression of pharmaceutical proteins in eukaryotic cells). In 1986, she founded a consulting firm to serve the biotechnology industry and the financial community. Dr Strimpel is the author of numerous articles on biotechnology matters and has lectured on various aspects of biotechnology to universities, colleges and schools.

Michele Svatos obtained a BA in Philosophy and Cognitive Sciences in 1988 at Tulane University (USA), followed by a PhD in Philosophy at the University of Arizona in 1994. Her areas of specialisation are moral philosophy, applied ethics, philosophy of law and social and political philosophy. From 1988 she was employed as Research Assistant and Graduate Teaching Instructor at the University of Arizona, and in 1993 she was appointed Assistant Professor at Iowa State University, teaching courses such as Bioethics & Biotechnology, Philosophy of Technology and Contemporary Moral Problems. Professor Svatos has presented a number of papers at various conferences and colloquia, including on topics such as the justification of intellectual property (biotechnology patents) in less developed countries, and the moral implications of the Human Genome Project. She has also published articles on these and other philosophical matters.

Christoph Then has studied veterinary medicine. Since 1992 he is involved in political work on issues of patenting. He is author and co-author of several oppositions against patents on plants and animals at the European Patent Office. From 1995 Dr Then is working for the Green Party in the Bavarian Parliament on issues of agriculture. He represents the organisation 'No patents on Life!', which was founded in 1992. Their office in München is mostly used as a service-point for spreading information, doing research on patenting, proposing lobby-campaigns and keeping the flow of information between the NGO's which are engaged in the 'No patents on Life!' campaign.

Michel Vandenbosch studied ethics at the Free University of Brussels (Belgium). He is the chairman of GAIA (Global Action in the Interest of Animals), and is involved in a variety of actions in defense of animal rights. Mr Vandenbosch has published on the subject of animal welfare and animal rights.

Dominique Vandergheynst obtained a Law degree at the Catholic University of Leuven (Belgium). He became Administrative Secretary and Consultant at the Office of Industrial Property which is connected to the Ministry of Economic Affairs and addresses legal issues surrounding patents at a national and European level. Since 1990 he is employed as Officer at the European Commission, studying documents regarding patents, specifically related to protection of biotechnological inventions. Mr Vandergheynst frequently participates in meetings in the European Parliament, the European Council and various other international organisations.

Luc Vankrunkelsven is the co-ordinator of WERVEL (Working group for a just and responsible agriculture) in Brussels. WERVEL tries to be a point of reference in the discussion between farmers and consumers' movements. This organisation places genetic engineering, patents, rBST-hormone and herbicide-resistance in the field of (dominant) industrial agriculture, as opposed to sustainable family farming; and defends the 'ethics of the liberation of life', as well as a safe food production and a sustainable agriculture.

Marc Van Montagu obtained a MSc in Organic Chemistry and a PhD in Organic Chemistry/Biochemistry at the University of Ghent (Belgium). He was appointed full Professor and Director at the Faculty of Sciences, Laboratory of Genetics. He accepted multiple invitations as visiting professor from all around the world and has organised a number of congresses, symposia, colloquia and workshops over the years, in Europe as well as outside Europe. Professor Van Montagu's major research area is the genetic engineering of plants. In 1982 he founded, with Professor Jozef Schell, the company Plant Genetic Systems Inc. Since 1982 he holds the position of Scientific Director in this company. He has received numerous scientific awards in Belgium and abroad. He has been given the title Doctor Honoris Causa at the University of Helsinki (Finland) as well as at the Université Technologique de Compiègne (France). Professor Van Montagu is member of a number of societies and organisations (USA, USSR, Japan, France, Sweden, Belgium etc.) as well as member of the editorial board of 14 scientific journals. He is the author of more than 600 publications in international and national journals and books.

Geertrui Van Overwalle obtained a Law degree at the Catholic University of Leuven (Belgium), followed by a doctorate with a thesis entitled 'Patentability of plant biotechnological inventions. A comparative study towards a justification of extending patent law to plants'. From 1985 she worked as Assistant and Researcher at the same university, and taught law at the Erasmuscollege in Brussels. Dr Van Overwalle has published a number of articles and papers on various law-subjects, but especially on biotechnology and patent law, and has given many lectures on the legal protection of biotechnological inventions. She is a member of several commissions regarding these matters.

Jan Van Rompaey obtained a Degree in Biology at the University of Antwerp (UIA, Belgium) in 1978. From 1978 until 1987 he was employed at the Antwerp State University (RUCA), mainly as Assistant. In 1990 he finished his Doctorate in Sciences at the UIA. Since 1988 Dr Van Rompaey holds the position of 'Patent Manager' (Manager Technology Protection) at Plant Genetic Systems (PGS) in Ghent.

Etienne Vermeersch graduated in Classical Philology and Philosophy at the University of Ghent (Belgium). He consequently obtained a Doctorate in Philosophy and was appointed full Professor in 1967. His teaching curriculum comprises subjects such as philosophy of science, contemporary philosophy and philosophical anthropology. In 1993 he was elected Vice-Chancellor of the same university, a position which he still holds. His scientific research relates to fundamental research in the field of social sciences, the philosophical aspects of research about informatics and artificial intelligence; more recently about general social and ethical problems, especially in the field of bio-ethics. Professor Vermeersch is Boardmember of numerous organisations and councils. To name just a few: Centre for Logics, Société Belge de Philosophie, Fund for Scientific Research, Flemish Board for Scientific Policy, Flemish Interuniversity Board, Flemish Interuniversity Institute for Biotechnology. He is also a member of the Supreme Council for the Co-ordination for Actions against AIDS, as well as a member of the Deontological Committee concerning animal testing. Etienne Vermeersch's commitment to contribute towards solutions of various social problems from an ethical viewpoint is also reflected in several publications, such as 4 books and more than 50 articles, published in national and international journals.

Foreword

Etienne Vermeersch

I write this preface a few days after reading the astounding report of the cloning of a sheep from the nucleus of an adult cell. Although the procedure is essentially different from the rDNA technique used in current biotechnology, this 'progress' and the reactions it stimulated remind us once more of the ever increasing importance of the ethical dimension in contemporary biological research.

A couple of months ago the importation of genetically modified soybean into several countries of the European Union provoked a discussion about the necessity of labelling such products, which would give individuals the right and the capability to choose the type of food they want to consume. This proves the importance of taking into account the public perception of the products of biotechnology.

The rejection (1995) of the first Draft Directive on the Legal Protection of Biotechnological Inventions proved that the members of the European Parliament were aware of feelings of uncertainty or unrest with the public concerning biotechnology in general and the patenting of living matter in particular.

For more than two decades the research on bioethics and environmental philosophy in our philosophy department has taken as a guiding principle, neither to dismiss these forms of concern as irrelevant, nor to let them deviate into antiscience, but to tackle them in a *rational* way.

The way in which the International Workshop on Biotechnology, Patents and Morality was conceived and realised, bears witness to this pursuit of rationality *and* responsibility. I am very grateful to our young collaborator Sigrid Sterckx for her skilled allegiance to the spirit of our department.

Preface

The Department of Philosophy and Moral Science and the Centre for Environmental Philosophy and Bio-Ethics of the University of Ghent are pleased to present the proceedings of the International Workshop 'Biotechnology, Patents and Morality: Towards a Consensus', which took place on 17, 18 and 19 January 1996 in Ghent.

At 'Het Pand' — the University's beautiful cultural centre, a former Dominican monastery — twenty-six invited speakers addressed the participants. These included, amongst others, professors, researchers and students in science and law; representatives of non-governmental organisations, of public and private research institutes, of the biotechnology industry and the European Patent Office (EPO); patent attorneys; social scientists; campaigners and policy makers at national and European levels.

The workshop was organised in the framework of the research project 'Biotechnology, Patents and Morality: Towards a Consensus', which started on 1 March 1995 and resulted in a final report that was delivered on 30 April 1996. This project was entirely funded by the Commission of the European Communities (DG XII - Science, Research and Development - Biotechnology Unit).

The purpose of the research project was to look into the problems involved in the patenting of inventions in the field of biotechnology. Of course the debate about the patenting of biotechnological inventions is intertwined with the debate about the moral permissibility of biotechnological inventions and techniques as such. The latter is being held in scientific and philosophical literature and in political circles. A 'new' branch of ethics has emerged over the past decades. This 'applied ethics' or 'practical ethics' is called upon to study these developments and to contribute to a solution of the problems resulting from them. A great deal of interesting work has already been carried out. Time is pressing, however, for many biotechnological inventions and techniques are being used and many more will follow. Consequently, the authorities on a national as well as on a European level have to take a position concerning these issues.

Questions concerning biotechnology arise particularly in the context of patents. A heated debate is on-going about the question whether it is morally permissible to patent living matter. Patent Offices have to judge on an increasing number of patent applications relating to all sorts of life forms, including humans (*cf* the patent application for germ line gene therapy that was filed in 1992 by the University of Pennsylvania). The 'Oncomouse case' is no doubt the most famous one — a patent application directed to a method for producing non-human mammalian animals, in the genome of which an oncogene sequence has been incorporated, and which are described by the patent applicant, Harvard University, as being useful tools for cancer research. This application was filed at the European Patent Office in June 1985. The patent was granted in 1992, but seventeen opponents, representing over two hundred organisations, filed notice of opposition to this patent. The opposition proceedings are still on-going, and after this stage there will almost certainly be an appeal, so this case is likely to drag on for several more years.

Another example of the complexity of the issues involved in the debate about biotechnology patents is the history of the *Directive on the legal protection of biotechnological inventions*. The Commission of the European Communities came up with a first draft in 1988, but the vote about this Directive in the European Parliament did not take place until 1 March 1995. The Directive was rejected. Ethical objections have played a very important part in this rejection. On 13 December 1995, the Commission of the European Communities announced a revised version of the Directive. However, many of the problems that surfaced during the previous debate remain unsolved and the new vote will most probably be preceded by another round of difficult discussions.

The European Patent Office has to decide on the granting of patents. After having examined a patent application, the Office ascertains the novelty and non-obviousness of the invention that constitutes the subject of the application. In addition, in order to be patentable an invention needs to be susceptible of industrial application and has to be described in sufficient detail. However, even when these criteria are met, there can be a decision against granting the patent on grounds of the so-called 'morality provision', reflected in article 53(a) of the European Patent Convention (EPC) of 1973. In this provision, it is stated that European patents shall not be granted in respect of 'inventions the publication or exploitation of which would be contrary to "ordre public" or morality, provided that the exploitation shall not be deemed to be so contrary merely because it is prohibited by law or regulation in some or all of the Contracting States'. This means that the patent examiners working at the European Patent Office have to consider whether the inventions underlying the patent applications they have in front of them are immoral.

The patenting of living matter is opposed by various groups like Greenpeace, 'No Patents on Life', animal welfare societies, organisations that defend farmers' interests, green parties, several religious groups, etc... A considerable part of their argumentation is based on ethical grounds. Besides these groups there are of course the national governments and the European Parliament who take great interest in these cases, both as a source of legislation and as a forum for discussing ethical issues. In addition to these questions, the debate about the patenting of biotechnological inventions has raised significant questions about the role of patent offices. Should, or can, a patent examiner — who is a technical expert — decide on the moral permissibility or non-permissibility of an invention?

The purpose of the international workshop was to gather round the table the diverse parties involved in the debate about biotechnology patents, to give them the opportunity to present their arguments to one another and to see to what extent we could get closer to a consensus or a compromise. Instead of focusing on the differences between the parties, we opted for a constructive approach and tried to find common grounds.

We have tried to reflect the diversity of the matter by the division of the programme into several headings. The first session of the first day was devoted to the scientific aspects of biotechnology. Here, the focus was on the nature of the developments in biotechnology: which kinds of research have proven to be 'beneficial to humankind' and what major breakthroughs can be expected in the near future?

The main question dealt with during the two afternoon sessions was: how does the public in general perceive biotechnology and to what extent do these perceptions relate to ethical, social and cultural factors? What is the attitude towards, for instance, plant genetic engineering as compared to human healthcare-related applications of biotechnology? In this second session, philosophers, psychologists and social scientists explored and tried to explain the attitudes of the public.

Of course, the main focus of attention was on the *patenting* of biotechnological inventions, rather than on the assessment of biotechnology as such. During the first morning session of the second day, the legal framework was laid out by several experts in the field of patent law, including people from the European Patent Office (Dr Schatz, Principal Director International Affairs, dedicated his entire lecture to the ethical aspects of patent law). The situation in the United States was described as well. The last morning session was devoted to the regulatory framework. Mr Dominique Vandergheynst, Principal Administrator at DG XV of the European Commission, the department that is responsible for the new proposal for a *Directive on the legal protection of biotechnological inventions*, clarified the modified text and commented on the differences between the old version and the new version (*cf* supra).

The two afternoon sessions of the second day were reserved for several presentations by opponents and advocates of the patenting of biotechnological inventions, as well as two rounds of questioning.

On the third day, a lecture was held on consensus formation and on the need to involve the public in the debate about biotechnology and biotechnology-patents. Afterwards, the discussions started.

Since the problems involved are extremely complex and the long term consequences of concrete policy options are in most cases hard to predict, it was often not possible to discuss in detail the many interesting views and suggestions expressed during the workshop. We are hopeful, however, that they will bear fruit by sparking off in-depth discussions in many forums, and stimulate exploration and evaluation in future research and writing. We would like to invite every reader to participate in the debate about the patenting of biotechnological inventions, for it is not only extremely captivating, but its outcome will have far-reaching consequences for our society at large.

Acknowledgements

First of all, I would like to thank Professor Etienne Vermeersch, who possesses all the qualities that are required to be a lot more to his collaborators than merely a professional supervisor. Although he must be the busiest man in Western Europe, he has managed to help me solve a variety of problems that have arisen during the preparation of this volume. Moreover, he made the publication possible in the first place, for he successfully defended my request for funding before the Research Council of our University. This allowed me to spend a few months full-time preparing the book. For all these reasons, and because day after day it comes home to me that he has offered me the most heavenly job on Earth, I dedicate this book to him.

Of course the international workshop would not have taken place if it was not for the generous funding provided by the Biotechnology Unit of DG XII of the Commission of the European Communities. I am very grateful that they selected the research proposal which constituted the framework of this workshop.

I am also indebted to all the people who assisted me in various ways while I was organising the workshop, as well as during the workshop itself: my colleagues Tom Claes, Dani Dewaele, Dirk Holemans and Johan Braeckman; Huguette, the 'chief' secretary of our Department; my brother Wouter, who made sure the recording of the proceedings went smoothly; and Geertrui Van Overwalle, for making many interesting suggestions concerning the lecturers. My mother provided fresh encouragement every time I needed it. Thanks to my father's professional advice, I acquired many organisational skills.

Siegfried supported me in every possible way one could dream of being supported. I owe him so much that it is impossible to put down in words how tremendously I appreciate all the things he has done for me.

Mrs Mariette Wouters, who took a great number of responsibilities off my shoulders during the workshop, also proved to be indispensable afterwards. She took care of the transcription of half the audiotapes, which was not an easy task for I had promised the lecturers that they would not have to prepare lengthy papers. While accomplishing the other half of the transcription job, I sometimes came close to regretting this... I could not have managed without Mariette.

xx

The fact that, at times, I wished the lecturers would have spoken a little more clearly or a bit closer to the microphone is of no significance whatsoever, compared to my gratitude towards all of them for having accepted my invitation to give a lecture. Numerous people congratulated me for succeeding in bringing together all these eminent specialists. The great number of interesting insights that were offered in the course of the workshop would not have been possible but for their enlightening lectures: they were responsible for the success of this workshop. I was also very impressed by the intense participation of many of the attendees.

From the moment I knew I would prepare this publication, I contemplated how interesting it would be if a few experts would write down some comments regarding the proceedings; to let them draw our attention to angles that were neglected in the debate, or react to particular statements that were made. Just a glimpse at the biographies of the commentators should suffice to understand how thrilled I was when I received positive answers from them. I gave 'carte blanche' for the content of their comments, for I had the feeling that even the mere suggestion of a title would be too restrictive. Of course, for some people — I am afraid I am one of them — being given 'carte blanche' is not a gift but an additional source of endless 'internal deliberation' about which road to take. I hope this 'profile' does not sound familiar to the commentators. I want to take this opportunity to express once again my enormous appreciation to them, for a job amazingly well done in addition to amazingly quickly. It is never very pleasant for an ethicist to burden people with such strict deadlines as the ones I had to deal with.

I am especially indebted to my editors Kate Hargreave and Rachel Hedges, not only for the indispensable guidance they provided but also for their encouragement. Finally, I want to express special thanks to my colleague Claudine Vanderborght, the superb proofreader of this publication. Indeed, superb, because the remaining inaccuracies are due to my 'wanting it my way' and occasionally ignoring her excellent advice. I need to add that, for more than eight years now, Claudine has been benefiting from her living together with Jeff, a native English speaker of the best kind. I, in turn, have been benefiting from the fact that Claudine's charms made Jeff become the second proofreader. Mentioning all the other responsibilities and irritating little and big tasks which Claudine took off my hands would take too long. Her professionalism has made a profound impression on me.

1 European patent law and biotechnological inventions

Sigrid Sterckx

Biotechnology

'Biotechnology' refers to a wide range of techniques that make use of living organisms. Although this term does not occur until the beginning of the 1970s, such techniques have been used for centuries. The making of cheese, the fermentation of wine and the breeding of plants and animals are examples of biotechnologies man has been applying for more than 5000 years. Until some two decades ago, however, the characteristics of living organisms could only be modified by means of gradual selection. The genes that had to cause a modification of a particular organism could only be chosen from the 'pool' of genes of the particular species this organism belonged to. At the beginning of the 1970s, a number of subsequent technological breakthroughs[1] created fundamentally new possibilities that marked the era of modern biotechnology.[2] When scientists unraveled the genetic code, they paved the way for genetic engineering. All living beings seemed to possess the same kind of hereditary 'building blocks', situated in the same molecule (DNA). Consequently, it had to be possible to 'exchange' this genetic material between different kinds of organisms. Indeed, in 1973 it was shown by Cohen and Boyer that DNA of *different* species could be assembled and subsequently inserted into a bacterium that served as a host cell. To 'cut out' particular pieces of DNA, they used restriction enzymes, which had been discovered earlier. This high-technological 'cutting and sticking', called 'recombinant DNA-technology' or 'genetic engineering', allows the combining of genes of totally different organisms with the purpose of introducing new properties into the receiving organisms. An example of the genetic engineering of animals is the modification[3] of certain genes of these animals in order to render them capable of producing particular substances which are normally not produced by these animals. Some animals, plants and micro-organisms are used as so-called 'bioreactors' for the production of rare pharmaceutical substances. Human insuline and growth hormone, for example, are being produced by micro-organisms for several years.

Organisms whose genetic material has been modified through these kind of techniques are called 'transgenic' organisms. Several transgenic plants and animals have already been created. The main commercial applications of genetic engineering are in the fields of health care, agriculture and the food industry

The widening of the debate

The debate about the legal, ethical and social problems connected with modern biotechnology has been on-going since this branch of technology came into being. One of the most persisting and important issues of the debate concerns the risks of genetic engineering to man and the environment. Since the organisation of the Asilomar conference in 1975, where scientists devised the first regulatory framework for the careful handling of genetically modified organisms (GMO's), a public debate about the assessment of these risks has developed.

In discussions about the possibilities of modern biotechnology, very often ethical considerations are brought into the picture, especially when techniques are concerned that are meant to be applied to human beings. Only too often, George Orwell and the Nazi's are brought into the debate. We agree with the assertion that, whenever possible, science should develop 'hand in hand' with ethics. However, we are opposed to an exaggeration of the possibilities of science with the purpose of causing fear and rejection of certain branches of science among the public. Unfortunately, such attempts to 'inform' the public are sometimes made with respect to biotechnology. Justified remarks concerning hazardous developments are used by some people to make biotechnology as a whole appear in a bad light. Biotechnology gives rise to very differing attitudes: some describe it as the ultimate means to solve all the major problems our society is struggling with. Others consider biotechnology to be the proof that humankind is digging its own grave.

Often, objections to modern biotechnology are confronted by the statement that there is no relevant difference between traditional biotechnology and modern biotechnology. Lord Howie, for example, who conducted an investigation in the United Kingdom in which he pleaded for a relaxation of the regulations in the field of biotechnology, stated that 'Biotechnology is no more than an enhancing of the evolutionary process.' Others consider biotechnology as an 'acceleration', instead of an 'enhancement' of natural processes:

> ... when Charles Darwin writes, of a certain Mr Wicking, that he took just "thirteen years to put a clean white head on an almond tumbler's [a variety of pigeon] body" ... what does the genetic engineer do but change the thirteen years that Darwin mentions to thirteen months?[4]

The great number of scientific advances which were achieved during the past decades in molecular and cellular biology have laid the foundations of the biotechnology industry as we know it today. Hundreds of companies have been set up, with activities in the fields of agricultural biotechnology, environmental biotechnology and biomedical applications of biotechnology. In environmental

2

biotechnology, micro-organisms are modified with the purpose of being used to clean up soil, water and air. Agricultural biotechnology applies recombinant DNA technology to develop new techniques of plant and animal breeding and to change the properties of plants and animals. Among the transgenic plants that have been developed up to now are transgenic tomatoes, potatoes, sugar beets and tobacco. Often, the reason why crops are genetically modified is to make them resistant to insects, fungi or herbicides. Whereas the production of insect- and fungal resistant crops appears to be a beneficial development from an ecological point of view, the genetic modification of plants to make them herbicide-resistant is considered by many as undesirable because they believe this will lead to an increase instead of a diminishment in the use of herbicides. Many transgenic animals have been developed as well. The technique most frequently applied in this respect consists in the micro-injection of DNA into newly fertilised eggs. The aim is to integrate the inserted DNA into the genome of the animal. If the integration is successful, the DNA will also be passed on to the progeny of the animal. This micro-injection has been done for the first time in 1980 with mice — in the mean time, hundreds of strains of transgenic mice have been developed. Mice are said to be cheap and easy to work with and the functioning of their genes is better known than that of higher mammals. The micro-injection technology has subsequently been utilised to produce among others transgenic amphibians, fish, rabbits, rats, poultry, cattle, goats, sheep and swine. A great number of transgenic mammals serve as models for the study of genetic diseases. Scientists apply the resulting insights to try to understand the development of diseases in humans and to design strategies for therapy. Possible treatments are tested on these animals. Other transgenic animal models have been generated for studying resistance to animal disease. Transgenic animals are also used as 'factories' for the production of valuable human and animal protein products which cannot be produced in bacteria (*cf* 'molecular farming', the use of animals as so-called 'bio-reactors').

In the domain of health care, biotechnology is used to produce medicines for diseases such as cystic fibrosis, various cancers and hemophilia. A wide gamut of diagnostic kits are also developed with the aid of biotechnology. An area that is still in its early stages, but considered by many as very promising, is the so-called 'gene therapy'. In this field, attempts are made to correct congenital disorders.

When reviewing critiques of biotechnology, it must be kept in mind that some critics condemn biotechnology as such, whereas others make nuances in their arguments, depending on the kind of organism involved. For instance, some people feel that the genetic modification of micro-organisms is acceptable, whereas interventions in the genome of plants, animals and human beings are not.

Critiques of biotechnology as such

A few examples of frequently mentioned critiques of biotechnology as such are that man should not tinker with genes because:

3

- the gene pool is the joint property of the entire human race (*cf* the concept of 'common heritage of mankind') and therefore it should be passed on unchanged to following generations

- in doing this, man is 'playing God'

- in doing this, man is playing 'Mother Nature'

Some critics state that the results achieved in the field of biotechnology are not relevant enough for society as a whole. In this respect, it is noteworthy that, very often, it is difficult to balance the positive effect of a biotechnological invention on society's well-being on the one hand and the risks of that invention to man and his environment on the other hand. The main reason for this difficulty is that, in many cases, the risks involved may only become apparent in the long term.

Another general critique is that, due to a lack of public control of the research in the domain of biotechnology, decisions as to which kinds of research are ethically acceptable are made by scientists, who are ill-suited for tackling ethical questions.

Critiques depending on the kind of organism involved

Some critics take into account the kind of organisms involved in biotechnological research. The arguments they use to assess, for example, genetic modification of plants, are different from the criteria applied to evaluate transgenic animals. They do not make statements about the merits and deficiencies of biotechnology as such.

Micro-organisms Regarding micro-organisms, attention is especially drawn to the risk of genetically modified micro-organisms 'escaping' from laboratories and spreading into our environment. Of course, some organisms are genetically modified with the very purpose of releasing them into the environment. For both these cases, regulations have been devised.[5]

Plants Regarding plants, a similar warning is expressed about the danger of disrupting ecological balances through the introduction and uncontrolled release of genetically modified plants into the environment.[6] In this respect, it is sometimes said that experts in the field of genetic engineering are inclined to reason in terms of the individual organism that results from the introduced genetic modification ('if the organism is harmless, there is nothing to worry about'). Some critics disagree with this way of reasoning: they assert that the effects of a transgene organism on its environment (including all the other organisms living in that environment) should be taken into account. They immediately add, however, that it is very difficult to assess those effects, for we are dealing with *new* organisms, i.e. organisms that did not exist before.

Another frequently heard objection to the genetic modification of plants is that it causes a reduction of biodiversity, for in the long term only plants with 'interesting' characteristics — i.e. characteristics that serve the interests of man — will remain. This kind of objection has been formulated, for instance, as follows:

> A further source of concern and possible constraint centres upon the cumulative effect of genetic engineering on genetic diversity. There are at least two ways in which genetic engineering might contribute to the loss of genetic diversity ... One is through the extensive use of cloning, where space which could have been occupied by a genetically distinct individual is occupied by an individual who is simply a genetic copy of another. The other way in which genetic diversity might be lost arises from the fact that genetic engineering will inevitably reflect human priorities, and human priorities are inevitably parochial. *Life-forms which are in any way serviceable to humans will be developed in those respects in which they are serviceable and, in general, genes which express such properties will be favoured at the expense of those which do not.*[7]

Representatives of the biotechnology industry reply to this objection by stating that, instead of reducing it, genetic engineering leads to *more* genetic diversity. In the context of the opposition that was filed by Greenpeace Ltd to a patent granted to the Belgian company Plant Genetic Systems, Greenpeace's statement that genetic engineering reduces biodiversity was answered as follows by the patent holder:

> ... making non-plant genes available in plants, can *increase* genetic diversity.[8]

Animals Part of the objections to the genetic modification of animals run parallel to the observations that are made regarding plants. Here too, attention is drawn to the danger of disruption of ecosystems caused by the deliberate introduction of transgenic animals into the environment or their escaping from laboratories. Here too, it is said that the effects of the interaction between these animals and the environment in which they are released are often disregarded. The objection concerning the reduction of genetic diversity — only animals with characteristics that are beneficial to man will remain — is a third example of a parallel critique. It is true that, in most cases, genetic engineering of animals is carried out with the purpose of improving their production characteristics and their resistance to disease, in view of the fact that disease results in financial loss for the breeder. This brings some people to the observation that the properties of these animals are changed in a way beneficial to man, but only in rare cases beneficial to the animal in question.

A specific critique — one that is not parallel to the objections raised against transgenic plants — concerns the problem of experimental animals: are animal experiments ethically acceptable? Are these kinds of experiments necessary? How many efforts should be made for research into alternative test models? In which cases is it justified to say that an animal suffers as a result of the experiment? Is it impermissible to conduct animal experiments for other purposes than the health and well-being of the animals themselves? Does it matter whether these experiments cause suffering or not? These are only a few examples of frequently asked questions.

The above-mentioned objection to the use of micro-organisms as 'bioreactors' is much stronger when animals are involved. Several questions arise: may animals *überhaupt* be reduced to bioreactors? Should this practice be allowed in some cases (for example, for the production of pharmaceuticals), while forbidden in other cases (for example, for the fabrication of cosmetics)? Should we accept the use of animals as bioreactors only for the production of pharmaceutical substances which are needed by a lot of patients, but cannot easily be produced otherwise?

Human beings   The most complex disputes, however, arise with respect to interventions into the genome of human beings. A few examples:

– Do we *own* our genetic material, or does it belong to 'society' or 'science'?

– Should any intervention into the human genome by definition be considered as an attack to 'human dignity'? Or should we say quite the contrary, viz. that precisely *not* delivering humankind from several serious hereditary diseases is an attack to human dignity?

– Are experiments on human genetic material allowable? Which guidelines should be taken into account? What about the principle of informed consent?

– Is somatic gene therapy ethically acceptable? Should germ line gene therapy be prohibited a priori?

Patents

A patent is a kind of intellectual property (other examples are copyrights and trademarks). An inventor who obtains a patent has the right to exclude others from the industrial or commercial exploitation of his invention. This right is limited in time: 20 years from the filing date.[9] The right conferred by a patent is also limited in space: a patent is only valid in the jurisdiction of the patent office by which it was granted — for instance, a patent granted by the United States Patent and Trademark Office is not valid in Europe.

Arguments concerning the justifiability of patents

Economical arguments have always played an important part in the justification of patent law. An often heard argumentation is the following: inventing something and subsequently developing that invention into a commercial product takes a great deal of time, work and money. These efforts would not be worth undertaking if the inventor would not be offered the possibility of recouping his investments. Without the guarantee that is provided by an intellectual property right, the inventor would be forced — by his competitors, who could immediately copy the invention — to

6

lower his price, so he would never be able to earn back the money he invested. According to this argumentation, exclusivity (the right to exclude others) is a necessary incentive for research and development.

A second kind of argument that is used in the justification of patents refers to their stimulating effect on innovation and industrial development through the promotion of the dissemination of technical knowledge. The achievement of the latter goal, it is said, is guaranteed because, by making the details of inventions public, other researchers can build upon this knowledge and thereby further advance the state of the technology. This way of reasoning is also reversed: if an area of technology is excluded from patentability, there is a disincentive for investment into research in this field and, to the extent that any research would still be carried out under these circumstances, also a serious disincentive for publication of the results of that research. If patents were to be taken away, it is said, the only remaining option for inventors is to keep the information about their inventions secret.

We would like to make the following comments on these arguments. Firstly, on patents as a 'necessary incentive'. On the one hand, it is said that the presence of patents is an extremely important tool to encourage people to carry out research. On the other hand, the very people who state this, often accuse individuals or organisations who oppose particular biotechnology patents on the ground of the 'morality article' of the European Patent Convention, of 'abusing patent law'. Opponents of biotechnology patents (to be precise, those that are no competitors of the patent holder, for instance NGO's) are advised to present their concerns to policy makers instead of patent authorities. The 'patent avenue' (i.e. attacking inventions at the stage of the patent procedure) is said to be a bad strategy, because the research will continue even if patents were abolished. In the Opposition proceedings in the Oncomouse case, for instance, Mr Bizley, the legal representative of the patent proprietor, said that 'opposing patents is a bad way to prevent a particular kind of technology'. In the context of his observations regarding the justifiability of the Oncomouse patent application under Art. 53(a) of the European Patent Convention — the so-called 'morality article', which states that European patents shall not be granted for inventions the exploitation of which would be against 'ordre public' or morality (cf infra) — Mr Bizley formulated the following question:

Does the existence of possible patent protection encourage such work [the research concerning the insertion of oncogenes into animals] to be done?

In general terms: what is the relationship between the availability of patent protection in a particular technical field and the carrying out of research in this field? Mr Bizley's answer was very disappointing:

In law this is not, I submit, a relevant question.

We believe that this question is extremely relevant. As we already mentioned, the encouraging effect of the availability of patent protection is always cited, by the advocates of the patent system, as the very 'raison d'être' of the patent system. Although Mr Bizley considers the question about the encouragement offered by patents to be irrelevant, he seemed to expect to be asked to answer it, for he added that 'there is not a shred of evidence that scientists would cease their incessant probing towards further and better understanding of the world in which we live and natural phenomena even if patents were to be abolished tomorrow!' In his lecture at the workshop, Mr Stephen Crespi expressed the same view: 'I have always believed that the human spirit in this area is so strong that people will invent and innovate, whether they have patents or not.' It is clear that to reconcile this kind of statements with the statement that patents are necessary to stimulate people to carry out research, is problematic. From Dr Schatz's interventions during the workshop, it became clear that he is aware of this: ' ... patenting is giving a strong incentive to research and development. So, if your objective is to slow down research and technical progress in biotechnology, you are right: you should stop patenting such inventions.' Of course, at the same time, Dr Schatz realises that we do not live in a vacuum, and changing patent policies in Europe would in no way prevent others from continuing the kinds of research which Europe would be discouraging through its change of procedure.

As to the argument that the taking away of patent protection, or even the mere tightening of the patentability requirements, leads to inventions being kept secret, we believe that this is an overstatement. Especially in the field of biotechnology, many kinds of technical information concerning inventions simply cannot be kept secret. Once *products* are brought to the market, clever competitors should not have a very hard time working out how to copy them. The details about the *processes* through which these products are made would of course be more difficult to find out, but advocates of biotechnology patents themselves assert that 'process protection' without 'product protection' is not very interesting from a commercial point of view.

'A patent does not confer the right to exploit the patented invention'

When a patent is opposed on moral grounds, a reply that is often heard is that a patent does not confer on its proprietor any right to exploit the patented invention. This was said, for instance, by the legal representative of the biotechnology company Plant Genetic Systems during the opposition proceedings in the case 'PGS versus Greenpeace'[10]:

> ... neither the patent at issue nor any other patent confers upon a patentee a right to exploit, let alone a "monopoly" right to exploit, his patented invention. Exploitation of a patented invention is always subject to governmental regulation and control in each of the Contracting States of the EPC. Nothing in the EPC diminishes or restricts the ability of the Contracting States to control or prohibit by law or regulation the exploitation of a patented invention. ... Hence, although a patentee has a right to exclude others from using his patented invention, a patentee does not have a positive right actually to exploit his invention.

8

Dr Ulrich Schatz, Principal Director International Affairs at the European Patent Office, phrased this idea as follows in the lecture he gave at the workshop:

> ... a patent does not *per se* amount to an authorization to exploit the invention. To be even more precise, we should say that *a patent gives no right whatsoever to exploit an invention*. The sole right which a patent does confer is the right of the inventor to *exclude others* from exploiting the invention.

It is true that a patent does not *automatically* give its proprietor the right to exploit his invention. It is also true that some sorts of inventions — among which there are many kinds of inventions in the field of biotechnology — may only be exploited after approval has been obtained from regulatory authorities (e.g. bodies comparable to the Food and Drug Administration, governmental bodies for the regulation of the use of genetically modified organisms, etc...). It is obvious that a patent holder may not exploit his invention if laws exist which prohibit the exploitation of that particular kind of invention. On the other hand, it should also be clear that exploiting of an invention is not at all dependent on whether that invention is patented or not. If an inventor chooses not to file a patent application, this has no influence on whether or not the exploitation of his invention is authorised. If he does choose to file an application and he obtains a patent, the same applies.

However, taking into account these very same facts, the Board of Appeal of the European Patent Office in the 'PGS versus Greenpeace' case (T 356/93, *cf* also infra) gave a different definition of a patent, one which does not define a patent as a merely *negative* right (a right to prevent others from exploiting of the invention), but as a *positive* and negative right at the same time, its 'positive part' (the right to exploit the invention oneself) not being unconditional:

> A patent confers on its owner(s) for a specified time *an exclusive right to exploit* the subject-matter of the claims, i.e. to manufacture, use and market it, *and to prevent others* from doing the same. ... *the right to exploit the invention is not unconditional*. On the contrary, the invention claimed in a patent may only be exploited within the framework defined by national laws and regulations regarding the use of the said invention. (§18.2 of the decision, our emphasis)

In practice, there is often a strong link between patenting and exploitation. First of all, obtaining a patent costs a lot of money (fees have to be paid to the European Patent Office), as well as a lot of time and effort (the specification and the claims have to be written). Nobody would undertake this effort if he did not, at least, have a reasonable expectation of obtaining the patent. In many cases, the act of filing a patent application indicates that the inventor intends to exploit the invention (to exploit it himself or to sell it to somebody else, who will exploit it). It is hard to understand what would be the point of trying to exclude others from the commercial exploitation of an invention if one does not want to reap the commercial fruits of the invention oneself. As we mentioned above, one of the most frequently used arguments

9

in favour of patenting is that, if patent protection were not available, there would be no incentive for companies or research institutions to engage in research, for they would have no prospect of recouping the money they invested in the development of their invention. It is hard to see how this money could ever be recouped if the invention were not exploited, whether by the patent proprietor himself or by somebody to whom the proprietor sold a license!

Secondly, and much more importantly, if patenting and exploitation would in no way have anything to do with each other, the drafters of the European Patent Convention would not have included Art. 53(a) in the Convention. This provision clearly establishes a link between the exploitation of an invention and the grant of a patent for that invention: if the *exploitation* of an invention would be contrary to morality, a *patent* shall not be granted. The wording 'would be' indicates, moreover, that it does not even matter whether an invention is actually exploited or not. If its exploitation *would be* contrary to morality, this is a sufficient ground for considering the invention non-patentable. However, it seems to be the case that every time an opponent to a biotechnology patent brings forward arguments under Art. 53(a) as to why the exploitation of the invention in question is contrary to morality, the patent applicant, as well as Opposition and Examining Divisions and Boards of Appeal of the European Patent Office, reply that a patent does not give the proprietor the right to exploit the invention. However, this reply is not only *irrelevant* in view of the wording of Art. 53(a); in addition, it makes one wonder whether there is any point at all in raising objections under Art. 53(a)! Is there only any use in doing so if the patent application contains a solemn oath, sworn by the applicant, saying 'I swear that I intend to exploit my invention and I will do everything within my capacities to obtain approval from the competent regulatory bodies'? If one really carries further the argument that patenting and exploitation are in no way interrelated, one should strongly disagree with the assertion, made in the *Guidelines for Examination in the European Patent Office*,[11] that a letter-bomb should be excluded from patentability on 'ordre public' and morality grounds. According to the afore-mentioned argument, granting a patent on a letter-bomb does not give the proprietor of the patent the right to exploit the letter-bomb. Well, then why not patent the letter-bomb? The example of germ line gene therapy is also a very good one, for exploitation of this kind of invention is regarded by many people as inconceivable.[12] It is even mentioned in the new proposal of the European Commission for a 'Directive on the legal protection of biotechnological inventions': paragraph (2) of Art. 9 lays down that germ line gene therapy shall be considered unpatentable. Since the first paragraph of Art. 9 of the Directive is the equivalent of the 'morality article' of the European Patent Convention, it is clear that the drafters of this Directive exclude germ line gene therapy because they consider its exploitation to be contrary to morality. Again, according to the afore-mentioned reasoning, granting a patent on germ line gene therapy does not give the proprietor of the patent the right to exploit germ line gene therapy. Well, then why not grant patents for germ line gene therapy?

We feel that those who hold the view that there is no link between patenting and exploitation should show some intellectual honesty by admitting that, in their view, not one single patent should ever be refused on grounds of 'ordre public' or morality. So they might as well admit that, if it were up to them, Art. 53(a) EPC might just as well be abolished.

Sometimes very confusing statements are made concerning the 'real' scope of the rights that are conferred by a patent. On the one hand, objections to the patenting of living organisms are often answered by the assertion that the existence of a patent covering a living organism does not mean that the patent holder 'owns' the organism in question. On the other hand, when the US Commissioner of Patents and Trademarks announced in 1987 that the US Patent and Trademark Office intended to grant patents on animals, it was added that human beings were exempt from patentability, because the 14th Amendment to the US Constitution prohibits 'ownership' of one person by another.

Patentability requirements in Europe

The patent law of most of the western industrialised countries dates from the 19th century. In 1963, the Strasbourg Convention came into effect. It was intended to strengthen the unification of patent law in Europe. The basic principles of the Strasbourg Convention were fully taken over in the European Patent Convention (EPC) of 1973, the 'bible' of the European Patent Office. The EPC made it possible to obtain a patent in some or all of the contracting states through one single application. In fact, a European Patent is a 'bundle' of national patents. The responsibility of judging whether the requirements an invention has to meet in order to be patentable are fulfilled, lies with patent offices. These exist on a national as well as on a European level. The European Patent Office's headquarters are located in Munich.

According to European patent law, *any* invention — be it an invention regarding 'living material' or a photocopier — that meets the patentability-requirements 'novelty', 'inventive step', 'susceptibility of industrial application' and 'sufficiency of disclosure' can be patented.

The general principles of patentability, the general requirements that should be satisfied in order for an invention to be patentable, are laid down in paragraph (1) of Art. 52 EPC: the invention has to be 'susceptible of industrial application', 'new' and has to involve an 'inventive step'.

The first condition is dealt with under Art. 57 EPC, according to which 'An invention shall be considered as susceptible of industrial application if it can be made or used in any kind of industry, including agriculture.' 'Industry' is understood as comprising any physical activity of a technical character. This requirement does not signify that, in order to be patentable, an invention must represent a technical progress, nor that it should have any useful effect.

Novelty is dealt with under Art. 54 EPC. This requirement is satisfied if 'the subject-matter does not form part of the state of the art'. The state of the art is considered to comprise 'everything made available to the public by means of a written or oral description, by use, or in any other way, before the date of filing of the

European patent application'. In other words, the meaning given to the concept of novelty in the context of patent law differs completely from the way this concept is normally understood, viz. as something new being something that was not there before. Here, the question is not whether it existed before, but whether somebody could have known about its existence.

Thirdly, the invention has to involve an inventive step. This condition, which is also often referred to as 'non-obviousness', is treated under Art. 56 EPC: 'An invention shall be considered as involving an inventive step if, having regard to the state of the art, it is not obvious to a person skilled in the art.' 'Having regard to the state of the art' means that the patent examiners must ask themselves whether, in view of the state of knowledge in the technical field in question at the time of the patent application, the invention would have been considered obvious by a skilled person[13].

Whereas the first paragraph of Art. 52 EPC lays down the general requirements inventions have to meet, the European Patent Convention does *not* define what is meant by the term 'invention'. However, the second paragraph of Art. 52 contains a list of things that shall not be regarded as inventions: (a) discoveries, scientific theories and mathematical methods; (b) aesthetic creations; (c) schemes, rules and methods for performing mental acts, playing games or doing business, and programs for computers; and (d) presentations of information. In this list, 'discoveries' are mentioned. The difference between a 'discovery' and an 'invention' is certainly not always clear when dealing with biotechnological inventions. An argument frequently used by opponents of patents on living organisms and parts of living organisms, e.g. genes, is that these can only be discovered, in other words that one can never speak of an 'invention' in this context. The *Guidelines for Examination in the European Patent Office* do not make us much wiser concerning this issue. In part C of the *Guidelines*, Chapter IV, 2.3, it is stated that, 'to find a substance freely occurring in nature'[14] is 'mere discovery and therefore unpatentable'.

However, if a substance found in nature has first to be isolated from its surroundings and a process for obtaining it is developed, that process is patentable. Moreover, if the substance can be properly characterised either by its structure, by the process by which it is obtained or by other parameters ... and it is "new" in the absolute sense of having no previously recognised existence, then the substance per se may be patentable.

Thus, if a method is found to isolate a substance from its surroundings, there is no longer a discovery, but an invention. If the substance in question is 'new' — i.e. if, up to then, it was not made available to the public — and its characteristics are described[15], the 'invention' becomes patentable. Many people have a hard time understanding the logic of this: why would the mere fact of isolating a substance from its natural environment, or purifying it, turn that substance from a discovery into an invention? This question poses itself even more strongly when realising that, due to the nature of the techniques for isolation and purification available to scientists today, these processes no longer require any inventivity. We will come back to this issue in the context of our comments on Art. 3 of the new proposal for a 'Directive on the legal protection of biotechnological inventions.'

The problem of the difference between an invention and a discovery also arises with respect to transgenic organisms. Take the example of a transgenic animal: does the insertion of one or a 'handful' of foreign genes into an animal possessing 50,000 to 100,000 genes justify the claim of having 'invented' the resulting transgenic animal? In the case of animals which are genetically modified to serve as models for carcinogenicity testing, such as the Harvard Oncomouse, we feel that the invention lies in the conception of the test model. 'But', it is said, 'the test model can only function inside the animal'. Although this is true, it does not change the fact that the animal as such has not been invented, conceived, by the inventor.

Finally, Art. 52 EPC also excludes from patentability 'methods for treatment of the human or animal body by surgery or therapy and diagnostic methods practised on the human or animal body', because these are not regarded as inventions which are susceptible of industrial application. This exclusion does *not* apply to products, in particular substances or compositions, that are used in these kinds of methods. In his lecture, Dr Schatz from the European Patent Office explained the reason for the exclusion from patentability of surgical, therapeutic and diagnostic methods:

> What is the reason for this general exclusion of medical treatment? The reason is that the patent system is a regulation of competition in industry and trade, whereas the medical art has to abide by medical deontology rather than by the rules of commercial competition. We do not consider that the exercise of medical art is a business which should be governed by the laws of commercial competition. Patent law should not apply to the medical profession.

At the beginning of this section, apart from the 'susceptibility of industrial application', 'novelty' and 'inventive step' requirements, we mentioned another condition that needs to be fulfilled in order for an invention to be patentable: 'sufficiency of disclosure'. However strange it may sound, this requirement is extremely important from the point of view of ethics. The relevant articles of the EPC are 83 and 84. Article 83 reads as follows:

> The European patent application must disclose the invention in a manner sufficiently clear and complete for it to be carried out by a person skilled in the art.

An important link exists between this provision and the requirement reflected in Art. 84 EPC, that:

> The claims shall define the matter for which protection is sought. They shall be clear and concise and be supported by the description.

A patent application contains a so-called 'specification' (description of the invention) and a set of 'claims', which determine the extent of the protection the applicant wants to obtain. According to Art. 84 EPC, a patent applicant cannot simply claim things that are not described in the specification. For a person to be able to repeat an invention, even if this person is skilled in the art, a sufficient amount of information is required. Depending on the interpretation that is given to the term 'sufficient' in this context, broader or more narrow patents will result. Patents that are too broad are extremely problematic, not only from the point of view of the law but also from an ethical viewpoint. We will deal with this issue in detail in the context of our comments on the explanation concerning Art. 3 that is given in the new proposal for a 'Directive on the legal protection of biotechnological inventions'. At this point, however, we also need to add that the requirement of 'sufficiency of disclosure' (Art. 83 EPC) and the condition of 'susceptibility of industrial application' (Art. 57 EPC) are interrelated: if an invention is not disclosed in sufficient detail by the patent applicant, i.e. if that invention cannot be repeated, this means that it cannot be regarded as applicable in industry.

Exceptions to the general principles of patentability

The notorious Art. 53 of the EPC mentions two exceptions to the general principles of patentability.[16]

Morality Paragraph (a) of Art. 53, the so-called 'morality provision', states (first half-sentence) that 'inventions the publication or exploitation of which would be contrary to "ordre public" or morality' are excluded from patentability. It is not said how the concepts of 'ordre public' and 'morality' should be understood. The only description that is provided is a negative one: according to the second half-sentence, the exploitation of an invention should not be deemed to be contrary to 'ordre public' or morality 'merely because it is prohibited by law or regulation in some or all of the Contracting States' of the EPC. In the *Guidelines for Examination in the European Patent Office*, the following instructions are provided about the way patent examiners need to determine whether an invention is contrary to 'ordre public' or morality:

> This provision is likely to be invoked only in rare and extreme cases. A fair test to apply is to consider whether it is probable that the public in general would regard the invention as so abhorrent that the grant of patent rights would be inconceivable.[17]

These instructions are far from clear, for it is not explained *how* the attitude of 'the public in general' towards a particular invention has to be 'discovered'. Should opinion polls be taken into account? It is not very likely that opinion polls will be available that show the attitude of the public in every Contracting State towards all kinds of inventions (we believe that, in the context of an EPC-provision, 'the public in general' should be interpreted as referring to all the Contracting States of the

14

EPC). Such polls are not even available on inventions that receive widespread attention in the press. Should the attitude of the governments in the Contracting States serve as a criterion? Or is it the viewpoint of the parliaments that should be taken into account? One thing we do know is that, concerning the patentability of animals, the European Parliament has adopted a resolution[18] which declares that 'transgenic animals engineered to suffer' in general, and the animals from the Harvard patent entitled 'Method for producing transgenic animals'[19] in particular, should be considered as *non*-patentable because the production of such animals is incompatible with public order (Art. 53a EPC) and contravenes the exclusion from patentability of plant and animal varieties (Art. 53b EPC, *cf* infra). The European Parliament has called upon the European Patent Office to revoke the Harvard patent and not to respond favourably to any further applications for animal patents until the legal uncertainties[20] have been clarified. We believe that this fact provides a reliable indication of the views of 'the public in general' concerning this particular range of inventions. Harvard's legal representative Mr Bizley, however, asserted during the Oral Proceedings in Opposition that the European Parliament is not representative of the European population.

At least one thing is clear about Art. 53(a): due to its vague formulation, it is not very useful for the evaluation of concrete inventions. The text of this provision has been taken over in Art. 9(1) of the proposal for a 'Directive on the legal protection of biotechnological inventions'. Some attempts have been made to create more clarity. Unfortunately, these cannot be regarded as successful. We will deal with the 'morality provision' in detail in the context of our comments on Art. 9 of the proposed Directive.

Plant and animal varieties and essentially biological processes for the production of plants or animals The other exception to the general principles of patentability is laid down in Art. 53(b) of the EPC:

> European patents shall not be granted in respect of plant or animal varieties or essentially biological processes for the production of plants or animals; this provision does not apply to microbiological processes or the products thereof.

There is no consensus about the legal or scientific definition of the concept of 'plant varieties', but the definition given in the UPOV Convention[21] is more or less internationally accepted. Regarding the exclusion of 'animal varieties', the lack of clarity is even worse. Additional confusion is caused by the fact that the European Patent Convention, which is drafted in English, French and German, contains three different words to designate the excluded subject-matter: 'animal varieties', 'races animales' and 'Tierarten'.[22] 'Tierart' means species, whereas 'animal variety' and 'race animal' are a sub-unit of a species. 'Varieties' constitute the lowest rank in the taxonomic classification, whereas 'species' are on a higher rank. This is a problem since, according to Art. 177(1) of the European Patent Convention, the English, French and German versions of the Convention are equally valid. In spite of the total lack of clarity concerning the scope of the exclusion from patentability of 'animal

15

varieties', the European Patent Office has already taken two decisions in the field of genetically engineered animals. The patent on the Harvard Onco-animals is the most famous one.

The *Guidelines for Examination in the European Patent Office* offer no clarification either. The only thing they communicate is that:

> Also excluded from patentability are "plant or animal varieties ...". One reason for this exclusion is that, at least for plant varieties, other means of obtaining legal protection are available in most countries.

'Other means' refers to the afore-mentioned convention on plant breeders' rights. Originally, the UPOV Convention contained a provision — Art. 2(1) — prescribing that double protection was to be excluded (in other words, that an invention should not be protected by both a plant breeders' right and a patent). In 1991 however, the UPOV Convention has been revised and it was decided that double protection is no longer prohibited. The *Guidelines* say that the availability of another system of protection is 'one reason' for the exclusion from patentability of plant varieties. No light is shed on the other reasons. In the field of animals, there is no comparable system of protection. 'Animal breeders' rights' do not exist. What then could be the reasons for the exclusion from patentability of animal varieties? The *Guidelines* remain silent on this issue. The Examining Division of the European Patent Office in the Oncomouse case stated that the reason behind this exclusion is that:

> ... animal varieties are not an appropriate subject-matter for patent protection[23]

The same Examining Division suggested that the exclusion from patentability of animal varieties may have originated from *ethical* considerations:

> It should ... not be neglected that the protection sought by the applicant does not "only" refer to mice but to all other non-human animals amongst which there are animals to which mankind historically developed a certain association and possibly affection. ... the specific exclusion in Article 53b may also have been caused by ethical consideration.

We will treat the issue of plant and animal varieties in detail when we discuss the new proposal for Directive, for this exclusion from patentability has been taken over in Art. 4(2) of the Directive.

Apart from plant and animal varieties, 'essentially biological processes for the production of plants or animals' are also excluded from patentability under Art. 53(b) EPC. Paragraph 3.4 of Chapter II, part C, of the *Guidelines* is devoted to the criterion for distinguishing essentially biological from non-essentially biological processes:

The question whether a process is "essentially biological" is one of degree depending on the extent to which there is a *technical intervention by man* in the process; if such intervention plays a *significant* part in determining or controlling the result it is desired to achieve, the process would not be excluded [for being essentially biological]. (our emphasis)

This means that, in the case of patent applications comprising a process for the production of plants or animals, the patent examiners at the European Patent Office have to consider whether there is a sufficient degree of 'technical intervention by the hand of man' involved in the process and whether this intervention has a significant impact on the final result. Of course these instructions are still rather vague. The exclusion from patentability of essentially biological processes for the production of plants or animals has been taken over in Art. 6 of the proposed Directive, so we will come back to it.

To conclude on Art. 53(b) EPC, it needs to be added that the second half-sentence of this article — 'this provision does not apply to microbiological processes or the products thereof'— lays down an exclusion to the exclusion that is expressed in the first half-sentence. This means, firstly, that 'microbiological processes' are patentable in any case, and, secondly, that 'products of microbiological processes' are also patentable in any case, *even* if they constitute plant or animal varieties. So, plant and animal varieties are not patentable, unless they result from a microbiological process. The patentability of microbiological processes and their products has been taken over in Art. 5 of the Directive (*cf* infra).

The patentability of the human body and its parts

Nothing is mentioned in the European Patent Convention about the patentability of human beings and parts of the human body.[24] It is often called a 'generally accepted principle' that they are non-patentable subject-matter. However, in this context, the expression 'parts of the human body' does not encompass the following: human genes, proteins and cells, provided that these are isolated from the human body. In the context of patentability, the question whether a particular isolated substance is derived from a micro-organism, a plant, an animal or a human being is considered to be irrelevant. Several patents have already been granted for isolated parts of the human body. This practice is sometimes called 'even more immoral' than the patenting of material derived from plants and animals. Very often, the people who oppose these kinds of patents refer to the European Patent Convention in support of their arguments. They say that genes can only be 'discovered', so nobody can 'invent' them, whereas Art. 52(2) EPC lays down that 'discoveries' are not patentable. Of course the question is what distinguishes inventions from discoveries. We have already mentioned this problem in the context of our comments on Art. 52(2). We will come back to it, for Art. 3 of the proposed Directive is concerned with the patentability of the human body and its elements.

The provisions governing American patent law are laid down in Title 35 of the US Code. For an invention to be patentable in the US, it needs to be novel, useful, and non-obvious. 'Usefulness', the so-called utility-requirement, is comparable to the European patentability requirement of 'susceptibility of industrial application' (Art. 57 EPC). 'Non-obviousness' means the same as the formulation 'involving an inventive step' in European patent law (Art. 56 EPC). Plant varieties, animal varieties and essentially biological processes for the production of plants or animals are not excluded under US patent law. Neither are inventions which are contrary to 'ordre public' or morality. The US patent law does not contain an equivalent of Art. 53(a) of the EPC, the so-called 'morality provision'.

In April 1988, the United States Patent and Trademark Office (USPTO) granted the first patent on an animal to the President and Fellows of Harvard College. Two researchers from Harvard, Philip Leder and Timothy Stewart, had produced transgenic mice whose germ cells and somatic cells contain an activated oncogene sequence, which was introduced into the animal at an embryonic stage. The result of this genetic modification of the mice was an increase of the probability of the development of neoplasms (particularly malignant tumors). The patent that is known as the 'Harvard Oncomouse'-patent covers in fact all non-human mammalian onco-animals, so this denomination is misleading. So is the title of the patent, 'Method for producing transgenic animals', for 'onco-animals' — animals in the genome of which an oncogene sequence has been inserted — are but one kind of transgenic animals.

The controversy over the granting of patents on life forms has been going on in the US since the so-called 'Chakrabarty case'. Ananda Chakrabarty, a microbiologist who was employed by the company General Electric, had genetically modified a strain of bacteria so that they were able to break down crude oil. He applied for a patent at the USPTO, but the application was rejected. In 1980, the case came before the US Supreme Court, which ruled in favor of Mr Chakrabarty. This decision is generally regarded as a landmark decision in US patent history, because the Supreme Court declared that 'anything under the sun that is made by man' is patentable. This means that whether an invention relates to living or to inanimate matter is considered to be irrelevant. In April 1987, the Board of Patent Appeals and Interferences of the USPTO ruled (*cf* In re Allen) that an invention relating to certain polyploid oysters was patentable, including the oysters *as such*. Four days later, the then Commissioner of Patents and Trademarks announced that the USPTO intended to grant patents on 'non-naturally occurring, non-human multicellular living organisms', including animals. As a result, the heated debate which had been going on for several years intensified. The idea of patenting animals was criticised by religious leaders, animal welfare organisations, farmer's interests advocates and environmentalists. At the request of a House Representative, Mr Kastenmeier, the USPTO agreed to an eight-month moratorium on animal patents. After this moratorium expired, in April 1988, the world's first patent on an animal was granted to the 'inventors' of the Oncomouse.

This short history shows that the frequently heard assertion, that the patenting of higher life forms has not caused any serious controversy in the US, is false. As a result of the controversy, several hearings were held in the *Subcommittee on Courts, Civil Liberties and the Administration of Justice* of the US House of Representatives. Moreover, legislative initiatives were taken. In August 1987, Democratic Representative Charles Rose introduced a House bill to place a two-year moratorium on the granting of patents for 'vertebrate or invertebrate animals, modified, altered or in any way changed through genetic engineering technology'. This bill also called for a revocation of any patent previously granted for such animals. The purpose of this initiative, according to Representative Rose, was to provide a secure legal basis for continuing congressional consideration of this issue. This bill did not pass. In February 1988, Republican Senator Mark Hatfield introduced a bill that called for an indefinite ban on the patenting of genetically altered or modified animals. This bill did not pass either. In August 1988, however, legislation was approved that created an infringement exemption for farmers using patented animals. This bill — the 'Transgenic Animal Patent Reform Act', which had been introduced by Democratic Representative Robert Kastenmeier — provides that on-farm reproduction of genetically engineered animals is not a patent infringement.[26]

The debate in the US about patenting in the field of biotechnology is not over yet. This is shown by the fact that the issue of human gene patenting is a priority on the agenda of the National Bioethics Advisory Commission, the members of which were named by President Clinton on 19 July 1996.[27]

A very special phenomenon

Within a period of nine months after the grant of a patent, any person, company or organisation can file an opposition to that patent, provided that the required administrative fees are paid to the European Patent Office. The purpose of filing an opposition is to convince the patent office that the patent should be revoked, in part or in its entirety.

As the European Patent Office began to receive patent applications for genetic material of micro-organisms, plants, animals and humans, as well as for micro-organisms, plants and animals *as such*, a very special phenomenon started to occur. The oppositions against these patents were filed not only by competitors of the patent holder — which is common practice; all kinds of patents are opposed by competitors all the time — but mainly by organisations like Greenpeace, 'No Patents on Life'-coalitions, animal welfare organisations, etc... In their objections, the EPC-provisions referred to most frequently are Art. 53(b) and especially Art. 53(a), the 'morality' provision. The latter is never used as an objection by lawyers who represent biotechnological companies. Very often, the moral grounds on which the patenting of inventions in the field of genetic engineering is opposed, include references to the detrimental effects these patents will cause to the environment. This observation has provoked the following, rather straightforward, comment from a Director at Directorate General II of the European Patent Office:

When Greenpeace UK says: "The short and long term effects of granting patents on living organisms on the environment, society and animal welfare have been inadequately debated and whether the public interest will be served by granting patents on life should be decided democratically, not by the free market." What do they mean by 'democratically'? ... What do we do here? Count heads? Count countries? In the context of the European Patent Office, it is often difficult enough for the basic 'jury' of three examiners to reach a verdict on the novelty or inventivity of an invention. If the examiners had to examine the issue of morality as well, their professional lives would become impossible. ... Examiners are pragmatic people. They would leave it to the public to raise the issue of morality and get on with their job, which is quite difficult enough as it is, thank you. The way is, and always was, open to the public to mount an opposition on grounds of lack of morality. The fact that the public has not chosen to do so till now, that those who have been so silent for so long have suddenly been stimulated to articulacy on the topic of patenting 'higher' life forms is not particularly meaningful, to my mind. I call it fashion. It is the obsession of our time. [28]

Discussion to be continued...

The European Commission's new proposal for a 'Directive on the legal protection of biotechnological inventions'

Introduction

The complete title of the new proposal for Directive is 'Proposal for a European Parliament and Council Directive on the legal protection of biotechnological inventions' [COM (95) 661 final]. The original proposal for Directive was published by the European Commission in October 1988 [COM (88) 496 final]. This proposal did not take into account the ethical dimension of the patenting of biotechnological inventions. On the basis of this document, negotiations started between the European Commission and the European Parliament, during which the Parliament undertook great effort to convince the Commission to modify the proposal in the sense that it would include some basic ethical principles, like the exclusion from patentability of the human body and its parts, the derogation for farmers and the exclusion from patentability of inventions that are contrary to 'ordre public' and morality. The mere mentioning of the fact that these negotiations lasted four years (until December 1992) should suffice to give an idea of the extent to which the opinions differed. In February 1994, the Council announced its position. To find ways out of the differences in opinion between the Council's views and those of the Parliament, negotiations were held in a Conciliation Committee — consisting of members of both legislative bodies — during its meetings in October 1994 and on 12 and 23 January 1995. A joint text was drawn up.

On 1 March 1995, about one month after the Conciliation Committee's last reunion, the joint text was rejected by the European Parliament. For the first time, the European Parliament rejected a Conciliation Committee project.

The rejection of the Directive was a serious defeat for the biotechnology industry. Ethical considerations have played an important part in determining the no-result of the vote. The new proposal, the one that will be discussed under this heading, was announced in December 1995. A second vote by the European Parliament will take place in the near future.[29]

If the Directive would be approved, only the EC-member states would be required to adjust their legislations accordingly. The European Patent Office would not have to do so (the European Patent Convention can only be altered through a diplomatic conference of the Contracting States). It is important to realise that laws adopted by the European Union are not binding to the European Patent Convention and the European Patent Office. However, as Dr Margaret Llewelyn, the Deputy Director of the Sheffield Institute of Biotechnological Law and Ethics, pointed out during a meeting held in Brussels on 29 October 1996:

Whilst the Directive will have no official effect on the practice of the EPO, it is possible that it might be used in the sense of providing unofficial guidelines for the EPO. This would have the effect of ensuring consistency of European patent practice.

Nearly all member states of the EPC are also member states of the EU and it would be difficult for the EPO to disregard totally the views of its EU member states as expressed in the Directive.

However, despite following the progression of the Directive closely, the EPO has not stated publicly that it will be influenced by the Directive.

It is pure speculation as to the extent to which the EPO will allow itself to be influenced by the Directive.

Equally if the Directive is rejected by the European Parliament then the EPO will not be bound to take into account the reasons for the rejection when making future decisions. It will remain free to carry on its policy of granting patents as before.

Personally, we suspect that — for indeed nearly all the contracting states of the EPC are also member states of the European Union — if the Directive were to be adopted, it is rather likely that it would be regarded by the European Patent Office as a set of unofficial guidelines.

Should the new proposal be adopted, this would indicate that the Members of the European Parliament changed their views on a number of the issues involved — as compared to the views they held in March 1995 — for the differences between the previous version and the new text are only minor ones. During the workshop, Mr Stephen Crespi made the following remark in the context of the lecture given by Mr Dominique Vandergheynst from the Commission:

21

... my first impression is that the Commission has stuck to its position on many of the points that the Parliament had problems with. ... So much is in common with the previous Directive text, that I am wondering if the Parliament has in the meantime been more educated to understand the issues that have been raised ...

As to whether a Directive is needed (regardless of its contents), whether a legal framework on the level of the European Union needs to be created for the patenting of biotechnological inventions, we would definitely take the affirmative view. A similar interpretation of patentability requirements throughout the member states of the EU should be aimed at. At present, as it is stated in the Explanatory Memorandum to the new proposal for Directive:

> ... it cannot be claimed that all European patents granted and entering the national stage in the designed Contracting States will be interpreted in the same way, regardless of the national court involved. Not only must a decision be taken as to whether an invention may be patented or not; the precise scope of the protection conferred by a patent must also be ascertained if the holder institutes infringement proceedings. In the absence of clear reference points, national courts may react differently.

It is important to realise that patents which have been granted by the European Patent Office may be revoked by the national courts of the member states of the EPC, if the subject-matter does not meet the general patentability requirements or falls under one of the categories of inventions that are excluded from patentability.[30]

Moreover, the existing concepts should be made clear. In the European Patent Convention from 1973, the provisions of the Strasbourg Convention[31] (1963) were incorporated. This means that many of the concepts that are now being applied to assess the patentability of biotechnological and other inventions were drafted more than thirty years ago. A very special characteristic of biotechnological inventions, as compared to inventions in other fields, is that they often involve self-reproducing material. This poses specific problems considering the scope of the legal protection that should be granted for inventions in this area. In this and other respects, the law should be made clear, in order to minimise the likelihood of divergent interpretations by different patent offices and national courts. This is not only in the interest of patent applicants, but of society at large.

Difference between discovery and invention / The patentability of parts of the human body (Art. 3)

Art. 3 of the new proposal reads as follows:

> (1) The human body and its elements in their natural state shall not be considered patentable inventions.

(2) Notwithstanding paragraph 1, the subject of an invention capable of industrial application which relates to an element isolated from the human body or otherwise produced by means of a technical process shall be patentable, even if the structure of that element is identical to that of a natural element.

The issue of the difference between an invention and a discovery was hotly discussed during the meetings of the Conciliation Committee, in its search for a joint proposal concerning the previous version of the Directive. As § 28 of the Explanatory Memorandum to the new version explains:

It had to be determined whether the words "as such" in point (a) of the second subparagraph of Article 2(3) differentiated sufficiently between a discovery and an invention as regards body elements of human origin: "*On this basis, the following inter alia shall be unpatentable: (a) the human body or parts of the human body as such...*"

The delegation — in the Conciliation Committee — from the European Parliament had serious problems with the words 'as such'. They felt that these had to be deleted, to create absolute clarity concerning the non-patentability of elements of the human body, but Parliament did not get its way. No doubt, this is one of the reasons why the previous proposal for Directive was rejected. In the new version, the Commission has deleted the words 'as such'. As stated in § 52 of the Explanatory Memorandum to the new proposal:

Parliament stressed that the words "*as such*" — the aim of which was to distinguish the natural elements of the human body to be excluded from patentability — gave the impression of making discoveries patentable, which they cannot be. Accordingly, in order to clarify the question of the patentability of elements of human origin, it appears sensible not to include the words "*as such*" in the present proposal.

However, the words 'as such' have been replaced by 'in their natural state' (*cf* supra, § 1 of Art. 3), which in this context of course means exactly the same. One might wonder, therefore, whether this time the drafters of the text have given a more convincing explanation of the difference between a discovery and an invention with reference to elements of the human body. The following is stated in the Considerantia:

(13) Whereas it should be specified that knowledge relating to the human body and to its elements in their natural state falls within the realm of scientific discovery and may not, therefore, be regarded as patentable inventions; whereas it follows from this that substantive patent law is not capable of prejudicing the basic ethical principle excluding all ownership of human beings;

(14) Whereas significant progress in the treatment of diseases has already been made thanks to medicinal products derived or otherwise produced from elements isolated from the human body, and medicinal products resulting from a technical process aimed at obtaining elements similar in structure to those existing naturally in the human body and whereas, consequently, the patent system should promote research aimed at obtaining such elements;

(15) Whereas, therefore, it should be made clear that an invention capable of industrial application and based on an element isolated from the human body or otherwise produced by means of a technical process is patentable, even where the structure of that element is identical to that of a natural element, since no patent may be interpreted as covering an element of the human body in its natural environment forming the basic subject of the invention;

(16) Whereas such an element isolated from the human body or otherwise produced may not be regarded as unpatentable in the same way as an element of the human body in its natural state, that is to say, may not be equated with a discovery, since the element isolated is the result of the technical processes used to identify, purify and classify it and to reproduce it outside the human body, techniques which human beings alone are capable of putting into practice and which Nature is incapable of accomplishing by itself;

Of course it cannot be denied that elements isolated from the human body have indeed enabled the development of many useful drugs, aimed at improving the quality of life of patients suffering from severe diseases. Therefore, we agree with the statement that 'the patent system should promote research aimed at obtaining such elements'. However, this does not permit to present as an 'inevitably following conclusion' the statement that those elements *themselves* are patentable. 'Research aimed at obtaining such elements' can perfectly be understood as research into technical processes for the isolation of such elements. So, one could say that, in order to promote research into this kind of processes, it has to be possible for researchers to get patents when they invent such processes. It is perfectly clear that — if these processes are new, involve an inventive step, are industrially applicable and described in sufficient detail — patents are available in this context.

But what about the elements *themselves*? In the proposal for Directive, it is said that, as long as they are inside the human body ('in their natural state') they are not patentable because they constitute 'discoveries'. However, the argument is used that, once these elements have been isolated from the human body or otherwise produced by means of a technical process, they turn from a discovery into an 'invention'.

The fact that it is *the nature of the process of isolation* which makes all the difference is very clearly shown by § 43 of the Explanatory Memorandum to the new proposal:

24

... an element of the human body that has not been obtained with the aid of a technological process, but simply detached, removed or collected, may not be regarded as a patentable invention ...

Thus, it is said that, if the process for obtaining the element is not patentable, the element *itself* is not patentable either. Yet, if the process of isolation is of a technical nature, this turns the isolated elements of the human body from natural elements into 'artificial' elements. According to § 51 of the Explanatory Memorandum:

Elements isolated from the human body by means of a technical process are artificial and thus qualify as inventions, since they are technical solutions invented by man in order to solve technical problems.

We believe it is justified to say that, when isolated elements from the human body are used to develop drugs and therapies, they are used as solutions to solve particular problems (the treatment of particular diseases). We do not consider it justified, however, to call the elements *themselves* 'artificial', merely because they have been isolated by an artificial process. Neither do we consider the fact that this *isolation process* is invented and carried out by man, and could not take place without human intervention, a sufficient ground for saying that the *elements themselves* are invented by man. Why would the mere fact of isolating a substance from its natural environment, or purifying it, by means of technical processes, turn that substance from a 'discovery' into an 'invention'? Even several eminent scientists have a problem with this line of reasoning. Four respected societies from the United Kingdom — the Clinical Genetics Society, the Clinical Molecular Genetics Society, the Association of Clinical Cytogeneticists and the Genetic Nurses and Social Workers Association — have presented a joint statement on patenting, in which they assert the following about the patentability of human gene sequences of known function and utility:

A natural gene sequence is part of the human body, and as such should not be patentable. The suggestion that such sequence might be patentable if it is "isolated in a pure form" or "isolated outside of the body" seems to us a sophistry, and should not be allowed.

It should be noted that this statement concerns the patenting of human gene sequences *of known function and utility*, so lack of industrial applicability is not the issue. For subject-matter to be patentable, being industrially applicable is not sufficient. The subject-matter needs to be an industrially applicable *invention*.

According to § 16 of the Considerantia (*cf* supra), the reason why elements isolated from the human body by technical means may not be equated with a 'discovery' is that:

... the element isolated is the result of the technical processes used to identify, purify and classify it and to reproduce it outside the human body, techniques which human beings alone are capable of putting into practice and which Nature is incapable of accomplishing by itself;

In our view, these are all very convincing arguments for saying that those processes for identification, purification, classification and reproduction are not to be equated with discoveries, but instead do constitute inventions. However, these arguments are not sufficient to consider the *elements themselves* as inventions. So, it may well be possible to give a detailed description of these elements (e.g. a description of the structure of a gene and its coding regions), and it may well be possible to point out how the element in question can be applied in industry (e.g. for the production of a drug for Alzheimer patients), but if the elements themselves are not inventions, the question whether they fulfill the requirements for patentability is irrelevant.

On the required extent of disclosure of inventions (expl. Art. 3)

In their explanation concerning Art. 3, the drafters of this proposal for Directive express their view on the extent to which an invention should be disclosed in order to be patentable. This is done in the context of a clarification of the requirement of industrial application. Earlier, we already mentioned the link between the requirements of sufficiency of disclosure and industrial applicability. On page 16 of the proposal for Directive, sub paragraph 61, it is stated that:

> ... in order to qualify for protection, the subject-matter must constitute a technical solution to a technical problem. It thus proves essential to stress the industrial application requirement. ... The industrial application of an invention is specified in the description that must be submitted when the patent application is filed. The description must be sufficiently clear and comprehensive for someone skilled in the art to be able to carry it out.

The last sentence is almost identical to the 'sufficiency of disclosure'-requirement that is laid down in Art. 83 of the European Patent Convention (*cf* 'Patentability requirements in Europe'). This requirement is interpreted as follows by the drafters of the proposed Directive:

> Accordingly, it [the description of the invention] must:
> – specify the technical field to which the invention relates;
> – indicate the previous state of the art;
> – explain the invention such that the technical problem and the solution can be understood;
> – specify in detail at least one way of making or doing the thing invented.

Since all patent applications are published, the above mentioned requirements are said to offer the guarantee that 'patent law places at the disposal of all interested parties the scientific information relating to the invention'. Obviously, everything depends on what is understood by 'the scientific information regarding the invention'. We believe that there is a serious problem with the fourth requirement, viz. that it is sufficient for the patent applicant to specify one way of making or doing the thing he invented. 'At least one way' means that one way is considered to be sufficient. At first sight, it seems that this way of interpretation is also expressed in the *Guidelines for Examination in the European Patent Office*. In Part C of the Guidelines, paragraph 4.9 of Chapter II, the following is said about sufficiency of disclosure:

> A *detailed description of at least one way of carrying out the invention must be given.* Since the application is addressed to the person skilled in the art it is neither necessary nor desirable that details of well-known ancillary features should be given, but the description must disclose any feature essential for carrying out the invention in sufficient detail to render it obvious to the skilled person how to put the invention into practice. (our emphasis)

However, it is added immediately that:

> In many cases a single example or single embodiment will suffice, but *where the claims cover a broad field* the description should not usually be regarded as satisfying the requirements of Art. 83 unless it gives a *number of examples* or describes *alternative embodiments or variations extending over the area protected by the claims.* (our emphasis)

Thus, in the case of patent applications which contain broad claims, it is not sufficient for the applicant to provide one single example or describe one single embodiment. However, still according to the *Guidelines*, even in cases of broad patent applications, sometimes one example can suffice:

> There are some instances where even a broad field is sufficiently exemplified by a limited number of examples or even one example.

Reference is made to part C, Chapter III, paragraph 6.3 of the *Guidelines.* Here the issue is illustrated with an example from the field of plants. It is said that, in those cases where only the application of the invention to one particular kind of plants is described in the specification, whereas the claims relate to all plants, the invention should *not* be considered as sufficiently disclosed, because 'plants vary widely in their characteristics'. Indeed, plants vary widely in their characteristics; and this holds as well for animals, human beings and living organisms in general. Therefore it should be clear that all this is especially relevant when reviewing patent applications for living organisms. These kinds of applications should definitely be supported by a detailed description, comprising not just one single example or embodiment. We fully subscribe to the view expressed in the *Guidelines*

The proposed Directive deals exclusively with patents on biotechnological inventions, including patents on living organisms, and yet it contains the assertion that it is sufficient to provide one example of making or doing the invention. This is a serious problem for two reasons. First, through this assertion, the drafters of the proposal for Directive create the wrongful impression that, at the level of the law, unanimity exists surrounding this issue. This is not the case. We would like to illustrate this with some arguments that were presented in the Oncomouse patent procedure. The claims of this patent are extremely broad and they are definitely not supported by the description, although this is required by Art. 84 EPC (*cf* 'Patentability requirements in Europe'). The *description* that was provided by the applicant, Harvard University, only contains information about a method to produce onco-*mice*, whereas the *claims* are directed to *all non-human mammalian onco-animals*. The Examining Division of the European Patent Office rightly observed that:

> In the present case it is not believable that the skilled man would have success with all kinds of animals by using all kinds of known oncogenes and promotors. Animals which have been used in the prior art are mainly mice and *no instructions are to be found in the specification as to how success could be achieved with other animals* or as to which oncogenes are suitable for which animals. (our emphasis)

In the specification of this patent, an extensive explanation is given of a method for the production of mice carrying an activated oncogene sequence. Concerning any other animals than mice, the specification only contains the following paragraph (the last paragraph of the specification):

> *Other embodiments are within the following claims.* For example, any species of transgenic animal can be employed. In some circumstances, for instance, it may be desirable to use a species, e.g., a primate such as the rhesus monkey, which is evolutionary closer to humans than mice. (our emphasis)

Indeed, any embodiments relating to animals other than mice appear in the claims, whereas they should be described in the specification, in order to enable a person skilled in the art to carry out the invention. The title of the invention is 'Method for producing transgenic animals', but the specification only comprises a 'Method for producing transgenic mice' (moreover, as we mentioned earlier, 'onco-animals' are but one kind of transgenic animals, so even the designation 'transgenic mice' is too broad).

During the Examination-phase of this patent the Examining Division, which had serious problems with the 'sufficiency of disclosure' of this invention, suggested the Applicant to limit his claims to rodents. After the rejection of the application by the Examining Division — insufficient disclosure was only one of the reasons for rejection — Harvard lodged an appeal.[32] The Board of Appeal of the European Patent Office which was competent in this case, made the following comment concerning the Examining Division's suggestion to limit the claims to rodents:

> The Examining Division's view that, if limited to rodents instead of mammals in general, the claims would be acceptable ... would seem to be based on the *arbitrary assumption that all rodents would behave in the same way as mice for the purpose of the invention.*[33]

We feel that the extrapolation made by the Applicant, from mice to all non-human mammalian animals, is based on the far more arbitrary assumption that all non-human mammalian animals would behave in the same way as mice for the purpose of this invention. In our opinion, this kind of extrapolation is not justified. Harvard University felt that it deserved 'general principle protection'. However, only onco-mice were produced. No other onco-animals were produced, let alone proven to be viable. Consequently, it is not clear whether the principle according to which the onco-mice were produced can indeed be considered as a 'general principle'. According to the patent specification, any oncogene and any species of transgenic animal can be used. However, not one single instruction is given on how to produce these other onco-animals.

In its Decision T 19/90, the Board of Appeal stated that:

> ... the mere fact that a claim is broad is not in itself a ground for considering the application as not complying with the requirement for sufficient disclosure under Article 83 EPC. Only if there are *serious doubts, substantiated by verifiable facts*, may an application be objected to for lack of sufficient disclosure. (our emphasis)

We believe that the Board was wrong when it concluded that, in this case, serious doubts are not justified by the facts. If this case should not give rise to serious doubts regarding sufficiency of disclosure, no case should.

Coming back to the point that, even on a legal level, there is no unanimity concerning the interpretation of the 'sufficiency of disclosure'-requirement; this becomes clear if one looks at the different interpretations which are given within the European Patent Office itself. In support of his arguments, Mr Bizley, Harvard's legal representative in the Oncomouse case, referred to a decision of one of the Boards of Appeal of the EPO (T 292/85), which states that an invention 'is sufficiently disclosed if at least one way is clearly indicated enabling the skilled person to carry out the invention'. The Examining Division in the Oncomouse case — to be more precise, first examiner Dr Christian Gugerell, Head of the Directorate 'Genetic Engineering' at the European Patent Office — supported his arguments by referring to another decision of one of the Boards of Appeal (T 226/85), stating that 'this [the 'sufficiency of disclosure'-requirement] means that substantially any embodiment of the invention, as defined in the broadest claim, must be capable of being realised on the basis of the disclosure'. The difference is, to say the least, significant: 'to indicate at least one way' versus 'to describe substantially any embodiment'. So there is definitely no unanimity about this issue on a legal level and therefore the assertion which is made by the drafters of the Directive — that one example is enough — is unjust.[34]

The second reason why this assertion poses a serious problem has to do with ethics, viz. with the ethics of the patent system in general. On behalf of Harvard, Mr Bizley said the following about the way the 'sufficiency of disclosure'-requirement should be interpreted:

... what is necessary is that the specification should be sufficient to enable the invention to be performed and that the breadth of claims should be such as to be a reasonable extrapolation from the specific content of the description and the particular work performed.

Of course the question is what is to be understood by 'a reasonable extrapolation'. During the proceedings in examination and opposition, Mr Bizley repeatedly referred to the so-called 'principle of fair protection'. He felt that, in view of this principle, the patent as claimed had to be granted since the extrapolation made by the Applicant (from onco-mice to all non-human onco-mammals) was reasonable. We recognise the importance of fair protection. We feel, however, that this principle should be interpreted as the need to balance the interests of the inventor with the interests of society, whereas Mr Bizley only takes into account the interests of the inventor. This is a wrong approach, for the purpose of the patent system is comparable to a contract: inventors can obtain a patent if, in return, they disclose their invention to society. This is the feature of the patent system that is always said to guarantee the stimulating effect of patents on research and development. Patent applications are published in order to bring information about inventions in the public domain, so other researchers can build on it and come up with new findings, in order to serve the public interest. Patents should not only benefit inventors, but also society as a whole. If a patent applicant wants to 'take a lot' (obtain a broad monopoly), while only 'giving a little' (disclosing only a minor amount of information about his invention), this is to the disadvantage of society as a whole. Therefore, broad patents pose not only legal but also serious ethical problems.

Furthermore, as we already mentioned, advocates of the patent system always assert that — due to the high costs involved in developing an invention — patents are necessary to stimulate creativity and inventivity, to promote the advancement of science and technology. Supposing that scientists indeed need the 'financial stimulus' offered by the availability of patent protection, then it also has to be true that scientists are not very inclined to conduct research into a field that is already 'covered' by somebody else's patent.[35] This would mean that, according to the reasoning of the advocates of the patent system, broad patents have a negative effect on the creativity and inventivity of the scientific community. Many people in patent circles say this is recognised as a serious problem and that attempts are made towards solutions.

It is very unfortunate that the drafters of the Directive have not taken the opportunity to prevent such broad patents. It is even more unfortunate that they have done quite the opposite.

The exclusion from patentability of plant and animal varieties (Art. 4.2)

Art. 4(2) of the new proposal for Directive reads as follows:

Biological material, including plants and animals, as well as elements of plants and animals obtained by means of a process not essentially biological, *except plant and animal varieties as such*, shall be patentable. (our emphasis)

Thus, plants and animals are patentable, whereas plant and animal varieties 'as such' are not. This provision appears to be in accordance with Art. 53(b) of the EPC,[36] except for the words 'as such', the inclusion of which makes an enormous difference and should be understood in the light of stormy discussions that have been going on inside and outside of the European Patent Convention, as a result of the decision of one of the EPO's Boards of Appeal, viz. in the case 'Plant Genetic Systems versus Greenpeace'. This decision is of extreme importance, not in the least because it constitutes the most recent case law in the field of genetically engineered plants. We will briefly discuss this decision, and afterwards we will explain why an exclusion from patentability of plant and animal varieties 'as such' is untenable.

In January 1987, the biotechnology company Plant Genetic Systems from Ghent (Belgium) filed a patent application at the EPO, entitled 'Plant cells resistant to glutamine synthetase inhibitors, made by genetic engineering'.[37] The grant of the patent was published in October 1990. Greenpeace Ltd filed an opposition to the grant of this patent, but the patent was maintained as granted. Consequently, Greenpeace lodged an appeal. Art. 53(b) EPC was not the only ground for appeal, but in the context of this discussion about the patentability of plant and animal varieties, we will focus solely on the 53(b)-arguments. The PGS invention was concerned with plants which had been genetically modified with the purpose of making them resistant to a class of herbicides (glutamine synthetase inhibitors). Claim 21 of this patent was directed to this kind of transgenic plants. The patent specification contained working examples carried out on known plant varieties of tobacco plants. The fertility of the genetically modified plants was normal and the subsequent generations of these plants were homozygous for the herbicide resistance gene. The Board of Appeal in this case decided that the plants claimed by claim 21 are plant varieties and are therefore excluded from patentability under Art. 53(b).

It is extremely important to realise that, neither the European Patent Convention, nor the *Guidelines for Examination in the European Patent Office*, contain a definition of the concept 'variety'. Thus, although 'plant and animal varieties' are excluded from patentability in the EPC, it is not made clear how this exclusion should be interpreted. It is equally important to notice that the new proposal for a

'Directive on the legal protection of biotechnological inventions' does not contain a definition of the concept 'variety' either[38] Taking into account the intensity of the discussions which have been going on surrounding this topic since the Board of Appeal's decision in the 'Greenpeace case' — decision T 356/93 of 21 February 1995 — this is, to say the least, remarkable. If the purpose of the Directive is to remove legal uncertainty — and this is stressed repeatedly in the Explanatory Memorandum — a definition of the concept 'variety' should definitely be added. In an official letter[39] from the applicant of the Oncomouse patent (Harvard) to the Examining Division, the assertion was made that:

> ... it would be perfectly proper for the Examining Division to decide that absence of clarity in the exception to patentability is alone sufficient to enable the Harvard claims to go forward to grant.

This amounts to saying: the meaning of the concept 'variety' being unclear, why bother at all about the evaluation of patents under Art. 53(b) EPC? Why not skip the evaluation under this provision? As a matter of fact, why not just do away with the exclusion from patentability of varieties? However, this cannot be what the drafters of the Directive aim at, for they have taken over this particular exclusion from the EPC.

As to Art. 53(b) EPC, there is not only a lack of clarity regarding the concept that is used to designate the excluded subject matter; in addition, it is not clear how the *scope* of the exclusion itself should be interpreted. In the new proposal for Directive, the latter is explicit. According to paragraph (17) of the Considerantia:

> ... in order to determine the extent to which plant and animal varieties are to be excluded from patentability, it should be specified that the exclusion concerns those varieties as such ...

First, we will say a few words about the concept 'variety'. Then we will try to explain that only one sound interpretation of the scope of the exclusion is possible, and this is not the one given in the proposed Directive.

The EPO Board of Appeal in the 'Greenpeace case' has provided the up to now most extensive definition, given by the EPO, of the concept 'plant variety'.[40] The concept 'animal variety' has not yet been defined by the EPO, although three decisions have already been taken in the field of animals.[41] However, since the same English term is used in Art. 53(b) for both plants and animals — 'plant or animal varieties' — we do not see why the Board of Appeal's definition of 'plant variety' could not be extrapolated to animals. It reads as follows:

> ... any *plant grouping* within a single botanical taxon of the lowest-known rank which ... is characterised by at least one single transmissible characteristic distinguishing it from other plant groupings and which is sufficiently homogeneous and stable in its relevant characteristics. ... *Plant cells* as such ... cannot

be considered to fall under the definition of a plant or of a plant variety. (§ 23 of the decision)

After having assessed the PGS patent in the light of this definition, the Board of Appeal came to the conclusion that the plants described in claim 21 were not allowable under Art. 53(b) EPC because the very process of genetic modification had turned these plants into a plant variety (the stably integrated characteristic distinguishing them from other plant groupings being their resistance to herbicides of the class 'glutamine synthetase inhibitors'). Accordingly, the non-allowable claims were deleted from the patent, which considerably limited its scope. In biotechnology industry circles this decision has caused a great deal of concern. Hardly surprising, for it comes down to saying that the very practice of genetic modification of plants and animals creates plant and animal varieties. Since plant and animal varieties are excluded under the EPC, this means that genetically modified plants and animals are excluded from patentability.[42] Of course the consequences to the biotechnology industry would be immense if a decision of this kind becomes established case law. Attempts have been made to have the decision revoked by the Enlarged Board of Appeal, the decision of which is binding on the Board of Appeal. However, the Enlarged Board did not revoke T 356/93, which thus remains the most recent case law of the European Patent Office regarding genetically modified plants.

So far on the concept 'variety'. As to the scope of the exclusion from patentability of plant and animal varieties, there are two possible interpretations:

- Either the exclusion refers *only* to cases where a *specifically designated* variety is claimed.

- Or the exclusion *also* refers to cases where varieties are *covered* by a claim.

If the claims of a patent application are not *directed* to one or more specific varieties, it is still possible that they *cover* varieties. If the claims in question are directed to a taxonomic classification unit higher than 'varieties', e.g. 'species', they do *cover* varieties, for higher taxonomic classifications cover lower ones. To make this point clear, one can draw a parallel with alcoholic beverages: claims to alcoholic beverages (taxonomically higher class) are not directed to beer or whisky (taxonomically lower classes), but they do cover beer and whisky. The question that arises concerning the interpretation of the scope of the exclusion from patentability of varieties is whether rejection of patent claims is only justified if these claims are directed to varieties, or whether the fact that claims cover varieties is a sufficient ground for rejection. The Board of Appeal in the 'Greenpeace case' stated the following, in § 24 of its decision T 356/93:

A product claim which *embraces* within its subject-matter "plant varieties" as just defined (*cf* point 23 supra) is *not patentable* under Article 53(b), first half-sentence ... (our emphasis)

'Embraces' has the same meaning as 'covers'. Consequently, in this decision — which, as we already mentioned, constitutes the most recent case law it is stated that, if claims cover (embrace) varieties, this is a sufficient ground for rejection.

This decision has provoked a lot of commotion in the biotechnology industry, for it considerably limits the allowable scope of claims in the field of plants — and, in our view, there is no reason whatsoever why it should not be interpreted as also applying to the field of animals, for it is a decision concerning the scope of the 53b-exclusion itself. The scale of the resulting commotion can be shown by the fact that the President himself of the European Patent Office referred the following question to the Enlarged Board of Appeal:

Does a claim which relates to plants or animals but wherein *specific plant or animal varieties are not individually claimed* contravene the prohibition on patenting in Article 53(b) EPC if it *embraces* plant or animal varieties? (our emphasis)

In his 'Statement of reasons' for this referral, the President argued in favour of an interpretation of the exclusion from patentability of plant varieties to the effect that it:

... is *only* intended to exclude from patentability a claim which *defines specific plant varieties individually.*[43] (our emphasis)

Thus, the President did not agree with the decision of the Board of Appeal in the 'Greenpeace case' and wanted the Enlarged Board to revoke it. According to Art. 112(1) EPC, the President of the EPO may refer a point of law to the Enlarged Board of Appeal where two Boards of Appeal have given different decisions on that question. In this case, the President argued that the Greenpeace decision, T 356/93, was in contradiction with two earlier decisions, T 49/83 (the first famous plant decision,[44] concerning a patent application by the company Ciba Geigy) on the one hand and T 19/90[45] (Harvard Oncomouse) on the other hand. For the details of the President's arguments about the existence of a contradiction, we would like to refer to the Summary of the Procedure, given by the Enlarged Board of Appeal in its decision (G 0003/95). Here, it is sufficient to say that the Enlarged Board decided that there was no contradiction. Thus, the Greenpeace decision remains the most recent case law.

The interpretation, given in the Greenpeace decision, of the scope of the exclusion from patentability of varieties — viz. that a claim which embraces (covers) varieties is *not* patentable — is in our view the only sound interpretation one can give to this provision. In the European Commission's proposed 'Directive on the legal protection of biotechnological inventions', however, the other interpretation is expressed: the words 'as such', added after 'plant and animal varieties', indicate that Art. 4(2) of this Directive intends to exclude only claims which are directed to specific varieties, i.e. claims in which specific plant or animal varieties are individually claimed. This interpretation is not tenable. To clarify our point, we would like, once more, to draw attention to the analogy with alcoholic beverages. Let us suppose that Art. 1 of the European Patent Convention would lay down that whisky and beer (taxonomically lower classes) are excluded from patentability. In these circumstances, would a claim to alcoholic beverages (taxonomically higher class) contravene the prohibition on patenting in Art.1? Of course in this claim, whisky and beer would not be individually claimed, but they would be embraced by this claim, for classes that rank taxonomically higher cover lower ranking classes. We believe the answer to this question is very simple: a claim to alcoholic beverages would indeed contravene the prohibition in Art.1 (non-patentability of beer and whisky).

Likewise, a claim that covers varieties, although it is not directed to any specific varieties, does contravene the exclusion from patentability of plant and animal varieties. This perfectly logical interpretation was already expressed during the Opposition Proceedings in the 'Greenpeace case'. According to the Minutes of the oral hearing in opposition, Greenpeace's legal representative, Mr Daniel Alexander, asserted that:

> ... it could not be the case that if a claim were drafted directly to a plant variety it would be contrary to Article 53(b) EPC, whereas if its scope were made larger ("dress up the claim") it would *not* contravene Article 53(b) EPC. (Minutes of the oral hearing, p. 7, our emphasis)

A similar view, viz. that claims are being 'dressed up' and given an interpretation which solely aims at contravening their exclusion from patentability, was expressed by Professor Michele Svatos, in her commentary on the proceedings (cf page 303):

> I find the argument that genetically engineered plants and animals constitute new species, or any taxonomic slot other than variety just so long as it's patentable, to be a bit of opportunistic and unconvincing reasoning, both biologically and philosophically.

We do not see the use of an interpretation of the non-patentability of varieties which considers *only* the claiming of individual varieties as a contravention of this provision. Surely, any patent attorney who would write an application which includes claims to excluded subject-matter (individual varieties) would be a stupid patent attorney. Any clever attorney who wants to obtain patent protection of varieties for his client, will formulate his claims in such a way as to make sure they cover varieties but do not define those varieties. If the legislator allows this, if he allows legal obstacles to be by-passed easily, he might as well do away with those obstacles. To interpret an exclusion narrowly (cf for example to lay down that plant cells do not fall under the concept of 'plant varieties') is one thing; to interpret an exclusion so narrowly that it loses all significance and might as well be abolished is quite another. If this narrow interpretation is what the drafters of the Directive had in mind, they should have deleted the non-patentability of varieties in their text. The present formulation of Art. 4(2) of the Directive, containing the words 'plant and animal varieties *as such*' is untenable.

The patentability of microbiological processes and their products (Art. 5)

Art. 5 of the proposed Directive lays down that:

Microbiological processes and products obtained by means of such processes shall be patentable.

This wording is essentially the same as the second half-sentence of Art. 53(b) EPC. However, the scope of this exclusion is entirely dependent on the interpretation of the concept 'microbiological process'. Art. 2(2) of the new proposal provides the following definition:

... any process involving or performed upon or resulting in microbiological material; a process consisting of a succession of steps shall be treated as a microbiological process if at least one essential step of the process is microbiological.

It is hard to compare this definition with the one that is given in the *Guidelines for Examination in the European Patent Office*, for paragraph 3.5, Chapter IV of part C, tells us only what a microbiological processes 'are not only' and what they 'also are':

The term "microbiological process" is to be interpreted as covering not only industrial processes using micro-organisms but also processes for producing new micro-organisms, e.g. by genetic engineering. ... The term micro-organism covers plasmids and viruses also.

This is by no means a satisfactory definition. It does not even clarify whether processes for producing new organisms (i.e. not just micro-organisms but also plants or animals) by genetic engineering are to be considered as microbiological processes.
    It is a lot easier to make a comparison of the definition that is provided in Art. 2(2) of the Directive and the one that is provided by the Board of Appeal in the Greenpeace decision. The concepts of 'microbiological processes' and 'the products thereof' are defined in § 36 of the decision. This definition is based on the following definitions of the terms 'micro-organism' (§ 34) and 'microbiological' (§ 35):

According to the current practice of the EPO, the term "microorganism" includes not only bacteria and yeasts, but also fungi, algae, protozoa and human, animal and plant cells, i.e. all generally unicellular organisms with dimensions beneath the limits of vision which can be propagated and manipulated in a laboratory. Plasmids and viruses are also considered to fall under this definition ...

Accordingly, the term "microbiological" is interpreted as qualifying technical activities in which direct use is made of microorganisms as defined above ... Therefore, *as an example, genetic engineering processes carried out on vegetable cells may be defined as 'microbiological processes' and their products, namely genetically-modified vegetable cells and their cultures, may be defined as 'the products thereof'.* (our emphasis)

... the concept of "microbiological processes" under Article 53(b) EPC, second half-sentence, refers to processes in which microorganisms as defined above ... or their parts, are used to make or to modify products or in which new microorganisms are developed for specific uses. Consequently, the concept of "the products thereof" ... encompasses products which are made or modified by microorganisms as well as new microorganisms as such.

None of the elements of these definitions appears to differ from the first half-sentence of Art. 2(2) of the proposal for Directive: 'any process involving or performed upon or resulting in microbiological material'. The second half-sentence of the Directive's definition, however, does pose a problem. As we mentioned already, it states that:

... a process consisting of a succession of steps shall be treated as a microbiological process if at least one essential step of the process is microbiological.

Of course it is clear that processes which consist of a succession of steps, of which at least one essential step is microbiological, cannot *as a whole* be considered as 'essentially biological processes'. But why should such processes be considered as a whole, in their entirety? In the case of the Oncomouse patent, the applicant claimed the process for genetically modifying the animals, the animals resulting from this process and the progeny of the latter. According to the Examining Division in the Oncomouse case, whereas the genetically engineered animals can be considered as the products of a *non-essentially biological* process,[46] the subsequent generations of onco-animals, however, are obtained through normal sexual reproduction, which is an *essentially biological* process. Hence, the Division stated inter alia the following in its 'Reasons for refusal of the patent':

... the product claims 17 and 18 ... contain two different process steps, namely the non-biological step ... and the breeding step ... in order to extend the claims to generations of animals which were not themselves genetically manipulated. The two steps result in two different products. Animals which were genetically manipulated themselves are products of a non-essentially biological process, whereas further generations are the product of sexual reproduction which is exclusively biological ... *The artificial connection of the two steps aims at circumventing the exclusion provision in Article 53(b) first part EPC.* (our emphasis)

37

We believe that, in cases of genetically modified plants and animals, it is correct to call every generation of plants and animals after the first generation 'the product of an essentially biological process', since these animals are indeed the product of sexual reproduction. However, it is incorrect to state that, as a consequence, these animals are excluded from patentability under the first half-sentence of Art. 53(b), for this provision excludes 'essentially biological processes for the production of plants or animals' and *not* 'the products of essentially biological processes for the production of plants or animals'.

According to the EPO Board of Appeal in the 'Greenpeace case', the first generation, i.e. those plants (and animals — again, there is no reason to limit the relevance of this decision to plants) who were themselves the subject of a genetic engineering process, can be considered as the products of a non-essentially biological proces, because the step of transforming cells or tissue with recombinant DNA is an essential technical step which has a decisive impact on the final result (*cf* criterion laid down in the *Guidelines*). The Board notes that the subsequent steps of the process, the regeneration and replication, make use of a 'natural' machinery, but the first step, the insertion of the relevant DNA sequence into the genome of the organism, is said to be the decisive step. Thus, genetically modified plants are considered as the products of a non-essentially biological process. In this respect, they are not excluded from patentability. However, as we explained under the previous heading, this very same Board of Appeal also ruled that the practice of genetic modification turns the resulting plants and animals into plant varieties and animal varieties. Plant and animal varieties are excluded from patentability, *unless* they are 'the products of a microbiological process'. The Board, however, came to the conclusion that the process of genetic engineering should not be regarded as a 'microbiological process'.

The process of genetic engineering was described by the Board as a multi-step process, the initial step of which is microbiological (the transformation of cells or tissue with recombinant DNA). This first step is said to have a decisive impact on the final result, because it is through this step that the organism acquires its characterising feature that is passed on to future generations. Nevertheless the subsequent steps — regeneration and replication — are said to have an important added value and to contribute to the final result as well. Hence, genetically engineered organisms should not be regarded as merely being the result of the initial microbiological step, but also of the steps that come afterwards. In paragraph 37 of its decision T 356/93, the Board says the following about multi-step processes such as genetic engineering:

As modern biotechnology often uses or develops *multi-step processes* for producing plants which include at least one microbiological process step (e.g. the transformation of cells with recombinant DNA), *it has to be decided whether such processes as a whole can be considered to represent "microbiological processes"* ... and whether, owing to this, the products of such processes (e.g. plants) may be regarded as being 'the products thereof' ... (our emphasis)

The Board immediately went on to answer this question:

... microbiological processes ... and technical processes comprising a succession of steps, wherein at least one essential step is of a microbiological nature, may not be considered to be of the same kind or similar ... Consequently, the concept of "microbiological processes" ... may not be extended to include all the steps of such technical processes. (§ 38)

... nor can the resulting final products ... be defined as 'products of microbiological processes' ... (§ 39)

Thus, according to the most recent case law in the European Patent Office, genetically engineered plants and animals are excluded from patentability for:

– they constitute varieties, which are excluded

– an exclusion to this exclusion would apply if genetically engineered plants and animals could be considered as 'the products of a microbiological process', but this is not the case.

The second half-sentence of Art. 2(2) of the proposed 'Directive on the legal protection of biotechnological inventions' tells a completely different story, for it declares that the multi-step processes of the kind mentioned by the Board *shall* be treated as 'microbiological processes' (and, consequently, as patentable). For a non-biotechnologist, it is very hard to assess which of these interpretations has the most merits, but our point in this section has been to show that, according to the proposed Directive, genetically engineered plants and animals are patentable, whereas according to the most recent case law of the European Patent Office they are not.

The non-patentability of essentially biological processes for the production of plants or animals (Art. 6)

Art. 6 of the new proposal for Directive reads as follows:

Essentially biological processes for the production of plants or animals shall not be patentable.

This provision is entirely in accordance with the first half-sentence of Art. 53(b) EPC. The explanation that is provided in the Considerantia is that:

... for the purposes of determining whether or not it is possible to patent essentially biological processes for obtaining plants or animals, human intervention and the effects of that intervention on the result obtained must be taken into account ...

39

This is exactly the same criterion as the one mentioned in the *Guidelines for Examination in the European Patent Office* (*cf* supra, 'Exceptions to the general principles of patentability'). The same interpretation is given in § 28 of T 356/93, the Greenpeace decision, which states that the following should *not* be considered as an 'essentially biological process':

> ... a process for the production of plants comprising at least one essential technical step, which cannot be carried out without human intervention and which has a decisive impact on the final result.[47]

Whereas the *Guidelines* and T 356/93 remain vague on the precise meaning of this criterion (starting from what point can one call the human intervention 'significant', 'decisive'?), the drafters of the new proposal for Directive have shed a lot more light on the way they want the concept of 'esentially biological processes for the production of plants or animals' to be interpreted [*cf* Art. 2(3)]:

> ... any process which, taken as a whole, exists in nature or is not more than a natural plant-breeding or animal-breeding process.

The wording 'taken as a whole' was probably included to make sure that processes consisting of a succession of steps (*cf* supra) — e.g. consisting of one microbiological step followed by several essentially biological steps — are not considered as falling under the heading 'essentially biological processes', which would make them unpatentable.

Inventions the exploitation of which would be contrary to public policy or morality (Art. 9)

The general exclusion on grounds of public policy or morality   The first paragraph of Art. 9 of the new proposal for Directive lays down that:

> Inventions shall be considered unpatentable where exploitation would be contrary to public policy or morality; however, exploitation shall not be deemed to be so contrary merely because it is prohibited by law or regulation.

This provision says essentially the same as the so-called 'morality article' contained in the European Patent Convention, Art. 53(a) (*cf* supra, 'Exceptions to the general principles of patentability').

Paragraphs 19-23 of the Considerantia provide some explanation concerning the general exclusion from patentability of inventions the exploitation of which would be contrary to public policy or morality. According to §§ 20 and 23:

> Whereas such a reference to public policy and morality should be included in the operative part of this Directive in order to bring out the fact that some

applications of biotechnological inventions, by virtue of some of their consequences or effects, are capable of offending against them;

Whereas such moral considerations should be given greater weight in appraising
the patentability of biotechnological inventions, both on account of the subject-
matter of this branch of science, namely living matter, and because of the often
far-reaching implications of the inventions to be examined; whereas these
considerations do not, however, change the nature of patent law as a primarily
technical body of law and are no substitute for the other legal checks which
biotechnological inventions are required to undergo from the start of their
development or at the marketing stage, particularly with regard to safety ...

We fully agree with these statements. All kinds of patent applications need to be
assessed under the heading of morality, but this assessment should be given greater
attention in the case of inventions in the field of biotechnology. This assessment
should not be regarded as a substitute for scientifical, legal and ethical checks in the
stages of research and introduction to the market, but as an *additional* assessment.

Of course, the main question is how this assessment should be carried out. In
Paragraph 21 of the Considerantia, the view is expressed that:

... it must be determined whether applications offend against public policy and
morality *in each specific case*, by means of an appraisal of the values involved,
whereby the benefit to be derived from the invention, on the one hand, is weighed
and evaluated against any risks associated therewith, and any objections based
on fundamental principles of law, on the other hand ... (our emphasis)

This so-called 'case-by-case' approach is also taken by the European Patent Office.
We believe that indeed the only possible option for patent examiners is to assess
each case on its merits, to look at the particular facts of each case. The reference to
a weighing up of benefits and risks is probably inspired by the decision of the Board
of Appeal in the Oncomouse case (T 19/90). Under the next heading, we will come
back to the 'balancing exercise'. We agree on the necessity of weighing up the benefits
and risks of inventions in order to assess their patentability in the light of public
policy and morality. We believe, however, that one of the most serious problems
involved in performing such a weighing up in the stage of a patent application is
that, in many cases, at this stage of development of inventions, only little information
is available about the benefits and disadvantages that would be produced by the
exploitation of the invention in question. Inventors try to apply for patents as soon
as possible; often the invention is still in its initial stages. In the case of the Opposition
proceedings in the Oncomouse case, this was not a problem for these proceedings
took place ten years after the animals had been produced, so the fact that no important
contribution to the development of cancer treatments had resulted from the invention
could be shown (the Examining Division nevertheless accepted the applicant's
assertion that the invention was beneficial to mankind when it granted the patent

four years ago, but the Opposition Division — which has not reached a decision yet — would have a much harder time justifying this). In most cases, however, the amount of information available to the 'evaluators' of a patent is not that impressing. We believe that this problem is manifested most clearly when it comes to the assessment of the 'risks to the environment' associated with an invention. Up to now, several patents have been objected to because these risks were said to be too high. People who oppose patents on this ground frequently use the argument that harmful effects on the environment often only become apparent in the long term. The argument that has been used repeatedly by EPO-divisions — that the evidence of the environmental risks submitted by the opponents is not convincing — is therefore no surprise. However, this means that the decisions that are taken by the EPO Opposition Divisions are always in the interest of the Patentee (proprietor), for the onus of proof in Opposition proceedings lies with the opponent. The Opposition Division in the Relaxin-case,[48] for example, expressed the following view on the possibility to assess the risks associated with genetic engineering:

The Opponent's inability to prove the extent of the risks ... is hardly surprising since experts all over the world have for at least the past fifteen years been intensively addressing themselves to the question of possible risks associated with genetic engineering and in particular with the release of genetically engineered organisms into the wild. Despite all this effort, there is still no agreement concerning the extent of these risks and the Opponent has indeed conceded that the risks are impossible to determine with certainty. ... it is difficult to see how examiners could ever be in a position to take a stand on such questions ... If examiners were to attempt to do so, the results could only be arbitrary and superficial and thus unfair to applicants. (§ 3.13. of the 'Reasons for the Decision')

The second paragraph of Art. 9   The second paragraph of Art. 9 reads as follows:

On the basis of paragraph 1, the following shall be considered unpatentable:
(a) methods of human treatment involving germ line gene therapy
(b) processes for modifying the genetic identity of animals which are likely to cause them suffering or physical handicaps without any substantial benefit to man or animal, and also animals resulting from such processes, whenever the suffering or physical handicaps inflicted on the animals concerned are disproportionate to the objective pursued.

The purpose of giving these examples is explained in § 22 of the Considerantia: this 'illustrative list' should serve as a general guide to national courts and patent offices, on how to interpret the reference to public policy and morality in Art. 9(1).

The provision of guidelines for interpreting the morality clause should be welcomed. As we mentioned earlier (*cf* 'Exceptions to the general principles of patentability'), the *Guidelines for Examination in the European Patent Office* certainly do not perform well in this respect. Apart from mentioning the example of the letter bomb and vaguely referring to the 'public in general',[49] they do not offer much clarification. The guidance that is provided in Art. 9(2) of the proposed Directive, however, poses problems as well. We will briefly comment on the 'proportionality principle' concerning inventions related to the genetic modification of animals (paragraph b). Next, we will discuss the exclusion of germ line gene therapy on humans (paragraph a), which will be treated in more detail.

The 'proportionality principle' concerning inventions relating to the genetic modification of animals   We welcome the inclusion of Art. 9(2)(b) in the proposal for Directive. However, it is still very vague. It is not clear what should be understood by the concepts of 'substantial benefit to man or animal' and the being 'dispropor-tionate to the objective pursued' of the suffering or physical handicaps inflicted on the animals.

The same problems arise in respect of the instructions that were given by the Board of Appeal in the 'Oncomouse' case. In Decision T 19/90, the view was expressed that the evaluation of the Harvard patent under Art. 53(a):

> ... would seem to depend mainly on a careful weighing up of the suffering of animals and possible risks to the environment on the one hand, and the invention's usefulness to mankind on the other ...

This direction, given by the Board of Appeal to the Examining Division, has become known as the 'balancing exercise'. Such a balancing exercise is a typical example of consequentialist moral reasoning, for it is said that, in order to determine the morality or immorality of the Harvard invention, the consequences that would be produced by the exploitation of the invention need to be evaluated. Three kinds of criteria have to be taken into account to evaluate these consequences:

- the amount of happiness of mankind that would be produced by the exploitation of the invention

- the amount of pain of animals that would be produced by the exploitation of the invention

- the amount of harm to the environment that would be produced by the exploitation of the invention

In our view, the instructions that were given by the Board of Appeal are by no means sufficient to enable the performance of the balancing exercise. If only one criterion has to be taken into account, for instance the general happiness of human beings (the happiness of all members of society, the 'well-being' of society), the most important question to be asked is how to 'measure' this happiness. The answering of this question is far from being easy (e.g. does a satisfying communication with another person make somebody happier than a delicious meal?). In the case of the balancing exercise devised by the Board of Appeal, however, not one but three criteria have to be taken into account. Consequently, first of all 'measurement-questions' arise with respect to any of the three criteria: how to measure the general happiness of mankind, how to measure the suffering of animals, how to measure harm to the environment? In addition to these questions, the issue rises of how to determine the relative weight of each of the three criteria. Most people would probably feel that, if the avoiding of human suffering can only be achieved through the causing of animal suffering, the 'weight' of the interests of human beings is greater than the 'weight' of the interests of animals. Even this is not all that easy to decide, for it is also important to know what *kind* of interests are at stake (how significant are they?). Determining the weight of the interests of the environment, in relation to the weight of the interests of humankind, is not easy either. If asked to do this, many people would probably be inclined to take into account the interests of the environment only insofar as the good 'functioning' of the the environment is necessary for the ability of man to live in that environment. In this case, the outweighing of the interests of the environment by the interests of mankind would be presupposed.

What if it would become apparent that the exploitation of a particular invention would produce a lot of animal suffering as well as a lot of harm to the environment, while at the same time producing a lot of benefit to mankind? How easy or how difficult would it be for the 'balancers', which of course belong to one of the parties involved, to decide that, in this case, the exploitation of the invention would be contrary to morality?

The Board of Appeal has not given any instructions as to which 'weighing factor' should be ascribed to each of the three criteria. For all these reasons, the Board's instructions do not provide sufficient information to allow the performance of the balancing exercise.

Although its shortcomings absolutely need to be remedied, we are convinced that a balancing exercise of this kind is necessary in the context of the evaluation of inventions under the heading of public policy and morality. Moreover, we do not agree with the view that such a balancing should *only* be performed in the case of patents in the field of animals. In our opinion, the application of the balancing exercise should be widened to the 'morality-evaluation' of all kinds of biotechnological inventions (as stated in § 21 of the Considerantia of the new proposal for Directive), and maybe even to all kinds of inventions *tout court*. Surely, biotechnological inventions are not the only inventions that could produce harm to society or to the environment.

However, consequentialist reasoning (the balancing exercise) should not be the only 'tool' to determine the (im)morality of an invention. We feel that, in view of the shortcomings of the former, deontological reasoning should play a part as well. Everybody can conceive of things that 'should not be done', even if doing them would serve our interests. The genetically engineering of animals with the purpose of making them develop a serious disease may well be one of those things. Personally, we believe that the 'case-by-case' approach should not be pushed to its extremes. Deontological reasoning could prove very valuable for the drawing up of certain general guidelines or principles. One such principle could be, for example, that the genetic modification of animals to make them develop serious diseases should always be considered immoral, unless *extremely convincing evidence* is provided of the utility of such an invention.

The non-patentability of methods of human treatment involving germ line gene therapy   As to the exclusion from patentability of methods of human treatment involving germ line gene therapy, it should first of all be observed that 'methods for treatment of the human or animal body by surgery or therapy and diagnostic methods practised on the human or animal body' are excluded from patentability under Art. 52(4) EPC, because they are not regarded as inventions which are susceptible of industrial application (*cf* supra). 'Methods of human treatment involving germ line gene therapy' clearly fall under this exclusion. It is not clear to us why Art. 52(4) EPC has not been taken over into the proposal for Directive.

In any case, in the proposal for Directive, the exclusion from patentability of germ line gene therapy is clearly grounded on reasons of 'public policy and morality'. We are aware of the fact that the practice of germ line gene therapy has a serious connotation of 'immorality' in the opinion of a great majority of the public. This can be perfectly understood in view of the fact that, when hearing 'germ line gene therapy', most people immediately think of eugenic interventions of the kind conceived by the Nazi's. However, this is but one aspect of germ line genetic engineering, the so-called 'enhancement' type of germ line genetic engineering. The second type is the 'correction' or 'repair' type of germ line genetic engineering, and many people are not even aware of this distinction.

There are four different types of gene therapy: the 'repair' type of somatic gene therapy, the 'enhancement' type of somatic gene therapy, the 'repair' type of germ line gene therapy and the 'enhancement' type of germ line gene therapy. Thus, firstly, a distinction needs to be made between the *type of cells* being the target of the gene transfer: somatic cells or germ cells. In the case of interventions in somatic cells (body cells), the purpose is to modify the genome of the individual who is the 'subject' of the intervention, whereas in the case of interventions into the germ cells (reproductive cells), the modification is passed on to the progeny of that individual. The second distinction which needs to be made, between repair and enhancement, has to do with the *purpose* of the gene transfer, either to 'repair' (cure or prevent) a genetically determined disease, or to 'enhance' (improve) certain characteristics which are genetically determined but have nothing to do with disease.

The 'repair' type of somatic gene therapy comes down to the insertion of genetic material into a human being with the sole purpose of correcting a gene defect, related to disease, in that patient. This is generally considered to be justified from an ethical point of view, provided that the following three conditions are fulfilled (which are comparable to the conditions that should be satisfied for the use a new drug or a therapeutic procedure): the new gene can be put into the correct target cells and will remain there long enough to be effective ('delivery'); the new gene will be expressed in the cells at an appropriate level ('expression') and the new gene will not harm the cell or, by extension, the animal ('safety'). Although the relevant techniques are still in their early stages, many people are very optimistic about the possibilities of somatic gene therapy to alleviate and cure various serious diseases. As yet, very little is known about the risks of somatic gene therapy. Hence, it is generally thought that the first candidates for gene therapy should be severe diseases, to find out whether the results are satisfactory.

As to the 'repair' type of germ line gene therapy, a question that is often asked is whether a treatment which, if it goes wrong, would pass on the mistakes and problems to future generations, is justified from an ethical point of view. Our knowledge of germ line gene therapy is extremely poor. According to the *Group of Advisers on Ethical Implications of Biotechnology* of the European Commission:

> ... there is no sufficient experience and monitoring of somatic gene therapy ... safety and effectiveness of its use ... much less, of its extension to the self-perpetuating germ-line;
> ... there are no sufficient animal studies on germ-line gene therapy applicable to humans;
> ... there is not yet any clinically useful and clearly proposed protocol for germ-line gene therapy.[50]

However, as to the benefits of germ line gene therapy, the Group quotes from a report from the XXIVth Round Table from the Council for International Organisations of Medical Sciences (Inuyama, 1990): as a consequence of the fact that no other technique exists to produce a genetic change in all the cells of an individual, germ line gene therapy might well be 'the only means of treating certain conditions, so continued discussion of both its technical and its ethical aspects is therefore essential'. We agree with this observation and we believe that too many people *a priori* reject any possible applications of germ line gene therapy. The 'prophylactic efficiency' of this therapy is also mentioned as a benefit. If germ line gene therapy would be applied, it would no longer be necessary to carry out somatic gene therapy in multiple subjects of successive generations. Concerning the risks, the Group rightly observes that, due to the present lack of knowledge, the risks of germ line gene therapy cannot even be evaluated, whereas any detrimental results would be passed on to future generations. We agree with the Group that this is a sufficient reason to prohibit any attempt of germ line gene therapy in humans. On this matter, a broad consensus exists. However, the Group makes the following extremely important observation:

> The differences of opinion begin when we ask whether this [the present lack of knowledge] is the *only* reason to oppose germ-line therapy. For some, the present uncertainty on risks and benefits is the only reason for prohibition or non authorization, while, for others ... *additional basic and non circumstantial objections* exist, which lead to the *condemnation of germ-line therapy not only in the present situation but per se*, even if one day sufficient safety guarantees were at hand.[51]

In our opinion, the presence of 'safety guarantees' — a situation which might occur in a few years — may be a sufficient reason to withdraw the prohibition of 'repair'-applications of germ line gene therapy (the applications which aim at curing or alleviating diseases), but it is certainly not a sufficient reason to remove the ban on 'enhancement'-germ line gene therapy (the applications which are aimed at modifying genes which have nothing to do with disease). However, it should be stressed that an 'enhancement' type of *somatic* gene therapy exists as well, and this is often not mentioned in the discussion. Neither did the drafters of the new proposal for Directive exclude it from patentability on grounds of morality, whereas they do exclude the 'repair' type of germ line gene therapy. The exclusion laid down in Art. 9(2)(a) of the proposed Directive is a general exclusion, which makes no distinction between 'repair' (correction) and 'enhancement' (improvement) applications.

The European Commission's Group of Advisers' final conclusion concerning the ethical allowability of germ line gene therapy is that there are no arguments which prove *a priori* that *any* kind of germ-line gene therapy would necessarily affect human dignity and justice. At present, the poor state of our knowledge necessitates a prohibition. By the time we know enough to assess the benefits and risks of this technique, an ethical re-evaluation should take place. We fully subscribe to this conclusion.

Objections to germ line gene therapy can, roughly speaking, be divided into two categories: safety (technical, state of the art-) objections and ethical objections. If the former would no longer hold because of the perfection of the relevant techniques, this would not mean that all grounds for the latter would be lost. Against both the 'repair' and the 'enhancement' applications of germ line gene therapy, safety- as well as ethical objections have been formulated. The safety-objections are likely to disappear in the (relatively) near future. Predictions concerning the ethical objections, however, are far more difficult.

One could be tempted to expect that, when these techniques have become safe and constitute the only or the best possible alternative for some patients, the current ethical objections to 'repair' applications of germ line gene therapy will no longer be raised. Personally, we believe that, in these circumstances, 'repair' applications of germ line gene therapy would be ethically acceptable. To expect the majority of the public to share this conviction may, however, be too optimistic, for three reasons: *Firstly*, some people are against *any* intervention in the human genome, regardless of its purpose (in other words, they feel that even if such an intervention were necessary to cure a life-threatening disease and the results of this intervention would not be passed on to the progeny of the patient, it should not be done). Whereas some people only hold this view as far as the human genome is concerned, others feel the same way about animals and/or plants and/or living organisms in general. Very often, advocates of these positions consider interfering with 'the building blocks of life' as a total lack of respect for 'creation' or 'nature'. We believe that, in most cases, this kind of views are held strongly. One could call them 'central attitudes', attitudes that are not very amenable to change, be it by changes in scientific and technological insights or by other developments. *Secondly*, according to some people only those

47

applications of gene therapy in which the introduced modifications are not passed on to future generations, i.e. techniques of *somatic* gene therapy, are ethically acceptable. Among these people, some approve only of 'repair' applications, whereas the others have no ethical objections to 'enhancement' somatic gene therapy either. *Germ line* gene therapy, however, they consider to be ethically unacceptable in any case, even if it would be used to 'repair'. An argument that is often heard in this respect is that, since one can never obtain the informed consent of future generations, one should never introduce deliberate modifications into their genome. *Thirdly*, some people have no objections of principle to 'repair' applications of germ line gene therapy, but fear that allowing these techniques to be carried out would open the door for 'enhancement' applications. They believe that the one would lead to the other and since they are strongly opposed to 'enhancement' modifications of the germ line, their conclusion is that the 'repairing' should be prohibited as well (as a kind of 'preventive strategy'). To put it simply, these people feel that it is impossible to 'draw a line'. This consequentialist argument is sometimes called the 'slippery slope'-argument. In this context, the European Commission's Group of Advisers on the Ethical Implications of Biotechnology made the following remark:

> ... [this] argument has been used in the past ... against somatic gene therapy and it was neither successful nor correct. *Somatic gene therapy* is being carried on with societal approval and considerable benefits, while no attempts are known of *somatic enhancement engineering*.[52]

The ethical objections to the 'enhancement' type of modifications of the germ line
The ethical objections to 'enhancement' applications of germ line gene therapy are even far less likely to disappear as the safety of the relevant techniques becomes established, than those brought forward against the 'repair' type of applications. For example, David Suzuki (an internationally renowned geneticist) and Peter Knudtson (a biologist) make the following remarks:

> ... the ethical dimensions of gene therapy shift the instant the target changes from somatic cells, with fleeting lifetimes, to germ, or reproductive, cells ... The fact that early proposals for future human gene therapy experiments have focused exclusively on somatic cells suggests that most scientists recognize the profound biological difference between somatic and germ-line cells in the human body. But, in part, this early reluctance to tinker with the genes of human germ cells genes can also be attributed to the imprecision of current gene therapy strategies. As the medical art of gene therapy becomes more refined, we must be on guard against those who may be inclined to apply these skills to the germ-line genetic engineering of our species. ... proposals for germ-line gene therapy are likely to be cloaked in the noblest of medical motives — the dream of saving lives in future generations and ultimately even eradicating hereditary disease ...[53]

The history of eugenics suggests that once a human characteristic — such as a particular skin colour or ... poor performance on IQ tests — has been labelled a genetic "defect", we can expect voices in society to eventually call for the systematic elimination of those traits in the name of genetic hygiene.[54]

According to some people, 'repair' applications of germ line gene therapy are not justified from an ethical point of view because they contravene the principle of informed consent. We find this argument most unconvincing. In the case of germ line gene therapy there simply cannot be any informed consent. At most, we can *presuppose* that, if they were to decide whether or not to 'remove' the diseases in question, our descendants would give their informed consent to the 'removal-option'. In our opinion, this presupposition is perfectly legitimate. Therefore, the fact that one acts without informed consent cannot constitute a sufficient argument for the unacceptability of 'repair'-germ line gene therapy.

Regarding 'enhancement'-germ line gene therapy, we have to ask ourselves the same questions. Would it be in the interest of future generations? Is it justified to presuppose that they would give their informed consent? These questions are far more difficult to answer than in the case of 'repair'-germ line gene therapy. Although we are aware of the fact that attitudes towards health and disease are to an important extent determined by cultural specificities, we believe that being free of serious diseases can be called 'generally desired' (with 'generally' also including future generations). As far as, e.g., esthetical and personality traits are concerned, it is impossible to indicate 'that which is generally desired'. To determine 'that which *should* be generally desired' is not so much that impossible — as we know, it has been done before several times in history and, definitely not always without less 'far-ranging' intentions, it is done frequently nowadays. We are convinced, however, that it is reprehensible, improper, wrong to think that one is justified in presupposing, e.g., that the people of future generations want to be beautiful, intelligent, good-humoured, mild-mannered, heterosexual, etc... This, we feel to be a sufficient argument for the unacceptability of 'enhancement'-germ line gene therapy.

It is true that 'slippery slopes' (*cf* supra) are far more likely to occur *within* the field of somatic gene therapy and *within* the field of germ line gene therapy, than *between* these two sorts of treatment. From a legislator's point of view, it is much easier:

– to *allow somatic* gene therapy and *prohibit germ line* gene therapy

than to:

– *allow 'repair'*-applications of both somatic- and germ line- gene therapy and *prohibit 'enhancement'*-applications of both somatic- and germ line- gene therapy.

Although we are aware of the fact that the 'dividing line' between repairing and enhancing is not always clear and it would take a considerable amount of reflection and discussion to arrive at workable guidelines in this context, we believe that, nevertheless, the criterion repair versus enhancement is a lot more relevant from an ethical point of view than the criterion somatic cells versus germ cells.

Notes

1   These developments were made possible by the availability of scientific data about the structure of DNA, which was revealed by Watson and Crick in 1953.

2   In this context, we use the term 'biotechnology' to refer to *modern* biotechnology.

3   Also referred to as 'manipulation', but we prefer the more neutral term 'modification'.

4   Source unknown.

5   *Cf* the Council Directive of 23 April 1990 (90/219/EEC) on the contained use of genetically modified organisms, 'contained use' meaning 'any operation in which micro-organisms are genetically modified or in which such genetically modified micro-organisms are cultured, stored, used, transported, destroyed or disposed of and for which physical barriers, or a combination of physical barriers together with chemical and/or biological barriers, are used to limit their contact with the general population and the environment'.

     Cf also the Council Directive of 23 April 1990 (90/220/EEC) on the deliberate release into the environment of genetically modified organisms, 'deliberate release' meaning 'any intentional introduction into the environment of a GMO or a combination of GMOs without provisions for containment such as physical barriers or a combination of physical barriers together with chemical and/or biological barriers used to limit their contact with the general population and the environment'.

6   Sometimes the risk of 'transboundary effects' is mentioned. These effects, it is asserted, may occur if 'unsafe' agricultural products — products that were obtained by genetic engineering in countries where little or no governmental regulation of these practices exists — are imported in countries that do have regulations. The term 'transboundary effects' is also used to refer to the spreading of pollen from one country to another. Pollen that is carried along with the wind does not halt for national borders. This way, pollen of maize that has been genetically modified in Great Britain can 'fly' to Belgium or Germany and pollinate 'Belgian' or 'German' maize.

7   Holland, Allan (1990), 'The biotic community: A philosophical critique of genetic engineering.', in Wheele, P. and McNally, R. (eds), *The Biorevolution. Cornucopia or Pandora's Box?*, Pluto Press: London, pp. 166-74, our emphasis.

8   Observations of Patentees in response to notice of Opposition, p. 5, our emphasis.

9   Until recently, the patent term in the US was 17 years from the grant of the patent, but the TRIPs Agreement (*cf* Chapter 32 'Conclusions') has imposed a 20 years' term from the filing date. In Europe, this term already applied.

10 European Patent No. 0 242 236, entitled 'Plant cells resistant to glutamine synthetase inhibitors, made by genetic engineering'. The grant of this patent was published on 10 October 1990.

11 These are addressed to the patent examiners working at the European Patent Office. The *Guidelines* have no force of law. They are regularly adjusted to keep up with the evolving case law.

12 Personally, we do not agree with those who say that germ line gene therapy should *a priori* be considered as immoral. We will come back to this in the context of our discussion of Art. 9 of the 'Directive on the legal protection of biotechnological inventions'.

13 According to the *Guidelines* (C, IV, 9.6), 'The person skilled in the art should be presumed to be an ordinary practitioner aware of what was common general knowledge in the art at the relevant date. He should also be presumed to have had access to everything in the "state of the art" ... and to have had at his disposal the normal means and capacity for routine work and experimentation ...'.

14 In this context, 'in nature' also means, for instance, 'in a plant', 'in an animal' or 'in a human being'.

15 We do not see how a description of the process that was used to obtain (isolate) a substance can be considered as a 'proper characterisation' of the substance itself. In our view, a description of the structure of the substance needs to be provided in order to have a 'proper characterisation'.

16 A guideline that is generally accepted within the European Patent Office is that these exceptions should be construed narrowly.

17 Part C, Chapter IV, 3.1.

18 Resolution B3-0199, 0220 and 0249/93, adopted on 11 February 1993; published in the Official Journal of the European Communities on 15 March 1993, No C 72/127.

19 The so-called 'Oncomouse patent', which covers in fact all non-human mammalian onco-animals, *cf* infra.

20 *Cf* the discussion of Art. 53(b) EPC.

21 The International Convention for the Protection of New Varieties of Plants (1961), which led to the development of national laws on plant breeders' rights.

22 For plants, identical designations are used.

23 Communication from the primary examiner of the Examining Division, 20 April 1988.

24 The Australian Patents Act of 1990 does lay down that 'Human beings, and the biological processes for their generation, are not patentable inventions.'

25 See also the commentary by Dr Harriet M. Strimpel.

26 The original text of the 'Transgenic Animal Patent Reform Act' offered two types of farmer exemptions. First, a broad exemption which was aimed at smaller farms, so-called 'family-farms', allowing these farmers to reproduce through breeding, use, or sell a patented animal. Secondly, a narrower exemption, applying to larger farms, which did not include the subsequent sale of a patented animal for further reproduction. Representative Kastenmeier's original bill also stated that engaging in research or experimentation on a patented animal is not a patent infringement if there is no commercial purpose. Finally, the bill called for an amendment of Section 101 of Title 35 of the US Code, to stipulate that human beings are not patentable subject matter. Two amendments were made to the original bill. First, the exemption for researchers was left out. In the second amendment, the exemption for farmers was reformulated: language was deleted that limited the exemption to farmers earning less than $500,000 or to farmers in a so-called 'single-family enterprise'.

27 Rhein, R. (1996), 'New Bioethics Panel Inherits Backlog of Hot-Button Issues', *The Journal of NIH Research*, Vol. 3, September, pp. 36-7.

28 Karet, Bryan (1993), 'Moral dilemmas in the history of patent legislation', in Cookson, Nowak and Thierbach (eds), *Eposium® 1992. Genetic Engineering - The New Challenge. Conference Proceedings and Essay Competition*, München: EPOscript [A series of publications from the European Patent Office , Vol. 1, p. 322.

29 By the time this book will be published, the result of that vote may be known.

30 Inventions the exploitation of which would be contrary to 'ordre public' or morality; plant varieties; animal varieties; essentially biological processes for the production of plants or animals.

31 The 'Convention on the unification of certain points of substantive law on patents for invention'.

32 The Board of Appeal decided that the Examining Division had to reconsider the patent application. The result of this reconsideration was that the patent was granted (in 1992). Seventeen parties filed opposition to the grant of this patent. The Opposition proceedings in this case are still on-going. After a decision is reached by the Opposition Division, there will probably be another appeal procedure. So the Oncomouse case, which has been going on since June 1985, is not coming to an end yet.

33 *Cf* Decision T 19/90, our emphasis.

34 In respect of the discussion about the 'sufficiency of disclosure' requirement, we would also like to refer the reader to the commentary by Dr Harriet M. Strimpel, who makes some very interesting comparative remarks concerning the policies of the American and the European Patent Office.

35 Of course something like cross-licensing exists, but in this case the inventor is dependent on the proprietor of the more comprehensive patent.

36 Some people assert that the intention of the drafters of the EPC, as far as animals are concerned, was to exclude not only animal varieties but animals in general. This was even said by the Examining Division in the Oncomouse case, on the

basis of the fact that different words are used to indicate the excluded subject-matter, which is not the case for plants:

'... if Article 53(b) EPC is considered in the three official languages it is apparent that with regard to plants the legislator used identical designations in all three languages whereas this is not the case with respect to animals. The German version "Tierarten" is definitely different from the English "animal varieties" and the French "races animales". This is a further indication that the legislator's intention was not the exclusion of some particular group of animals but rather the exclusion of animals in general'.

Harvard's reply was that, if the drafters of the EPC would have wanted to draw a distinction between 'plant varieties' and 'animal varieties', they would have used a different term. We tend to agree with this argument.

37 The so-called 'PGS/BASTA' case.

38 Art. 2, the 'definitions-article' of the new proposal, only provides definitions of 'biological material', 'microbiological process' and 'essentially biological process for the production of plants or animals'.

39 Dated 16 April 1991.

40 This definition is, in part, based on the definition of 'plant varieties' given in decision T 49/83, the first famous decision in the field of plants.

41 The Oncomouse decision is of course the most famous one. In Decision T 19/90, the only thing that was said about the meaning of the terms 'animal variety', 'race animale' and 'Tierart' is that they do not refer to animals as such.

42 Unless they could be considered as 'the products of a microbiological process' (the exclusion to the exclusion), but the Board ruled that this is not the case (cf infra).

43 Opinion of the Enlarged Board of Appeal (given on 27 November 1995), Case number G 0003/95, Summary of the Procedure, p. 7.

44 This decision was published in the Official Journal of the EPO, 1984, 112.

45 Cf Official Journal of the EPO, 1990, 476.

46 It should be noted that the Examining Division does not say 'microbiological' process.

47 This definition refers to the production of plants (animals are not mentioned), but there is no indication whatsoever that its relevance is limited to plants.

48 European Patent Application 83307533.4, entitled 'Molecular cloning and characterisation of a further gene sequence coding for human relaxin', granted on 10 April 1991 to the Howard Florey Institute of Experimental Physiology and Medicine, c/o the University of Melbourne.

49 Cf supra: according to the Guidelines, 'This provision is likely to be invoked only in rare and extreme cases. A fair test to apply is to consider whether it is probable that the public in general would regard the invention as so abhorrent that the grant of patent rights would be inconceivable.'

50 European Commission, Secretariat-General SG-C-1 (Secretariat of the Group of Advisers on the Ethical Implications of Biotechnology) (1994), The ethical

aspects of gene therapy. Press Dossier relative to the Opinion from the Group of Advisers on Ethical Implications of Biotechnology, 13 December.

51 Ibid., p. 26, our emphasis.

52 European Commission, Secretariat-General, SG-C-1 (Secretariat of the Group of Advisers on the Ethical Implications of Biotechnology), op. cit., p. 27, our emphasis.

53 Suzuki, David and Knudtson, Peter (1990), *Genethics. The ethics of engineering life*, Stoddart: Toronto, pp. 183-84, our emphasis.

54 Ibid., p. 188.

Part One
BIOTECHNOLOGY: SCIENTIFIC ASPECTS

2 Plant biotechnology: historical perspective, recent developments and future possibilities

Marc Van Montagu

First of all, allow me to thank the organisers and congratulate them for the initiative to have this workshop. I really think it is extremely important that this dialogue is held and that understanding finally wins. If there are new scientific developments which the general public does not understand, does not accept, we really have a problem in our society. In history there are very many examples of this. I do not want to dramatise and talk about the period when the position of the planets in the solar system was a point of discussion. We have seen how the attitudes have evolved. You might think I am exaggerating, but I am afraid that fifty years from now people will look back at our period and feel that we lost a tremendous amount of time in scientific and technological development through the current delays due to the incomprehension of biotechnology. The very logical attitude of society is not to progress immediately because we do need to have this type of consensus.

Why do I consider this issue so important? Because of the major problem of the overpopulation of our planet. Biologists in the fifties and the early sixties have regularly pointed out that this overpopulation is causing an enormous destruction of the environment. Fortunately, further in the sixties and in the last 25 years, a major movement, called the 'green movement', has taken over and has pointed out all the dangers that are accumulating. It is good to make an inventory of the problems, but it is extremely important to see how we can solve these problems. So I will try to show what is considered important from the point of view of plant biotechnology and which contributions can be made already or are in the making. This way you will be better informed to evaluate this progress.

My entire talk, and I think many of the talks here, should be understood against the background of the population curve. For many, many hundreds and hundreds of years, the population was below a billion. Suddenly, in the beginning of our century, it took off. And it took off not with an exponential curve, but with a hyperbolic curve, which means that at a finite time, you reach infinity. Of course biologists know that you will not reach infinity but the problem is when irreversibility occurs. It is impressive to see how fast this increase has been in the last fifty years. All our attitudes, the rules of behaviour we have developed, the ways in which religious leaders, philosophers, politicians, etc. think about this problem have been generated by many hundreds of years of reflection about the nature of society and human beings and about our place in society. We have to react to this problem very urgently.

If there is overpopulation, it is logical that there is a problem of food supplies. In the beginning, people said 'well, the green revolution in the sixties helped us and showed that you can have a good increase of food supplies'. Remember that this green revolution came about because of a very intensive use of agrochemicals, fertilisers and pesticides. Suddenly, we realised that a lot of our ecological problems are caused by this new type of agriculture, this agricultural revolution. So we actually have to rethink and see what can be done to avoid this intensive use of chemicals and still proceed with the increase of the production. Everybody knows that, with this increase in population, there cannot be peace on earth — the traditional term is 'global equilibrium' — if 80 per cent of the population is extremely poor. We have to make sure that the discrepancies in the level of welfare are not as they are at the moment. So we will have to find solutions. This means that in our kind of economy we must create more industrialisation. But if we have more industrialisation, what kind of industry will we develop? If we develop the same kind of industry as fifty years ago, we know there will be an enormous problem of pollution. At the same time, everybody knows that a higher population-density causes disease problems. How can these disease problems be solved? In the case of viruses, it is not easy at all. We have the example of AIDS. So our medicine needs to be ready for all catastrophes which can occur at that moment. In this context, I think plant biotechnology can play a major part.

We should remember some basics about our agriculture. All the crops we use today — and there are not that many — exist through the efforts of breeding. I will not go into this, you have read it in your textbooks at school: you have several varieties with certain properties, you breed them together and each time you try to select the ones you want. Through recursive breeding you can try to obtain a final product. For basic crops which grow fast this can take ten to fifteen years. If we look at other crops important to our society, such as trees, it is of course even more complicated. Poplar trees, for example, flower every ten years, so the number of crosses you can make is rather limited. I take the case of poplar trees first because our group in Ghent is working on it, and close to Ghent there is a very good poplar breeding station. The director of this station, who has retired now, started in the early fifties with breeding trees that are mould-resistant. The director of this station was fortunate to obtain this result rapidly, but for a lot of other trees it would be equally interesting to know 'why is this tree more resistant to fungi?' We have to identify this and see which ways there are through genetic engineering to help protect other trees from other varieties against mould. To repeat the results with the poplar trees could take fifty to hundred years of breeding.

One of the techniques which have been developed is what we might call a kind of DNA-fingerprint. It looks a bit technical, but it is worthwhile to look into for a minute, so that you can catch the value of the concept. On the top left, you can see blue and yellow lines, representing the two parents. The blue line has the trait, represents the genome that possesses the trait you want to have, let us say fungal resistance. So there is a little segment but you do not know where — it can be on this chromosome or on another. You would like to identify this blue segment which gives you this property. Then you do crosses and there is a lot of exchange of material during these crosses. You have an insertion of a lot of the blue in the yellow. What is done then is called 'bulk-segregation'. You bring together all the trees that have the trait, all the trees that come out of these crosses and that are represented by this mixed yellow-blue. You bring these mould-resistant trees together, *cf* the red line you can see in this kind of bar-code. So on the first line the parent has it, but you do not know where it is, and then you will have a group — the left group of what is called 'Bulk 1' — that does not have the trait. The others have it. The red bar divides them. What has to be done now is a digest of DNA, a kind of bar-code, and see whether you can identify a fragment that correlates with the trait.

The method which is used at the moment was developed by Mark Zabeau, who was a graduate student in our lab and is now Director of an institute in Wageningen (The Netherlands) called Keygene. You extract the DNA of the tree and then you separate DNA-fragments. There are different ways to cut it up. Then each time, in the bulk of the trees that have the trait, you look to see whether a fragment can be found which is absent in the others. This might take some months. Then it becomes the basis of in-breeding. You can now, immediately when the shoot emerges, take a DNA-fingerprint and see whether this fragment is present. If this is the case you know the tree will be mould-resistant and you can accelerate breeding. Of course, it is also important to find out — through the identification of the information on this fragment of DNA — how other trees can be genetically engineered in the future. This work is now being done in biotechnology for all kinds of crops. A lot of these crops could not be studied if it were not for this method. This type of DNA-marker-assisted breeding is crucial and is one of the first major results of biotechnology. These tools become available and they are used to help with breeding and to identify new DNA-fragments.

People have, through cell cultures, sometimes regenerated what they call 'mutant plants'. Here you have a fodder crop, mostly used in Canada. It is a leguminous plant. At the right side, you have the normal plant and at the left side you see a mutant that grows very well, but is sterile. We do not know which mutation occurred, but it is interesting that, with a minimum of mutation, you go from the plant at the right to the plant at the left. So it is worthwhile to know what this is and to know whether it is possible to regenerate new plants in the laboratory with a much higher production-yield than a fodder crop. I think we can go through a whole list of plants for which it would be extremely important to society to have the tools to do this type of work.

If we think about developing countries, for example in the Sahel area, Caseuorini trees are extremely important for having some wood in these areas. Nobody will invest money to do breeding on Caseuorini trees and, even if somebody would, with the traditional methods it would take hundreds of years to make good progress. With the DNA-fingerprint technique, however, and with some funding from international organisations and ministries for co-operation and development, this type of work can now be started. We can try to develop trees with a better nutrient uptake. A lot of improvements can be made.

We can go further. At the Max Planck Institute in Köln, a group is working on the resurrection-plant. At the left, you see the plant when it is dry. It has shrunk completely. If water comes, it blooms again, so this plant has an enormous capacity to dry out and still resurrect. If we knew the molecular basis of this process, if we knew the genes with this property, it is clear that that there could be enormous progress. Only DNA-work can make these kinds of solutions possible.

In some deserts, you see some plants growing under circumstances that would make any growth unexpected. Which properties make this growth possible? Again, this case seems a little fantastic, but it has never been studied. Biologists and botanists know it exists but nobody has the money to start a study.

Let us go back to the tropical forest. Everybody is very emotionally sensitised by the destruction of the tropical forests, and rightly so. This destruction is caused by economic factors and population pressure: people need the soil and the timber. We have to try to convince those governments of the enormous value that is present there. We know this value is ethical-biological, but that is not an argument in our society at the moment. Therefore we have to try to convince the local governments that maybe having more information on the trees will allow exploitation of them in a better way; they could create new kinds of plantations and as a result, a lot of new products would be made possible. How does one study a tropical forest practically? There are about 1,200 trees in a typical Amazonian forest. If you walk around in such a forest and you ask the specialists, they will tell you that maybe 20 of them, or maybe only 15, have been completely anatomically described, meaning that if they see a shoot emerging they know what it is. Twenty of the 1,200. For all the others, they have to wait 20, 30 or 50 years to know what kind of tree it is. You all know that a DNA-fingerprint would immediately give an answer to this question. At this moment, however, there are no institutes there to carry it out. If you see a very valuable tree and you see another one 500 meters further and yet another one 200 meters in another direction and you ask, 'how is the pollination? are these trees related?', the answer does not exist. DNA-fingerprints could give the answer. These kinds of studies should be carried out urgently and they require biotechnological techniques.

When you walk around in the forest, you sometimes find fruit trees that are extremely valuable, but the fruits cannot be transported easily, so nobody exploits them in an economic way. But when we consider the Calgene tomato, we see that it is possible to delay ripening in fruits. Maybe with some little intervention it would be possible to have valuable new fruit trees that could be used by the local population on their plantations. That could improve the agriculture in these areas. This is the type of work that becomes possible in tropical agriculture through biotechnology.

At the moment, all this sounds a bit like a dream, in part because we have such a lack of information. I am convinced, considering the very fast progress in plant biotechnology, that we will come to a much better understanding of the molecular basis of plant growth and development, because in plants there are only about 30,000 or 40,000 genes and the techniques to identify them and to understand their function are developing rapidly. Sometimes work with plants is much cheaper and much more efficient than work with other organisms, so I am convinced that in the next 20 years we will see progress in all these nearly science fiction topics which I have mentioned.

I will now give a quick overview of what has been done here in Ghent at Plant Genetic Systems in the last ten years. Much of it also looked like science fiction around 1980, but now it has become real. The three examples I would like to present are examples of gene engineering: isolating genes and expressing them in plants and in that way creating other plants, each time by introducing a bacterial gene and expressing this in the plant. When this work started, in the beginning of the 1980s, plant molecular biology was really nowhere. There was no knowledge about plant genes and the only genes easily available within the technology of that period were bacterial genes. One learned to express these bacterial genes in plants and give them a new trait. That of course is possible, since DNA from one organism can be expressed in another organism. This is easily said but of course to have a good yield of expression and to obtain some stability, a lot of technical knowledge had to be built up. This technical knowledge has been covered by many patents. Many of you have heard this story but it is always worthwhile to go back to it. It is about making plants that produce their own insecticidal protein. This is very interesting from an ecological point of view, because these proteins — that come from the bacterium Bacillus thuringiensis (Bt) — are extremely insect-specific. They are not like the chemical insecticides we know, which are active against very broad categories of insects but also at the same time against vertebrates, fishes, birds and mammals, unfortunately. The proteins are extremely specific against a particular insect, sometimes even too specific, for example in agriculture, but that is another story: then you have to introduce several genes. Already in 1984, Plant Genetic Systems was able to show tobacco plants that produced a high enough level of this protein to kill some of the larvae of, in this case, Manduca Sexa, a moth whose larvae destroy a lot of plants. Since then, Plant Genetic Systems learned to express much higher levels and began to attack a lot of other pests. Some of these pests are extremely important to our economy, like the European cornborer. This is one of the products we will see on the market in the future because in the United States it is an enormous problem. It is an insect that develops, as the name infers, inside the plant, so it cannot be reached by chemicals, unless you carry out a very complicated strategy of repeated spraying. It will be a benediction for the environment if this type of protection can replace the use of insecticides. This can be done for the protection of the plant itself and it might also be interesting for trees, although there are always discussions among biologists about integrated pest management: if resistance is built up, would that be a complication? Of course you always have other Bt's and people are now cloning the receptors of the Bt, so that, through protein engineering, it will be possible to make new Bt genes. I do not think this will be a major problem and the results show that, if worse comes to the worst, you can always go back to the chemical insecticides, but we can have this discussion later if you want.

There is not only the protection of the plant itself. You have seen the historic photograph of the tobacco plant with larvae of insects; in twelve days time, the normal tobacco plant at the left side is completely eaten, while the one that expresses this protein, even at a low level, is still very well protected. That has been repeated for quite a few crops and products are already made available by Monsanto for cotton. In the United States, as I mentioned, it has been done for the cornborer. Sometimes it is also possible for storage diseases, like the tubor moth, that attacks potatoes in storage. If you express the protein in these potatoes, they are absolutely protected. The advantage of this protein is that there are track records from 25 to 30 years that by spraying the bacilla, there is not the slightest toxicity. So we know that the receptor for this protein is only present in the larvae of these insects and that is why we know that it is absolutely non-toxic for vertebrates. This is one of the rare cases in which these many data are available. This is an example, I think, of a very important protection that came about through biotechnology.

Another example that sometimes generates a lot of misunderstanding is engineering plants that are herbicide-resistant. People think that this will lead to an increase in the use of herbicides and to much more problems in the field. I think it is clear that, if we make plants that are herbicide-resistant, it is to allow us to make what we call 'the new classes of herbicides'. We have to realise that modern agriculture is absolutely impossible without herbicides. It is economically impossible to remove weeds manually. It is sometimes done that way in tropical areas, but in our countries it is absolutely impossible. For many years herbicides are being used that actually should be forbidden because they are toxic, they are a danger for ground water and they create real problems. But as there are no alternatives and our economy has to go on, compromises are made in negotiations. Now there are some new herbicides that rapidly degrade in the soil and have a minimum of toxicity. They have been specifically studied for that because otherwise a herbicide had to be developed that was active against weeds but not active against a given crop. Thus it was possible to develop herbicides which were absolutely not toxic and left no residue in the soil. In some weeks' time, the herbicide had been degraded by the micro-organisms in the soil. But at that moment of course, if you develop this type of herbicide it will have lost its specificity, otherwise it would be a miracle compound. This would be really asking too much of chemistry. So this herbicide will then also be toxic for the crop itself, not only for the weeds. The solution is to make your crop resistant to the herbicide. Just like you incorporate an enzyme that would destroy or modify a molecule and would give protection against a toxic compound, there are also, from this soil, bacteria that can degrade. You isolate genes that make enzymes that modify or degrade the herbicide, or have it entering a pathway where it will later be degraded. Express that in the plant and as a result you obtain plants that are resistant against the herbicide. In Ghent, Plant Genetic Systems has worked on the herbicide 'Basta' and has engineered this resistance in an enormous variety of crops — 'Basta' is the old name we still use; Hoechst calls it 'Liberty Link'. Each time you spray with the herbicide you see that the original plant, a leguminous fodder crop, is completely destroyed while the engineered one that is resistant, survives perfectly. There is a very good detoxification and this herbicide-resistant variety will now be used in the United States. Hoechst has the registration authorisation coming up for a variety of crops. This is only done now because the price of the herbicide was high and to be commercially acceptable Hoechst had to develop new ways of producing the herbicide at an acceptable price for agriculture. We will see a lot of applications of this herbicide-resistance in the future and breeders already use it extensively, because if you bring in a gene through genetic engineering you can add this gene in all possible positions. Sometimes it can be linked to a trait that you want to follow, so you could call it 'tag a trait'. You have a certain property in breeding that you want to bring into progeny and you have this herbicide-resistance inserted in the close neighbourhood in your chromosome. At that moment you have tagged your gene. In breeding you can follow your trait together with the herbicide and accelerate your breeding programs. A lot of corn breeding in the United States done at the moment by the major corn companies uses this trait as a market trait in their breeding programmes, because they have saturated distributions of the gene inserted at different locations.

The following might be a bit more complicated technically but I think it is still understandable for the non-specialist. In the 1930s, it was realised that for corn one could really exploit what is called 'hybrid vigour'. You see two corn varieties and you see at the top indicated the dimensions of the ears. Now, for certain crosses the outcome or the progeny of your cross — the one you use as a male, the other as a female — the plant that comes out has a much higher yield. Why this happens, molecular biologists cannot explain — the molecular base of plant growth and development is still in its early stages, we are only building up our knowledge at the moment — but it has been used very intensively since the 1930s. At the same time the plant is more vigorous, more disease resistant and this is called 'hybrid vigour'. This hybrid vigour is valuable to many plants but only in corn it has already been used since the 1930s. This was because at the top of the corn plant you have the male part of the ear, the floret, where the pollen is, and in the middle you see that the ear will develop where the female part is, so the pollination goes from top to bottom. The top part can be manually removed — this is called 'manual emasculation'. Then you have only cross pollination, so the pollen from one will go on the female parts and find there the egg cells of the other plant. In the corn fields in the United States you see two rows of pollinators, five rows of emasculated plants which will be the receptor of the pollen and in that way hybrid seed is made. At the moment for corn world-wide all the seed is hybrid seed, because it has a much higher yield and for the farmer this is important. He cannot plant back, he must buy the seed, that is true, but for him it is much better than just planting back because the improvement in yield is so much better. It can be done for corn, but if you want to do it for other plants when you have the male and the female part in the same flower, then of course you are in trouble.

Fundamental research collaboration between the University of California (Los Angeles), Bob Goldberg and Plant Genetic Systems has made it possible to express a gene in a particular part of the plant. Here you have a section through the anthers and that is the blue line surrounding what is called the future pollen chamber, where the pollen grains will be. In the blue line in this area you expect an enzyme, an arenase, an enzyme that is absolutely not toxic. All organisms are full of arenases but these can be very well inhibited and their expression can be controlled extremely well. The result is that the pollen grains are formed but the pollen is sterile. So through gene engineering a male sterile plant is made. This plant grows normally as a female plant and you can do the cross pollination. This finding has brought an enormous revolution in the way plant breeding will be done in the future. Plant Genetic Systems has developed it for rape seed. At the moment in Canada it seems, after field trials during many years, that the yield increase can be 120 per cent to 125 per cent, sometimes 130 per cent in different soil conditions. It has been tested in all possible conditions and now it is ready for commercialisation. So this is the situation for rape seed. It will be applied to corn, to vegetables, and it will become important for all types of agriculture at the moment, because yield increase — whatever the European Union may say about overproduction — is always what the farmers want, it is economically interesting. Of course, it will be possible to use the same principle for tropical plants also; we will see a lot of applications there as well.

Apart from these achievements of Plant Genetic Systems, of course many other groups have been working on gene engineering in plants. Groups like Monsanto have engineered virus resistant plants. They first tried out expression of some genes that make virus resistance in tobacco and then in a lot of other crops. Actually it is one of the first plants largely used in the world because China has taken that over; China even produces cigarettes from tobacco engineered for virus resistance. There you have hundreds of thousands of hectares of engineered tobacco since at least two or three years. All these examples that I mentioned are not yet commercialised, because there was an enormous delay due to complications with regulations, and the authorities did not know how to handle the situation very well. The only product which has been commercialised since last year is what is often called the 'Calgene tomato' or the 'flavr savr tomato'. There is not only Calgene, but also several other groups who have engineered tomatoes with delayed ripening. Why is this important? If you delay the ripening, you can use varieties that otherwise are very difficult to transport. It is easy to understand that if you have very flavourful tomatoes but you can only grow them for you and your neighbours and you cannot transport them, you do not have a commercial product. Today if you have a product, you have to be able to produce it on a large scale and transport it over a long distance. Thus it is important to have selections for tomatoes that can be transported and this is what the breeding did. If you do selection for two or three traits that are important for transportation and disease resistance, well, flavour is not a major consideration. You then end up with the type of tomatoes and vegetables that can be found all over the United States: they look wonderful but have no taste at all, it is like eating cellulose and water. So it is possible to make for our agriculture new types of crops with old varieties but rapidly bring in the trait that makes them transportable. It is a gene involved in ripening that cuts down an enzyme; several crops each took another type of enzyme. At the moment Zeneca from ICI in the UK has this type of tomato. Right now, there is only one product on the market, but we see now that — with all

64

the field trials and the scale-ups that are allowed in the US — 20 to 30 different new products are on the way, so we will see an enormous acceleration in the future.

We could also talk about engineering plants for nematode-resistance, a type of research that is going on in our department. Nematocides are used as fumigation against nematodes, worms in the soil that attack potatoes particularly. These nematocides are extremely toxic and indeed should be forbidden. Accidents happen regularly. Nevertheless, this fumigation is still used because we have to fight the nematodes. There are other ways to try, through gene engineering, to inhibit the kind of structures which are made when nematodes infect and in the future we will probably see quite some progress in this area as well.

I have given examples in the field of agriculture but very important to our economy is developing new, less polluting, chemical industries. We like to hope that these chemical industries would come from plant products. It is still wishful thinking, because no work has been done on this yet, but we know that plants produce very important pharmaceuticals and thousands and thousands have been exploited. There are a lot of agrochemicals which can be used directly, there is a lot of construction material which can be used. There are lots of products which could come from plants but are not used. Because the plants as such, as we find them in nature, are not well enough yielding, it is not economically feasible to do it. However, if we know more about growth and development of plants in molecular terms, biotechnologists will be able to make new varieties of plants that can make these products with a much higher yield. At that moment maybe we can have new types of industries. Furthermore, we should not forget that with this type of plant, its energy to grow comes from the sun. It might be interesting to use solar energy for a chemical industry.

One thing is clear: to develop this concept, and to come so far, an enormous amount of research has to be done. Universities cannot really extract enough value out of these findings because doing research, training people and developing a product is ten times more expensive than doing scientific research. The only way out is to set up 'prototype development'; this is done more and more in the United States and in Japan, where it is done 'in-house' in the major big companies. So either a small company does it or it is done inside a very big company. If you have your prototype, you have to test it out under all possible circumstances in the field — and that is the extremely expensive part. Next you go to production, you have to hassle with the registration and all the legal authorities, then set up the complicated structure for doing the commercialisation. In our society these are parts which are terribly expensive and which the university cannot undertake. That is the reason why many of these steps are protected by patents and an attempt is made to make a good balance between what the society needs and what the industry needs to be able to function in the type of society that we have created.

As a conclusion, I would like to return to the slogan we started with: there is an enormous amount of things which can be done by scientists with respect to the needs we have. We just will not find the solutions in nature; we will really have to create them. And that is the reason why we need this plant biotechnology and other types of biotechnology as well.

3 Plant biotechnology and its future impact on science, agriculture and the environment

Jozef Schell

First of all I would like to say that I am extremely pleased to be able to participate in this workshop, because I join the organisers, Dr Sigrid Sterckx, and Professor Vermeersch, the philosopher who happens to be also the Vice-Chancellor, in their intentions to have a meeting different from what I have been used to in the last decade and a half. After Belgium I am now living and working in Germany, where the idea is that you discuss biotechnology — a very controversial topic as we all realise — so that it becomes a media spectacle. The animation is usually achieved by having two camps that have to discuss controversially. It will be no surprise to most of you that I am usually classified in the 'pro' camp, which I do resent. I think that as a scientist, although I agree with the presentation of Marc Van Montagu that the method has a lot of potential, I am nevertheless capable of thinking also of the contra's. So this classification is not at all to my liking. Instead of that I understand that the purpose of this meeting will be to see whether we can, after some debate, come to a consensus. This is idealistic. I do not believe that this will be possible. Furthermore I read that the intention is that, maybe more realistically, we would at least come up with a compromise. It is true that all this is very urgent, and not only because the European Patent Office has a few major decisions to make. It is also urgent because I think it is true to say that not resolving these issues is going to have the consequence that Europe will not be part of what I consider to be a major development in evolved societies, relative to less developed societies, in order to improve the conditions in our society in particular and world-wide in general. I believe that, for the developed part of the world to which we belong, it would be morally unjust not to participate in this effort. That is why we should think of the proposal made here as urgent and I would like to contribute in two ways.

The strategy is that one should make proposals for discussion which would allow minimal consensus. The first proposal on which I think we could reach a minimal consensus, is that biotechnology should be used to improve society, to deal with the problems which face society in a positive way. The area in which I am a specialist is the same as the one of Marc Van Montagu for good reasons — we started here together — so I fully subscribe to everything he said. I will also argue from the point of view of agriculture and environment, a major problem which this technology can help solve. I will also argue about what Marc has told us in other ways about this topic. Furthermore I must admit one thing in your presence — and that is another reason why I am really happy to be here. Starting all of this in the end of the 1970s and early 1980s, my biggest astonishment and frustration has been that exactly those groups in our society, well represented here, whose declared mission is to try to solve societal problems, are the ones who have opposed the principle of this technology the most. We know that technology per se is not the proper solution to all problems. However, I do not think it can be argued that a genetically engineered organism is an organism that intrinsically, by the fact that it has been made with a certain technology, is more dangerous. I think the opposite is true, as we now know. I am not saying that it was wrong to consider it that way, but having considered it one should at least move forward and draw the conclusions. And one of the conclusions is that indeed, if anything, the method of breeding by a combination of methods including molecular biology is safer and more predictable than doing it in the traditional way.

One way to reach a consensus is to not take oneself too seriously. And therefore I show this slide to introduce the people working at the Max Planck Institute in Cologne: my colleagues Francesco Salamini, whose group does work on the resurrection plants which Marc described, Heinz Saedler who works on transposable elements, Klaus Hahlbrock, a biochemist, and finally myself a few years ago, when my beard was not white yet. The important point to make about this picture is that it appeared on the cover of a journal which is the German equivalent of 'Scientific American'. People from this journal had spent a whole week in our Institute trying to describe what we are doing. Therefore it is not at all correct that they showed this, although it was a nice example of good humour. Anyway, it was an example of good humour to illustrate the Institute in that way, but it was also an example of good business to talk about gene technology in Germany. Not to show 'monsters' is supposed to be a mistake if you intend to sell a large number of copies of your journal. That is why it came out this way. One thing I can tell you is: what we are not trying to do is to make monstrous potatoes. We are using potatoes as a model system to try to understand disease resistance. We are working on potatoes to try to modify their carbohydrate content, maybe with the purpose of finding out whether one could produce biodegradable plastics in potatoes. That is something the scientific community is working on, but that's about it. It is certainly not monstrous.

I mentioned that I was going to use a slide coming from the work of Keygene in Wageningen, The Netherlands. It demonstrates that if anything, the breeding and the possibilities achieved by gene transfer are leading to products which are less questionable than the ones obtained by traditional breeding. Here you see for instance a tomato that was bred in a classical way to become resistant against a nematode. Marc, in one of his last slides, showed how detrimental nematodes — little worms in the soil — can be to tomatoes. Well, this is a traditional variety obtained by crossing a tomato in a traditional way with a wild species. Now one knows that by the methods described by Marc one can identify the amount of foreign DNA from that wild tomato species which has not gone through the long process of trial and error resulting in the crop plants we presently grow, but has been taken out of the wild somewhere. You can see here that two per cent of the total DNA in that tomato variety was in fact foreign DNA. Then they improved this variety by certain traditional means, also crosses, trying to remove some of the foreign DNA by more inbreeding and backcrosses, resulting in a new variety with only one per cent foreign DNA. Now, with the new methods Marc has described, it is possible to obtain yet a newer variety where the figure is down to 0.6 per cent. If one would do the same with transgenic approaches, as Marc described, it would be even less. So this illustrates that the argument that introducing foreign DNA is creating an unknown situation that nature is not used to, is independent of the technology used to create a given plant variety. In nature gene transfer can occur. The Agrobacterium system is a natural system. The fact of the matter is that you do not get very far in thinking that everything that is natural and that we have always been doing, is safe, and everything that scientists come up with is intrinsically unsafe.

You have been given a wealth of examples by Marc Van Montagu, but I am not going to continue in the same way. I am going to describe some of the work we are doing in Cologne, because it is important to this discussion. It is also my contribution to another expectation you may want to aspire to here, namely the hope that in society there will no longer be pro's and contra's, in the sense that one party is supposed to argue in one way and the other in another way, so that one is right and the other is wrong. I believe, instead of doing that, what one should do is have a dialogue with one another and say where we are going and what should be done. What one should do is take from the contra's the very positive aspect that society may enter into the desirable situation in which there will be a discussion to measure what the consequences might be before introducing new technologies. With this in mind, I am going to give you some examples of work that has been done in Cologne.

Here you see a plant. It is a tobacco plant which we use only as a model, not because we think tobacco is important for the third world or the developed world, but because it is a convenient model. This plant has been attacked by a major cause of plant diseases, namely fungi. One of the environmental impacts we suffer from is that farmers, to prevent these plants from dying because of moulds (i.e. fungi), need to use chemicals as protection, so that even in the presence of the pathogen (the fungus) the plant would grow normally. In this tobacco a protein has been introduced and expressed, in a way that could not be achieved by sexual crossing. This protein, called RIP for Ribosome Inhibiting Protein, is normally present in the seeds of barley. Barley has proteins in its seeds which are toxic to fungi. The idea was to see whether it would be possible to use that evolutionary property of some grain cereals and introduce it in other plants. Indeed, you see here that, when this was done, the tobacco plant became quite tolerant to infecting fungi. Then the question arose — a question of ecology in fact — that if all crops would have this anti-fungal property, soon there would be fungi which manage to survive in spite of this protein, just as they have developed to be not very sensitive to the chemicals produced by the agrochemical industry. This has led to the fact that more and more of fungicidal chemicals were used. Thus some of the chemical fungicides may end up in the groundwater and cause ecological problems. We would certainly like to avoid such a development. Therefore, we have done the following. We knew there was another protein, a chitinase, with antifungal activity. If you put tobacco seedlings into the soil and you let the seedlings germinate, many will grow to small plants or not at all as a result of the spores of the fungus in the soil. But if we introduce a chitinase-gene into tobacco seedlings, practically all of them grow out to full-sized plants. So we get a dramatic increase in tolerance and that is what the farmers are interested in: an increase in yield. The idea was as follows: what if one would combine the two principles? I showed you results with the RIP-gene and I showed you the result with the chitinase-gene. The new possibility is to combine both principles in the same crop plant. Imagine that we get a changed fungus that becomes tolerant to RIP. This pathogen would still be killed by the chitinase. Similarly, one that would become tolerant to chitinase would still be killed by RIP. So the idea was to find out what happens if you combine both. Would you then have something that will be environmentally superior? The answer was most definitely yes.

The tobacco which is not infected is of course not diseased. With normal tobacco, when infected, not all plants are diseased but you have the maximal disease level (index) you will obtain under these circumstances. If you introduce either RIP or chitinase alone, you see that the disease index is reduced. But if you make combinations of both, you see that the disease index is reduced more than proportionally. Not only will you have the hoped for result that the probability of resistance in the population which would require you to use more and more of the anti-fungal product is low. You also have a better resistance. You have less disease so you have a better crop. That is a point which is environmentally important, namely that this technology offers the possibility to combine different principles.

Let us take an example with grapevine. I happen to like wine, and I also happen to know that in the region of Bordeaux — where they have some very good wines, as you know — they are forced to argue, in all kinds of commissions whose responsibility it is to protect human health as well as the environment, that it is essential to continue the use of some products like copper derivatives or cyanide derivatives, which are so toxic that we should not use them anymore. In fact, the European regulatory bodies have said 'you cannot use them anymore', but they argued that an exception should be made because not using them right now would mean that there would be no Appellation Controlée Bordeaux wine anymore. Of course, when they made up the rules for Appellation Controlée and Cru Bourgeois or whatever it is, they did not know about molecular biology or gene technology. Recently also grapevine plants have been transformed. Thus all the examples Marc has given about tobacco plants can also be applied to grapevines. Therefore I have some hope that, when the Bordeaux people have to replant grapevines in their region, they will use transgenic grapevines, which will allow my children to enjoy the same Chateaux which I enjoy.

Rapeseed is a very European plant. It grows well in the northern regions of Canada, but also in Europe. Consider the possibility that farmers — and I think more and more environmentalists too — would rather like to take care of the land by growing something on it that would be useful to society and not contribute to overproduction instead of having rules in Europe that control overproduction by having land not in use set aside. That is how the notion of 'non-food-plants' originated: agriculture for other needs than food. The major candidate in Europe for this is certainly rapeseed. For instance, it is already a source of oil, but as one can see by using a gas chromatogram, there are different fatty acids in the oils of rapeseed. It would be useful for industry to have fatty acids that have either longer or shorter chains. There is a plant in nature named Cufea lanceolata which — when we do the same analysis — has far less C18's, but it has C16, C14, C12, even quite a bit of C10, and even C8. So it has shorter fatty acids. Therefore, it has been attempted to develop this plant as an agricultural plant, because for detergents short fatty acids are important. But here comes the problem. This plant is a wild plant and you cannot begin to grow it because it has not been adapted to become a cultured plant. Farmers do not know how to use it, so there are agronomical problems. On top of that, there is the major problem that this plant has not been bred to have high yields, nor has it been bred to become tolerant to disease, and so forth. So you cannot grow them continuously. If you grow them, they lose their properties, they develop all kinds of diseases, they look horrible. In Germany, people have tried to do this for years, and have not succeeded. So now the idea has come up to use genetic engineering to try to retrieve the genetic information which allows Cufea lanceolata to make short fatty acids, transfer it to rapeseed and then maybe make new varieties of rapeseed with predetermined fatty acid length. I want to show you that there was initial success with this approach.

I should mention that success was first realised with a similar plant in the United States by a company called Calgene. Depending on which gene one introduces, the proportional amounts of fatty acids will change. So the notion that, for our farmers in Europe and world-wide, one has alternatives to a set aside policy is a realistic one. Therefore, we can hope that farmers will continue their traditional and cultural role of taking care of our land. What kind of nature do we want? A nature of five years ago, ten years ago, twenty years ago, a hundred years ago? A fact in most of Europe is that there is no nature left fit for agriculture. It is all cultured land, land made by farmers. Now the question is 'what do we want ?'. Do we want farmers to grow things that will be acceptable and maintain the environment without polluting it, and on top of that will give them an income which will allow them to continue their traditional role of maintaining our culture? Or do we oppose this on the basis that we think it is unnatural? I must admit that I do not know what natural is. I think the nature of man is to exploit his environment. I do not know how to work with the argument that 'natural is good and unnatural is bad', because I do not know where we start. Concerning this debate about what is natural and what is not, I have two pieces of information. One comes from what is done in the Common Market. There is now, as you know, a new organ to advise the Commission of the EU on what research is essential for progress. It is called ESTA, the European Science and Technology Assembly, in which people from industry, breeders, agrochemical industry and some academics as well, participate. They discuss what should be done at a European level in terms of support of science. Since I am a member of this group,

I thought I could share with you what is on their programme. Number one is biotechnology and molecular biology. There is the traditional human genome project, but there is also marine biotechnology and many other things, including combinatorial chemistry. This is not biotechnology but the alternative to biotechnology proposed by chemical industries. Chemical industries want to improve their capacity to synthesise more products and test them. Combinatorial chemistry is very powerful and allows the chemical industry to speak about 'gram pro hectare-activities'. Whether that is a dream or a reality is a different matter. There will be a discussion on how we risk, by taking position against this technology, to put Europe out of the possibility of using it.

The type of plants presently worked on by industry are: corn, canola (the equivalent of rapeseed), cotton, tomato, potato, soy bean and zucchini. In the USA three different organisations have to decide on whether or not to allow the commercialisation of transgenic crops. I am not talking about field tests anymore, but about the permission to bring the product to the market. For the different companies, the United States Department of Agriculture has given allowances in 1994 and 1995 for the Calgene tomato (the first product allowed by the USDA); the second organisation has given allowance in 1994; EPA has given allowance in May 1995. Cotton resistant to herbicide has been allowed in May of last year. Other plants needed permission by two of the agencies, not necessarily by the third one. Not all of them need the third one, the Environmental Protection Agency. If you compare the US situation with what happens in other countries, like Canada, you see that companies such as PGS, Agrevo, Calgene, DuPont and Monsanto have received permissions in 1994 and 1995. Interestingly we often say that the country which is most advanced is probably the US or Japan. That is true in terms of the numbers, but in terms of dates the truth is that it is China. In China, two plants — tobacco and tomato, in both cases resistant to viruses — have been authorised for commercialisation in 1992 and 1993. Compared to that, Europe has rather few examples. You see some examples of international companies such as Ciba, Monsanto, PGS from Ghent with its rapeseed male sterility which is 'en attente' (this situation is a rather frustrating one), and then Ciba and Zeneca. So the situation in Europe is bad. And then, in terms of patents, I would like to refer to two articles which recently appeared in Germany. I found them quite striking. One is by an ecologist, Professor Markle from Konstanz, who wrote in 'Der Spiegel' — which is not really known for its pro-gene technology-attitude — something like: 'It is our duty as humans to fight against what many people think is nature'. I am not going to repeat what he said, but it comes down to what I said about the nature of humans which is to defend their own species through interfering with nature. If you have doubts, you should read the article, it appeared in 'Der Spiegel', in an issue of December 1995, on page 48. The other information comes from somebody who is now the president of the Deutsche Forschungs Gemeinschaft. He is in fact a German linguist; so the person responsible for German science policy is a linguist. His name is Wolfgang Fruhwald. He discusses 'Patente auf Leben' and says that he does not see any ethical problems. He only has an ethical problem with modifying the human germ line. I will end here by simply advising that you read those articles.

4 Biomedical applications of biotechnology

Désiré Collen

I would like to start with congratulating the organisers on this important initiative. What I would like to do is to discuss a couple of aspects, mostly from the point of view of an academic scientist who has been involved in patent business to a certain extent, and in the conversion of university research into commercial products. I will not speak as a scholar in this matter because I have no real education in it, but I do have some 15 to 20 years of practical experience which covers being involved in a successful patent on the one hand, and several patents not worth the paper on which they are printed on the other hand.

I would like to cover three aspects. First, I will briefly make some one-line statements about the patent system. Some pro's and con's as I see them from my personal experience. Secondly, I would like to present to you two case studies on the development of drugs for pharmaceutical use coming from university research in which both the patent availability and the use of experimental animals and pilot studies in patients were very important and actually necessary. Lastly, if time permits, I would like to give you a few examples of how complicated the patent system can be and how sometimes it can go to excesses. When I talk about patents, I am not talking as a person who has been frustrated with the system, because by Belgian standards I have done extremely well with the patent system, as well as my laboratory and university.

Table 4.1 shows a few one-liners, statements that I will briefly review with you. In our field of pharmaceutical development, patents are absolutely necessary. The reasons clearly are, first of all, that most of the pharmaceutical development is done within the industry. Industry is a for-profit environment. The development of a new drug nowadays costs five to ten billion Belgian francs. It is clear that without any protection and monopoly, the pharmaceutical industry would just not be able to provide that money. Secondly, even from the perspective of the university, research laboratories with more than 100 scientists, with budgets that approach between five and ten million dollars a year, cannot be funded with government support. So we need additional income from connections with the pharmaceutical or agricultural industry. It is necessary for the survival of some of this research that it is potentially applicable. Therefore I am afraid that patents are needed.

Table 4.1
Patents versus Science

ONE-LINERS

1. Patents are necessary
2. Patents are 'for-profit'
3. The patent system is made 'by lawyers, for lawyers'
4. The European patent system hampers academic research
5. The patent system needs more 'peer-review'

I have however mixed feelings about patents. I do not think they are perfect and that is why I have this second one-liner: submitting a patent application is an act for-profit, by which you try to get as much as possible. If you have a good patent agent, the first claim of your patent will ask half the world, and the following claims will progressively restrict this. So this is not a purely scientific endeavour; it is a legal document that is produced. Therefore I have this third statement: the patent system, as it has evolved in my opinion — and again, I am only talking from practical experience and not from any specific study — is a system that is made by lawyers and for lawyers. That means that it is a legal document which can be challenged in court. The claims as worded are usually written by lawyers for legal purposes. The meaning of a claim is basically the meaning that is given to it in a court, and not necessarily what the scientist either intended or wrote down or was advised so to do. So it is really a mixed system, in which the science is but one aspect.

The fourth one-liner I would like to mention is that in the European patent system, when you submit an application it is made public 18 months later and when the patent is issued several years later, there is an opposition period of nine months. This system is very much to the disadvantage of academic research. When a patent application is submitted from industry, usually the know-how is kept as a tradesecret within the company. When we discover something in our laboratories, it is usually done by young people who are there to make a career, to publish papers, to write a PhD thesis. It is in our interest and it is our duty to publish results as soon as possible. My personal attitude has always been, when we make contracts with the industry, to insist that we can publish and communicate our results as soon as possible, meaning usually not more than one month after the submission of a patent application. This means that, if you have something important, significant and of commercial value, the whole industry will be much better informed when it is published and when the patent is issued. Thus if you have something valuable there is a good chance that you will find yourself in an opposition hearing in which the industry is trying to invalidate your patent. In that respect, I think the American system, where the patent is secret until it is issued and then you can go to court, is more fair in the sense that it does not put this academic attitude at a disadvantage to the industrial attitude.

The last one-liner I would like to present to you is that I have great respect for patent examiners. They have to cover enormous ground. My research in cardiovascular disease or cardiovascular biology, covers probably about one per cent of what is going on in this field. I have all the trouble in the world to keep up-to-date with what is going on. The poor patent agent working for a patent office has to work through these things and probably covers a field a hundred times bigger than mine. Frankly speaking, I do not understand how they can do it as well as they actually do. The patent system is totally different from the scientific literature in the sense that, when there is a challenge, it will end up in court, with lawyers. A trial lawyer qualified or experienced in patent issues surely will, if he is confronted with two qualified scientists and if he gets enough time, usually succeed in making those scientists say totally opposite things. That is because there is formulation and there is interpretation. When I compare this whole system to the peer review system of scientific papers, the standards of scientific literature are clearly higher. In my personal opinion, this is lacking in the patent literature and in this respect maybe the system could be improved to some extent.

Now I would like to briefly discuss with you from practical experience two cases dealing with how academic research has been or is being converted into pharmaceutical products and what the role has been of patents and experimental animals to do this kind of work (cf Table 4.2). I will talk about recombinant tissue type plasminogen activator or tPA and about recombinant Staphylokinase. Those two drugs are used for treatment of myocardial infarction. A myocardial infarction or a heart attack is what you get when in your heart the blood vessels or the arteries that supply the heart muscle get clogged or occluded. So you have no blood supply to a specific region of the heart. That usually gives a very typical clinical syndrome of pain due to lack of oxygen, which will lead to necrosis of part of the muscle, loss of pump function and, in about ten per cent of patients that reach the hospital, to death of the patient. Of course a quarter already died before getting there, but those who survive long enough to reach the hospital still have a mortality of about ten per cent. In countries like Belgium, with our lifestyle and our culture, almost 20 per cent of the people die from a heart attack, so it is a very serious disease. If one could do something about the mortality, it would really have an impact on health.

Table 4.2
Two examples of development of results of academic research into pharmaceutical products for the treatment of heart attacks

– Recombinant tissue-type plasminogen activator (alteplase, activase[R], Actilyse[R])
– Recombinant Staphylokinase

The mechanism, the pathogenesis as we call it, which I have just briefly explained to you, has not been known for a very long time (cf Table 4.3). Actually it took until 1980 before an American group of investigators, cardiovascular surgeons headed by Dr De Wood, proved that when they studied patients with an acute infarction, immediately put them into a cardiac catheterisation unit, put a catheter or a little tube up an artery to the origin of the aorta where the coronary arteries originate which supply the heart with blood, put the tube into one of these arteries and put in a contrast dye, that the artery was blocked. They also saw that, as time went by, there was some spontaneous dissolution of the blood clot, but then the damage was usually done. A blood clot occurs in patients with arteriosclerotic disease of the vessel wall. In 1979 a German investigator called Rentrop found that if through that catheterisation tube, a clot-dissolving drug — Streptokinase — was injected instead of a radiographic contrast medium, he could recanalise the artery. Thereby he hoped that he could influence the outcome of the infarct. In parallel with these investigations we studied how this blood clot dissolving system, which is normally present in our blood but has to be activated in these situations for therapeutic purposes, is regulated. It is the understanding of that biochemical regulation that has fuelled the hope that we could try to make better drugs than those available in 1979.

Table 4.3
Milestones in the development of fibrin-selective thrombolytic therapy of acute myocardial infarction

- Thrombosis as a trigger of AMI (De Wood et al, 1980)
- Coronary artery recanalisation with streptokinase (Rentrop, 1979)
- Molecular mechanism of physiological fibrinolysis
  —> concept of fibrin-selective thrombolysis
- Discovery of fibrin-affinity of melanoma PA and subsequent identification as tissue-type plasminogen activator (t-PA)

The advent of biotechnology and recombinant DNA technology made it possible to make human t-PA — which normally is only present in our body in an extremely low concentration — at very high production yields so that it could be used for thrombolytic treatment. Thrombolytic treatment means the treatment of a heart attack aiming at dissolving the blood clot. By serendipity, we found a tumour cell line which, by a genetic deregulation, overexpressed t-PA. Thereby we had access to the gene and could study it in collaboration with an American company, Genentech.

How does the system work? (cf Table 4.4) We have a system in our blood that, when activated, dissolves fibrin, the solid component of a blood clot, into soluble degradation products. Haemostasis forms fibrin on the vessel wall which, once the wound heals, has to be removed. This occurs by a fibrin-dissolving enzyme, plasmin,

Table 4.4
How does the system work?

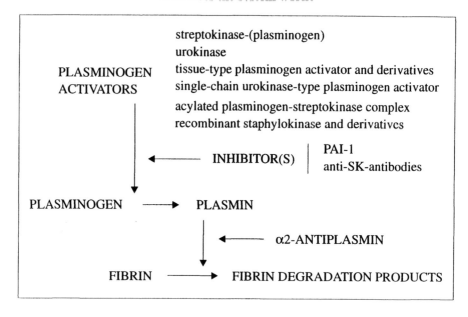

which normally does not circulate in the blood as such, but as a precursor, plasminogen. Activation of the fibrinolytic system, as we call this, consists in the conversion of plasminogen to plasmin by so-called plasminogen activators. Normally, we have two plasminogen activators in our body. One is the tissue type plasminogen activator or t-PA, which originates from the vessel wall, the inner lining which we call the endothelium. It is secreted in the blood and there it is provided to keep this blood circulation system free from clots. There is another one, which is called single chain urokinase, an activator which probably plays a lesser role in the removal of blood clots, but rather in other phenomena such as cell migration. Over the years, other proteins have been developed which can be used for therapeutic applications. The oldest one known, discovered in the 1930s, is Streptokinase, a bacterial protein. Staphylokinase, which was discovered in the 1940s, was abandoned for a while. This probably happened because of the simple mistake that when it was studied in animal models, they used dogs. This is the wrong species because it is an exception, in which Staphylokinase does not work well. If you look at this scheme, you might say, 'well, if plasmin dissolves the blood clot, then the best drug presumably is the one that makes most plasmin out of plasminogen', or in other words which has the highest catalytic efficiency. However, this is not the case. It turns out that this plasmin, if you generate it in purified systems, is a very potent enzyme but not very specific, so it will digest a lot of proteins, not only fibrin. Yet in our body normally and even under adequate therapeutic conditions, the action of plasmin is highly oriented

76

towards its physiological substrate fibrin. Understanding of the regulation of the fibrin-selectivity of plasmin is important. How can an aspecific enzyme be highly oriented to a specific substrate? Secondly, in our blood, there is a very high level of a potent inhibitor, alpha-2-antiplasmin, discovered in 1976. The rate of inhibition is so high that if plasmin were freely formed in the blood, its half life, the time to inhibit 50 per cent, would be one-tenth of a second. It is obvious that if plasmin has to exert a physiological function, namely to remove fibrin from the vessel tree, that it has to be protected from this inhibitor. Otherwise you can hardly imagine that it could exert a physiological function. It is these biochemical interactions studied in the early 1980s which have then provided a scientific background for the evolution of 'fibrin-specific drugs', as we call them now.

How does this work? (cf Figure 4.1) Although t-PA is a very specific activator for plasminogen, the affinity is extremely low. Clearly, nature has not developed t-PA to activate plasminogen in the circulating blood, because if it had, generated plasmin would immediately be neutralised by antiplasmin to make an inactive complex. What really happens when you have fibrin in the blood is that t-PA binds to the fibrin via structures that are reasonably well defined, and that the plasminogen binds to the fibrin via the same sites which are necessary for its rapid inhibition. Thereby local assembly is achieved and activation of plasminogen to plasmin initiated. Generated plasmin is protected from the inhibitor and digests the fibrin. That understanding of the physiological regulation of course has significant consequences for therapeutic development.

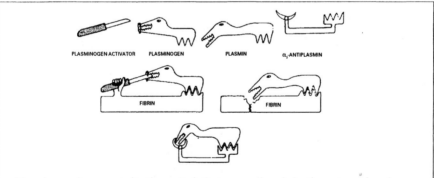

Plasminogen is converted to the proteolytic enzyme plasmin by tissue-type plasminogen activator, but this conversion occurs efficiently only on the fibrin surface, where activator and plasminogen are 'assembled'. Free plasmin in the blood is very rapidly inactivated by α_2-antiplasmin, but plasmin generated at the fibrin surface is partially protected from inactivation. The lysine-binding sites in plasminogen (represented as the 'legs' of the animal) are important for the interaction between plasmin(ogen) and fibrin and between plasmin and α_2-antiplasmin.

Figure 4.1 Schematic visualisation of the molecular interactions regulating physiological fibrinolysis

The concept of 'fibrin-specificity', first proposed in 1981, can now be derived (*cf* Table 4.5). If you consider a patient with an infarct and a coronary artery with a clot to be removed, theoretically one can distinguish within this patient's bloodstream two compartments. One is the circulating blood, and the other the solid phase of the clot. If you inject a plasminogen activator into the patient's blood stream, the following happens. If the plasminogen activator does not specifically distinguish between plasminogen in the circulation and plasminogen in the clot, most of the action will be against the plasminogen in the circulation. You will generate plasmin, freely circulating. The plasmin will be immediately inhibited by alpha-2-antiplasmin, and there is no therapeutic effect. On the other hand, an activator that specifically recognises clot associated plasminogen, will generate plasmin on the clot surface — that is now protected from inhibition and will digest fibrin to degradation products. The idea, in 1981 or 1982, to design plasminogen activators which would recognise the clot associated pool of plasminogen preferentially over the circulating pool, was something that fuelled the hope that we could make better drugs.

Table 4.5
Fibrin-specificity

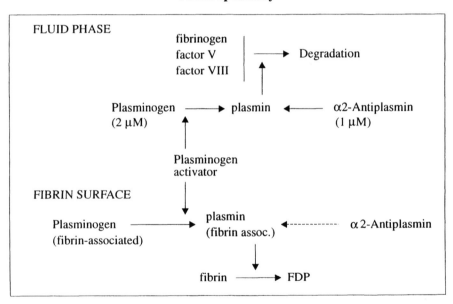

The way this has been done is very simple (*cf* Table 4.6). What happened is that first a good source of t-PA was needed. T-PA, the physiological plasminogen activator, has been known since the 1940s. The problem is that its concentration in the blood is so low — in the order of a couple of nanograms per ml — that, in order to get it out of the blood you would have to purify it ten million times, which would be very difficult. Secondly, the concentration in the blood is so low, that in order to get a

Table 4.6
Chronology of KU Leuven-Genentech collaboration on t-PA

Nov 1978	Acquisition of the Bowes melanoma cell line in Leuven
Feb 1979	Discovery of fibrin affinity of melanoma PA
Oct 1979	Purification of melanoma PA and characterisation as t-PA (with D.C. Rijken)
Jan-May 1980	Demonstration of thrombolytic activity of t-PA (with O. Matsuo)
Jun 10, 1980	Patent application via LRD
Jun 13, 1980	Communication of results in Malmö, Sweden Contact with Genentech via Dr D. Pennica
Sep 1980	Cooperative letter of agreement LRD/Genentech
May 1981	First patient treated with melanoma t-PA (W. Weimar, Rotterdam)
Sep 1981	Collaboration agreement for two years, with non-exclusive license
May 1982	t-PA cloned and expressed at Genentech
Mar 18, 1983	Exclusive licensing agreement with two year collaboration Collaborations with B. Sobel, H.K. Gold and F. Van de Werf
Feb 1984	First patient treated with recombinant t-PA (rt-PA)
Nov 1987	Approval of rt PA in the USA
Mar 1993	Demonstration in GUSTO of superiority of t-PA over SK
Worldwide sales in 1995 between 400 and 500 million US dollars	

milligram you would probably need a thousand litres of blood taken from a couple of thousand donors. Clearly, that is not a therapeutic option. The solution to the problem was recombinant DNA technology. The way this has been developed is illustrative of how academic research depends on patents. We discovered a good cell line that produced a lot of t-PA so it was easy to isolate it in relatively large amounts. When we had done this in 1979 with Dr D. Rijken, a collaborator from The Netherlands, we showed that this malignant cell derived t-PA was similar to the t-PA in blood. On 10 June we filed a patent application via Leuven Research and Development, which is the technology transfer arm of our university. Three days later we communicated our findings in Malmö, Sweden. A young Doctor in molecular biology from Genentech, Dr D. Pennica, was there. She was looking around for an interesting project. After concluding an agreement with Genentech, we very quickly supplied them with the cell line, pure protein and antiserum. They isolated the t-PA message and translated it back to DNA. In those days, for a protein of 70,000 containing 530 building blocks, the biggest one ever made, the anticipation was that that this was going to take years, but in less than a year and a half Genentech succeeded in cloning t-PA. The consequence of this productive collaboration with industry was that everything went very quickly, because basically the connection was made three days after the patent application. The downside was that, during the next seven years, until the patent was issued in 1988, we have talked and published about it so much that after our patent was issued by the European Patent Office, during the opposition period, many pharmaceutical companies were present. They tried to get this patent invalidated and they actually succeeded. There were two reasons why the patent was invalidated. Firstly, the cell line was not officially deposited in the registry so that it was not officially available to everybody, even though all the companies sitting around the table had received it directly from me; however, they said that they had no guaranteed access to it. The second reason had something to do with calibration.

Our early and detailed communications have had a clear negative effect on Genentech's business. We have published about 150 papers about this subject. It was our main research programme for many years.

With the work on Staphylokinase (cf Table 4.7), we may get a second chance. Again by mere serendipity, in 1989 a previous collaborator — a Japanese MD named Osamu Matsuo — who had been working with us on the first experiments in 1980 in Leuven, said that with the bacterial protein Staphylokinase he was able to obtain some in vitro results that looked similar to t-PA, more than to Streptokinase. We collaborated together for a while to demonstrate that this old protein, known since the 1940s, had been studied since the 1950s and 1960s in the wrong species: the dog. In the dog it does not work, because the dog has a somewhat unique fibrinolytic system for this application. In June 1991 we submitted a patent application on the use of Staphylokinase. We started treating patients in 1992. The patent was issued in the United States in 1994, and in Europe in 1996. I anticipate that, since we have published 35 papers, including clinical pilot trials in patients, we will probably have to go through an opposition proceeding.

Table 4.7
Efforts to develop staphylokinase via a spin-off company

Sep 1989	Discussion with O. Matsuo in Tokyo on fibrin-selectivity of Sak
1989-1990	Collaboration on biochemical mechanisms (with H.R. Lijnen and O. Matsuo)
1991	Cloning and production of Sak in Leuven Demonstration of thrombolytic potency in vivo
Jun 1991	Patent application on use of Sak, issued in USA in 1994
Jun 1992	First patients treated with recombinant Sak in Leuven
1992	Constitution of Thromb X NV by LRD and D. Collen
1993	Licensing of technology from IMB Jena via Medac GmbH
1994	Development of GMP production in Hans Knoll Institute, Jena, approved in Sep 1995
Oct 1995	Publication of first randomised trial in patients with AMI

How do you develop this into a commercial drug? If, as an academic research laboratory, you have a patent application or an issued patent and you go to the industry for an exclusive license — the only option you have to make it commercially valuable — you basically lose all control over it. In the case of t-PA, Genentech has paid us a lot of royalties over the years and probably will keep paying this for a number of years to come; I have no problems with the bills and neither do they. However, the problem is that they decide on the price-setting, on the commercialisation policy, on everything. It is basically their drug. This is the reason why we are trying to do this on our own as long as we can, so we set up a spin-off company from the University of Leuven, Leuven Research and Development and myself, and we are trying at least to take this a little further. The only problem is that, to develop such a drug, you probably need at least one billion Belgian francs. So sooner or later, I guess we will lose control of it anyway.

These are two examples of transfer. In all these cases, without patents or patent applications, you would have nothing. A company would not be interested without patents, that is one thing. A second thing of course is that it takes an enormous investment to conduct such clinical trials, to see if this is really superior to existing treatment. I will give you an example of what this involves. You have to treat patients with myocardial infarction. The patients who are available for study have a 10 per cent risk of dying without thrombolytic treatment. Since we now know that thrombolytic treatment saves lives, it has become unethical to treat patients in a control group without thrombolytic drugs. So what you give is the accepted, the best drug, and then you have to compare with an experimental drug. This means that one out of fifteen of your patients treated with standard treatment will die. Now you have to compare and see whether an improved drug or what you think is an improved drug has a mortality which is significantly lower. You can calculate, with biostatistical tables, that for such a study you would need at least 20,000 patients to have any reasonable statistical power.

Of course before we go to these patients, we need animal models. We can do things in cell culture, we can study the biochemistry of a purified product, but before you go to sick people you clearly need to have animal models where you can make clots, where you can dissolve the clots, where you can determine dose-response, where you can determine toxicology of the drugs, side-effects, pharmacokinetics and so on. This kind of research clearly has been shown to save people's lives — there are presently several thousands of people walking around in the world who would not be walking around without this treatment with t-PA or with similar somewhat less effective drugs like Streptokinase. So I would like to make a case that both the use of animals and the value of patents are of necessity for this kind of research, which can produce life-saving drugs.

Part Two
PUBLIC PERCEPTION OF BIOTECHNOLOGY

5 Adolescents' opinions about genetic risk and genetic testing

Gerry Evers-Kiebooms

First of all I would like to thank the organisers of this workshop for the invitation to give a lecture about public perception in the field of human genetics. It is a presentation of a different nature than the presentations in the first session, but nevertheless I hope that you will appreciate the content.

Introduction

During the last decades the field of human genetics has been characterised by an exponential growth of knowledge. The technology available to human geneticists for the diagnosis of an ever increasing number of inheritable disorders and traits has evolved over the last 25 years to a real arsenal of genetic tests. Predictive DNA-testing for late-onset diseases and carrier testing have become a reality for several diseases. Moreover, prenatal testing has become available for an increasing number of genetic disorders. As a consequence of the growth of knowledge and of an increasing number of tests, more and more individuals, couples and families will be confronted with decision making about genetic tests. Are they prepared to make this type of decisions?

The main part of my presentation today deals with the perception and the opinions of adolescents about genetic risk and genetic testing in Flanders. They are indeed the adults and parents of the future and the potential users of the new genetic tests. Some of them will be the future clients of genetic clinics, where they can benefit from genetic counselling, at least if those who need it find their way to these centres and if there are not too many other barriers, for instance to understand the genetic information, to cope with it and to use it in subsequent decision making.

Information about genetic risks, mostly given during one or more counselling sessions in a university hospital, and the availability of prenatal and other genetic tests do not only increase the control and responsibility of families, but they also entail the burden of difficult and/or painful decisions. This is true for decisions concerning further pregnancies when there is an increased risk to have a child with a hereditary disease. This is also the case for couples who are deciding whether they want to make use of prenatal diagnosis (either by amniocentesis or by chorion villi sampling). It is even more true after the discovery of a fetal defect, when the very painful decision to abort or not is required. At that time the wish to avoid the anticipated burden of bearing a defective child may be stronger than the desire for a child and may lead a couple to interrupt pregnancy, even in the face of pre-existent moral or religious objections to abortion.

Before presenting the findings of the study in adolescents, with their views and attitudes about some of these topics, I would first like to discuss very concisely the impact of genetic counselling, as it was measured in the past.

It always has been assumed that genetic counselling contributes to the families' better understanding of the recurrence risk, as well as to free informed responsible decision making concerning further pregnancies. Studies have shown that genetic counselling definitely improves the knowledge about risk of occurrence of a specific disease. They also reveal how difficult it is for many people to grasp the exact meaning of probabilistic information and how the emotional context of genetic counselling may also be an obstacle to coping with the cognitive facts. Many emotional factors (feelings of shame, guilt, fears, anxieties) are involved in genetic counselling. 'Genetic disorders strike to the heart of a person's self system since much of our self-image and self-esteem is bound up in our capacity for health and for producing healthy progeny'. Within this emotional context people may also be very disappointed by the probabilistic nature of the information. From their perspective risk information may not seem precise because it is only a probability and they want certainty!

Genetic counselling clearly has an effect on subsequent reproductive decision making. The effect can consist in a complete change of reproductive plans after genetic counselling, in a change from indecision to a clear-cut decision to have or not to have a pregnancy, as well as in a reinforcement of a decision that was made prior to counselling. Hereby the information about the numerical risk and the subjective evaluation of the risk level usually play a part, as well as the availability of prenatal diagnosis to prevent the birth of an(other) affected child. However, there are many other factors influencing the decision making process: the personality of the prospective parents, their relation, the perceived burden of the disease, the presence of living children, the strength of the reproductive drive, ...

It is the conviction of many professionals in the field that the poor knowledge of the general public impedes a correct understanding of the information given during genetic counselling and is an obstacle for free informed decision making. A lack of awareness about genetics is also a major reason why families seek information too late or not at all. In this context, professionals and patient organisations agree that a special effort should be made to improve school education about genetics, not limited to technical aspects. To get more insight in adolescents' spontaneous beliefs and opinions, as a starting point for an educational campaign, a study was carried out aimed at evaluating adolescents' attitudes about genetic disease and genetic testing. I will present a selection of results about the following three topics:

86

- Perceived severity of a handicap or a genetic disease

- Perception and subjective evaluation of genetic risks

- Preventive attitudes in the context of genetic risk

Methodology and sample study

The data were collected in close co-operation with the Medical School Health Service and the Department of Youth Health Care of the University of Leuven. Alternating self report questionnaires and standardised face to face interviews were used.

The sample was selected at random out of a population of adolescents meeting the following criteria: fifth grade high school students following general education or technical education and undergoing a medical check up in the Medical Health School Service in Leuven in the period January to May 1993. We refer to Decruyenaere et al. (1995) for more details about the methodology used and to Table 5.1 for a sociodemographic description of the sample.

Table 5.1
Sociodemographic characteristics of the sample (N=166)

Age:	16-20 years (mean age 16.7 years)
Sex:	55% boys
	45% girls
Type of Education:	53% General Education
	47% Technical Education

It is also important, because of the topics that will be studied, to give you an idea of the regularity of religious practice of this sample. 27 per cent was not religious, 39 per cent religious but rarely attending church, the others were irregularly attending church, rather regularly or very regularly attending church. This regularity of religious practice is not very different from the Flemish population of adolescents of that age.

Results

Perceived severity of a handicap or a genetic disease

What is the adolescents' idea about the severity of a number of handicaps? Hereby we distinguish the burden adolescents see for the parents of an affected child and the feelings attributed to a child or a person with a handicap.

87

Let us start with the burden for the parents (*cf* Table 5.2) This was measured on a seven point scale, from not serious to very serious. A child with a serious mental handicap is seen as having the highest burden for the parents, higher than a child with a serious genetic disease and than a child with a serious physical handicap.

Table 5.2
How do adolescents perceive the burden of the birth of a child with a handicap or a genetic disease for the parents?
Measured on a seven point scale (1 = not serious, 7 = very serious)

A child with 'a serious mental handicap'	6.14
A child with 'a serious genetic disease'	5.86
A child with 'a serious physical handicap'	5.48

Why do adolescents perceive the burden for the parents in this way? (*cf* Table 5.3) We looked at the reasons they have given to explain the severity and here I want to draw your attention to a result about genetic disease. It is very striking that 44 per cent of the adolescents gave feelings of guilt as one of the reasons why they perceive the burden of genetic disease so high. This is very important in the context of information about genetics to adolescents. The second way to know something about the severity of handicaps was asking adolescents what feelings they attribute to a person with a serious genetic disease, a serious physical handicap and a serious mental handicap. Genetic disease is considered as more negative for five of the adjectives, with a significant difference from the two other types of handicaps. This is true for 'feeling bad', 'feeling afraid', 'feeling shocked', 'feeling angry' and 'feeling sick'. So here too, we see that genetic disease has rather negative connotations and I think this is an important result.

Table 5.3
Reasons to explain the perceived burden of the birth of a child with a mental handicap, a physical handicap or a genetic disease

	Mental handicap	Physical handicap	Genetic disease
Shock, unexpected	46%	10%	10%
Necessary attention and care (by the parents)	32%	17%	13%
No normal life possible for the parents	29%	9%	(0%)
Not functioning like other children	19%	14%	(0%)
Fear for negative reactions of the environment	10%	(9%)	(0%)
Feelings of guilt	(9%)	(5%)	44%
Afraid of the risk for subsequent children	(0%)	(0%)	16%

We also tried to give an answer to the question whether adolescents think that the birth of a child with a mental handicap, physical handicap or genetic disease is a reason for not having subsequent pregnancies (cf Table 5.4). When we look at genetic disease, we see that 58 per cent of the adolescents think that genetic disease is often a reason for not having subsequent pregnancies.

Table 5.4

Do adolescents think that the birth of a child with a mental handicap, physical handicap or genetic disease is a reason for not having subsequent pregnancies?

	Mental handicap	Physical handicap	Genetic disease
rarely or never a reason	21%	32%	7%
sometimes a reason	52%	54%	35%
often a reason	27%	14%	58%

The perception and subjective evaluation of genetic risk

Table 5.5

Adolescents' numerical estimation of the risk of giving birth to a child with a genetic disease or congenital malformation

	Risk of random couple	Own risk
SPECIFIC ANSWER	N= 134	N = 130
Less than 1%	7%	21%
1-2%	22%	28%
3-5%	31%	25%
6-10%	23%	14%
11-20%	12%	5%
25%	2%	2%
50%	3%	4%
NO ANSWER	N= 32	N= 36

The perceived susceptibility to genetic disease (cf Table 5.5) is very important if we want to start an educational campaign about genetics. Here we have given a multiple choice question with seven numerical answer categories to the adolescents, and also the possibility of answering 'I don't know'. If we look at the last column, their idea about their own risk of giving birth to a child with genetic disease or congenital malformation, we see that one out of five adolescents immediately answer that they do not know. If we look at the answers of the others and we know that three to five per cent is the correct answer, we see that more adolescents underestimate the risk than overestimate the risk. So they think their risk is lower than it really is. If we look at their answers for the risk of a random couple, we find a very interesting phenomenon. We see that adolescents think that the risk of a random couple is on average higher than their own risk. This phenomenon is called 'unrealistic optimism'. So there is a shift to lower risk categories when you have to give an answer about your own risk compared to the risk of a random couple. This is illustrated in Figure 5.1, where you see a clear shift to lower risks for own risk, as compared to risk of a random couple.

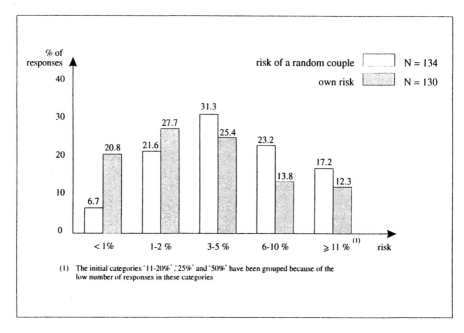

Figure 5.1 Unrealistic optimism when estimating the risk of giving birth to a child with a genetic disease or a congenital malformation

We also asked questions regarding the subjective evaluation of the risk (*cf* Table 5.6). Here we see that about half the adolescents feel that their own risk is very low or rather low. It is important to know this subjective evaluation.

Table 5.6
Subjective evaluation of the risk of giving birth to a child with a genetic disease or a child with a congenital malformation

	Risk of a random couple	Own risk
SPECIFIC ANSWER	N=155	N= 154
Very low	11%	19%
Rather low	39%	31%
Medium	33%	36%
Rather high	13%	10%
Very high	4%	4%
NO ANSWER	N=11	N=12

The preventive attitudes in the context of genetic risks

Now we go to the last topic, the preventive attitudes in the context of genetic risks. First of all we asked whether adolescents thought that the birth of a child with a genetic disease is preventable (*cf* Table 5.7). Here we see that in their spontaneous belief, about two thirds think that it is not or probably not.

Table 5.7
Do adolescents think that the birth of a child with a genetic disease is preventable?

Certainly not	18%
Probably not	45%
I do not know	16%
Probably	19%
Certainly	2%

A very important question is: 'Suppose that you want children in the future, would you ask a doctor for information about increased genetic risks before pregnancy?' (*cf* Table 5.8). Here we see, in the last two rows, that slightly more than a half would seek information about a possible increased risk before getting pregnant.

Table 5.8

Suppose that you want children in the future. Would you ask a doctor for information about increased genetic risks BEFORE PREGNANCY?

Certainly not	3%
Probably not	26%
I don't know	15%
Probably	39%
Certainly	17%

If we look at the arguments they give to seek information before pregnancy (*cf* Table 5.9), we see that about one-third would do so to make informed decisions about reproduction and that 20 per cent would do so to be better emotionally prepared for the birth of an affected child.

Table 5.9

Reasons for or against information seeking BEFORE PREGNANCY

(Reasons or comments given by more than 5% of the adolescents)

Arguments pro	
Making informed decisions about reproduction	33%
Being better emotionally prepared for the birth of an affected child	20%
Arguments against	
Strong desire to have children, regardless of possible genetic risks	11%
Not necessary - no genetic diseaese in the family	8%
Not thinking about risk in the context of family planning	7%
Other comments	
Dependent on the presence of genetic disease in the own family or in the partner's family	13%

Table 5.10 refers to the question about the use of prenatal testing to detect serious diseases in the child. If we look at the last two rows, we see that slightly more than half report that they would use testing. If we look at the arguments pro here, we make a rather surprising finding (*cf* Table 5.11). The main reason that is mentioned to ask for prenatal testing is 'being better emotionally prepared for the birth of an affected child'. So in adolescents' spontaneous belief, there is no direct link or not a very pronounced link between prenatal testing and decision making about the continuation of pregnancy. It is mainly to be prepared for the birth of an affected child.

Table 5.10
Would you use testing during pregnancy to detect serious diseases in the child?

Certainly not	6%
Probably not	20%
I don't know	11%
Probably	45%
Certainly	18%

Table 5.11
Reasons for or against prenatal testing in pregnancy
(Reasons or comments given by more than 5% of the adolescents)

Arguments pro	
Being better emotionally prepared for the birth of an affected child	40%
Making informed decisions about abortion	21%
To start special treatment in time	7%
Arguments against	
Reluctance about abortion	21%
Information is too worrisome	5%
Other comments	
Depending on the presence of genetic disease in the own family or in the partner's family	11%

Then we presented a number of hypothetical situations and asked, if such a type of situation would occur, whether they would terminate the pregnancy (*cf* Table 5.12). In the first situation, if the child would be mentally normal but would have a serious physical handicap, nine per cent would terminate the pregnancy. If the child would be seriously mentally handicapped, 18 per cent would terminate the pregnancy. I want to draw your attention to the high number of 'don't know' answers. That is of course also explained by the lack of life experience at the age of adolescents, but we absolutely wanted to know their spontaneous beliefs at that age, in order to start an educational campaign in the last year of the schools. In the third situation, if the child would have an incurable disease, associated with serious consequences for the rest of his life, 32 per cent would terminate the pregnancy. If the child would die soon after birth, 51 per cent would do so.

Table 5.12

Suppose that in the future you and your partner are in one of the following situations. Would you decide to terminate the pregnancy?

If the child would be mentally normal, but would have a serious physical handicap

Yes	9%
I don't know	29%
No	62%

If the child would be seriously mentally handicapped

Yes	18%
I don't know	37%
No	45%

If the child would have an incurable disease associated with serious consequences for the rest of his life

Yes	32%
I don't know	44%
No	24%

If the child would die soon after birth

Yes	51%
I don't know	30%
No	19%

Conclusions

Genetic disease clearly has negative connotations. Underestimation of the genetic risk also occurs much more often than overestimation. Moreover, adolescents have the feeling that their own risk is lower than the risk of a random couple ('it will not happen to me, it will happen to others'). This is the phenomenon of unrealistic optimism. Before pregnancy, slightly more than half of the adolescents would ask a doctor for information about an increased risk, one third to make informed reproductive decisions and one fifth to be better emotionally prepared for the birth of an affected child. During pregnancy slightly more than half of the adolescents would make use of prenatal testing. The main arguments in favour of prenatal testing were being better emotionally prepared for the birth of an affected child, for 40 per cent, and decision making about continuation or termination of pregnancy, for 21 per cent. The attitudes towards abortion in hypothetical situations of serious disease detected during pregnancy clearly show a reluctance. Adolescents clearly have a critical attitude as a function of the nature of the handicap. An important consequence of these findings is that too strong an emphasis on pregnancy termination could induce an aversion towards genetic information and genetic counselling. This should

94

be kept in mind to reach the main goal in educational campaigns: providing balanced information about genetic risk as a starting point for free informed decision making. To conclude my lecture, I just want to say that last year, some of the findings of this study have played an important part during the realisation and production of an informative video film about genetics for the general public, but with special attention paid to adolescents. The video film is ready and is entitled 'Genes, generations and society: about human heredity'. It is a production of the Centre for Human Genetics in Leuven and the audio-visual unit of the University of Leuven and it has been made with the support of the Flemish Action Program for Biotechnology. I just wanted to illustrate how important it is to have a view on the spontaneous beliefs and the knowledge of adolescents as the starting point of an educational campaign that is seen as very important, as well by professionals in genetic centres as by patient organisations.

References

Decruyenaere, M., Evers-Kiebooms, G., Welkenhuysen, M., Bande-Knops, J., Van Gerven, V. and Van den Berghe, H. (1995), 'Adolescents' opinions about genetic risk information, prenatal diagnosis, and pregnancy termination', *Journal of Medical Genetics*, Vol. 32, pp. 799-804.
Evers-Kiebooms, G., Denayer, L., Decruyenaere, M. and Van den Berghe, H. (1993), 'Community attitudes towards prenatal testing for congenital handicap', *Journal of Reproduction and Infant Psychology*, Vol. 11, pp. 21-30.
Garver, K.L. and LeChien, K. (1992), 'Geneticists' responsibility to other health care professionals and to the lay public', *American Journal of Human Genetics*, Vol. 51, pp. 922-3.
Green, J., Snowdon, C. and Statham, H. (1993), 'Pregnant women's attitudes to abortion and prenatal screening', *Journal of Reproduction and Infant Psychology*, Vol. 11, pp. 31-9.
Watson, E.K., Williamson, R. and Chapple, J. (1991), 'Attitudes to carrier screening for cystic fibrosis: a survey of health care professionals, relatives of sufferers and other members of the public', *British Journal of Genetic Practice*, Vol. 41, pp. 237-40.
Weinstein, N.D. (1987), 'Unrealistic optimism about ilness susceptibility: conclusions from a community-wide sample', *Journal of Behavioral Medicine*, Vol. 10, pp. 481-500.
Welkenhuysen, M., Evers-Kiebooms, G., Decruyenaere, M. and Van den Berghe, H. (in press), 'Unrealistic optimism and genetic risk', *Psychology and Health*.
Welkenhuysen, M., Decruyenaere, M., Evers-Kiebooms, G. and Van den Berghe, H. (in press), *Jongeren en Erfelijkheid*, Garant: Leuven.

6 Intrinsic value of balanced views and policies on biotechnology

Rio D. Praaning

May I be permitted to make one observation on the group as we are here? Many sciences and expertises are represented here, but I miss patients organisations. I do know that people representing animal welfare groups are present, but perhaps it could be useful to consider inviting patients organisations next time. They have different views every now and then.

Perhaps to provoke some discussion, I would like to start with a number of statements and I will very briefly explain why I make them. The first statement is 'There are lies, damned lies, and statistics'. To demonstrate the value of that statement, I will very quickly take you to a random number of opinion polls held throughout the past few years, both in the US, in Europe and in Asia. Much of what I refer to has been published in *Biotechnology* in the September 1994 issue, or otherwise by Eurobarometer or Gallup. 12 per cent of citizens in the European Union member states know about biotechnology in a certain degree of detail; 16 per cent in the US. Eight per cent in the US have heard a lot about biotechnology and 30 per cent in the US have heard something about biotechnology. In Japan it is 64 per cent who say they have heard something about biotechnology, 32 per cent in Japan claim they understand somewhat about biotechnology, 26 per cent in Japan say 'We can explain what biotechnology is'. Just as a measure of comparison, 6 per cent in China know what DNA is. We asked a trick-question in Germany a couple of years ago. During an opinion poll people were asked whether chewing gum or something close to it was probably or not a result of biotechnology, and whether bread was. The majority of the people said that chewing gum was, and the people were much less convinced that bread was. Only a week ago, there was an interesting conference of the European Commission in Brussels. Many members of the European Commission were taking part and amongst them was the Commissioner responsible for DG III. He said 'acceptance' does not imply 'total acceptance' and 'rejection' of biotechnology or genetic engineering does not imply 'total rejection'. To prove that statement — because I think there is a lot of wisdom in it — I will quote another set of numbers. 73 per cent of science policy leaders in the US fear that monsters could be created, or had ecological concerns over the further progress of biotechnology. In the same poll however, 61 per cent said they thought that the benefits of biotechnology and genetic engineering were much greater than the risks, and only 6 per cent said it was the other way round. In Europe, 85 per cent wish to counterbalance perceived high degree of risk with strong regulations and control, six per cent want a total ban of certain forms of biotechnology. In the US, in 1986 — I simply continue with a couple of numbers and I will point out later why I am using all these numbers roughly — 25 per cent say genetic engineering is morally wrong. In 1984, two years before, 80 per cent said it was compatible with their philosophical and religious values. In 1990, a similar question was asked in Europe; 20 per cent said it was unethical and 32 per cent said it was ethical to continue with genetic engineering. Finally, in Europe, the poll-results were as follows: according to Eurobarometer 74 per cent accepted gene therapy in 1991; 72.5 per cent accepted gene therapy two years later; 16 per cent were against in 1991 and 19 per cent were against two years later. One last number on gene therapy: in Japan 74 per cent of the Japanese are strongly against gene therapy, but 54 per cent would accept the technology if it would be used to cure their own fatal disease.

I come to the second statement. When I met a good friend of mine in England just a few weeks ago, I asked 'How is your wife?', and he replied 'Compared to what?'. In other words, expressions of public opinion in all these quotes and numbers are only really relevant if they are compared to each other and if you compare the way you ask questions and to whom you ask them, where you ask them, after which event you ask them, in which order you ask them. All these factors are extremely relevant. If you discount those questions, then you are bound to draw the wrong conclusions. Let me give you a small example. I refer to Switzerland, with a population clearly split mainly into Romanic Swiss and Germanic Swiss. They were also put to task on the issue of biotechnology and genetic engineering. The results were that 50 per cent of the Romanic Swiss said they would accept genetic engineering, compared to 33 per cent of the Germanic Swiss. Now, does this information say something about biotechnology and genetic engineering or does it say something about Switzerland?

My third and final statement is 'Political ideology, by its very nature, will be selective in its information and arguments. It is therefore important to understand the difference between isolated moral and ethical considerations concerning biotechnology on the one hand, and views of the organisation and management of societies on the other.' Now, if there is only one truth, are there not many ways of expressing that truth, wouldn't it be rather careless to presume that one expression is more valuable or correct than the other? If one would look back in time from the year 2020 for just one brief moment, what would biotechnology and genetic engineering look like in that period of time? How will we have reached that point? What will have survived in terms of research, development and production and how will people consider the benefits or non-benefits? If these questions, wise or not, are applied to biotechnology and genetic engineering, one may come to the conclusion that there is not one truth but rather a continuum of development, reflecting all the motion in thought, potential, understanding, application and rejection and acceptance, which is so typical for the development of any technology in a mature society. This very continuum makes it difficult to draw clear technological lines in the sand. Even if such lines could be drawn, any fresh wind could lead to change.

In other words, to turn back to the patenting issue, if you want to discuss pro or contra patenting, one might be on the wrong track or at the wrong point of the discussion because if you want to be against certain forms of research, development or technology, in that case it is obviously better to try to stop that research and not to discuss the patenting of it. In general, if public opinion is not led by a set of clear and understandable indicators and considerations, it is bound to reflect uncertainty, insecurity, fear, hesitation and sometimes outright rejection. This can be applied to a large number of fields: to the use of force, to military intervention, to a single European currency, EU-membership or technological development. So the dilemmas which face the technology we are discussing are not unique. In response to those dilemmas clear boundaries and limits are needed, as well as openness and transparency. These words are heard on both sides of the divide. But if applied, they may be just as simple as misleading, even though they may be helpful for whatever public understanding and support is sought. This leaves much room for manipulation and simplification, which can only truly be exposed by time itself. However, for the time being, all sides are apparently bound to use these methods and simplifications, since governments are relatively short-lived and tend by their very nature to deny that continua exist. When governments are mentioned, one automatically refers to the word 'control'. That is an important word, because if there is hesitation and if there is very careful progress in whatever technological development, people want to have control. Who should control? Obviously an authority should, and logically that would be a government. Now how does a government control? Let us look at another example: the development of chemical weapons. I was privileged to be close to the negotiations to ban chemical weapons. It is very hard to reach a point where you can effectively control, and where you can effectively be sure that somebody is not doing what he says he is not doing, or doing what he says he is doing. In fact, there has been a century of negotiations before reaching the position to agree on a convention. Since it is so difficult to control, I suppose we should try to use every method available to have control, openness and transparency in the area which we are discussing. If that is true, patenting could be an extremely important factor, because patenting drives industries and also the economic world to openness. After all, you must deliver your documentation, your papers, to a forum where everybody can see, read and know what you have done and what you plan to do; more or less, because of course you cannot be too explicit over what you do with a certain invention. Thus, patenting might be an extremely important tool to bring about openness and transparency, which would otherwise be lacking.

The effort to capture technological development in a legally, ethically and socially acceptable box seems to make progressive political groups conservative, and conservative industries progressive. That is a reversal of roles, which can also be seen in other fields. As an example, I refer to the peace movement. Some peace-movements were pleading for military intervention in the former Yugoslavia. This happened while we were looking at it. Given my very practical remarks on the biological and chemical warfare issue, I could well imagine that such a reversal of roles could take place here as well after a while. Anyway, such a temporary box cannot prevent the technological continuum from developing further. More reversal of roles may, and will be forthcoming. This process presumably will nor can be stopped. It may well be possible to delay or to stimulate certain parts of it, but unless there are universal brain operations an indefinite stop may be quite illusive. It may be precisely that fact, rather than any group in particular, which sows the seed of distrust among the participating parties in this dialogue. If handled badly by a national or international authority or by any of the other partners in the dialogue, the seed may flourish and cause conflict. This should not be too worrisome, because the conflict itself will only be a temporary state of affairs. If handled well, i.e. if authorities and partners in the dialogue finally would manage to avoid conflict, the effort to capture the continuum may include the following. Firstly, the marriage of ethical, social and legal aspects in a regulatory framework which reflects the broadest possible variety of cultural, religious and political views. Secondly, the explicit involvement of user groups, i.e. consumer- and patient-organisations, the latter usually with a different set of ethical considerations than their healthy counterparts. Thirdly, the explicit wish to inform and educate the public in a balanced, careful and somewhat distanced manner, which implies that the tendency to involve secondary and even primary schools must be viewed with considerable scepticism. It seems necessary to avoid that schools are seen as just another battlefield where victories can be more easily scored. Fourthly, the explicit reflection of flexibility and vision, which is imperative for open and progressive societies.

7 Biotechnology-related consensus conferences in Denmark

Lars Klüver

I will begin to thank the organisers for the invitation to come here. I think this will be as John Cleese says 'Now for something completely different'.

I will talk about procedures, rather than about biotechnology and patenting, although there will be a little substance in the end. First a little about who I am and what kind of organisation I come from, just to let you know who is speaking. The Danish Board of Technology was set up in 1986 by the Danish Parliament to perform parliamentary technology assessment. The purpose of the Board is to follow technological development, to carry out comprehensive assessments of the possibilities and consequences of technology for society and for the citizen, and finally to communicate the results to the Danish Parliament, to other political decision-makers in society and to the Danish population in order to initiate and support a public debate. It was quite a job to find out how we should do all this and we ended up with a list of four points about our mode of work. These four points are as follows. Danish technology assessment has to include the wisdom, experience and visions of the citizens, the insight and tools of the experts, the needs and working conditions of the decision makers, and finally the democratic traditions in Denmark. We usually use the phrase that the job we do is building bridges between the scientific, the public and the political scene. To do that huge job we have little budget. In international terms, I think we are quite a large technology assessment organisation, but as an organisation as such, we are small. We have 10,000,000 DKr, which is approximately 50,000,000 BEF, or 1.4 million ECU. That is not very much. We are around 13 people, among them eight academics plus different project consultants. We have a lot of people working for us for free, e.g. people from the academic society, who want to have us as some kind of tool to be able to talk to politicians. So I guess we involve around 100 persons directly in our projects during one year. We use a series of different methods, which can be split into three groups. Surveys or research carried out by one expert are what we call 'expert-oriented methods'. They are not used very much, and in that respect we are special in technology assessment. The reason we do not use these methods very much is political. I am sorry to say so, but researchers do not have very much credibility politically. Especially when you talk about social research, I would say, broadly, that counts. The reason of course is that, if you ask one researcher at one time and another at another time, you will get two different answers. That does not establish credibility. So what we do more often is to create a mix of experts and make them shout at each other. That is what we call 'cross-discipline expert groups'. I think that we are mostly known for our 'participatory methods'. We use 'scenario workshops', which are 'vision-producing workshops', where we use the lay people, the politicians, the scientists themselves, and we make them provoke each other and create visions. And we use 'consensus conferences', on which I will focus here. But we have to use public debates too. We rely heavily on the Danish tradition of local debate. Therefore we simply fund local initiatives for debate.

Now you have an idea of what we are. So, one of the things we do is 'consensus conferences'. We have two panels in a consensus conference. The heart of a consensus conference is the lay panel. They define the agenda, they put the main questions and sub-questions which the conference has to answer. They also make their own conclusions at the end of the conference. In order to get insight and overview, they put the questions to a panel of experts. We use a very broad concept of expertise: we say that anybody who can give us a special insight is an expert. Thus, in a medical technology assessment project, it could be a patient. Or it could be a mother of a sick child or somebody like that. It could be anybody who can give us a special insight.

The procedure is the following. We set up two preparatory weekends with the lay panel. During these weekends they get a basic education, preferably given by high school teachers. We do not go too deep into the matter, we have the conference for that later. We give them enough to show that they do not have to be afraid of the difficult words or of the way experts talk. We simply try to make them feel secure with the issue. During this weekend, they brainstorm issues and concerns and work a lot with the material created during the brainstorming. They group the issues into main questions and subquestions. Why do we do this very thorough preparatory work? All of you have been at some kind of seminar where you get five questions. You go into a group room, you sit there and you have to come out with some answers to those questions. Each group gets the same questions and yet they come up with different answers. That is simply because they understand the questions differently, each of them. We cannot afford to have that at such a conference. Therefore first of all we simply want the lay people to express their own concerns and let the conference be about their concerns. However, we also want them to work with the issues so they actually understand their own questions. Since they are going to draw conclusions, it is very important that they understand the questions.

The conference in itself is a three- to four-day event. The first day is a long day with presentations from the experts. The second day we have half a day of cross-examination. The lay people simply put further questions to the experts. Afterwards the lay people withdraw for the rest of the day, or a day and a half, and write the final document. The last half day of the conference is used for presentation of the final document, debate, and press-conference.

Not every topic can be taken up by means of consensus conferences. It has to be foreseen that this will be a future debate topic, that it will be on the public agenda some time in the near future. It has to be important for the public in general. It should not be too abstract. It has to be possible to delimit, so you cannot have a very academically defined problem. It has to be a problem which lay people can identify with. However, it might simply be a headline, like gene therapy or something like that, because they can identify with that. And they will very soon find out what they want to know about it. It has to contain conflicts. The lay people need to have something to work with, they have to have something to cut through — a cake to cut through. It has to call for clarification of attitudes, it requires expert-input and there has to be expertise available.

What do these conferences do? The conferences I will present here are not about patenting and some of them are not about biotechnology, but just to give an idea of what they can end up with, I cover some of them here.

One conference in 1987 was about gene technology in industry and agriculture; the Parliament decided afterwards not to fund animal gene technology in a research programme.

Irradiation of fruits in 1989: the Parliament decided on a policy against irradiation of foods, except for dry spices, and Denmark has stuck to that policy since then.

Mapping the human genome in 1989: the Parliament decided to make a law against use of gene tests in the areas of employment and insurance.

The future of private motoring in 1993: the Minister of Environment expressed that he wanted petrol prices to be increased fourfold and he is actually still working on that.

Childlessness in 1993: the Minister of Health has taken up a discussion on the economic gap between free infertility treatment in the Danish hospital sector and very expensive international adoptions.

Information technology in traffic: the Ministry of Internal Affairs expressed that they would change policy, so safety measures achieved priority over logistics.

Integrated agriculture in 1994: the Danish Council of Agriculture has, only three days ago, put forward a plan which supports introduction of integrated production in Danish agriculture overall.

So these conferences do have direct results, but maybe the diffuse and immeasurable results are the most significant in the end. Because they start a process in Danish society, they are what I usually call 'markers-in-time'. When a discussion is held two years later, someone will say 'well, at the consensus conference, the lay panel said that'. So they simply become a symbol of legal discussion around that theme.

Why do they have this impact and what makes these consensus conferences relevant? I will run very quickly through some arguments related to the political characteristics of lay people. First of all, they are motivated, they know that they and their children have to live with the thing we are dealing with. Here we have fourteen motivated people. They mean what they are saying. Second, they have relevant expertise, they know what it is to live in a society where decisions are taken by everybody else but them. So they have expertise in what it is to be in a society which is politically regulated and regulated by power-bases. Third, they have legitimacy, they have the right to speak as free individuals. Fourth, they are independent. They are non-stakeholders, only representing themselves. Fifth, they are neutral: they have no reason to harm anyone. Thus, what they are saying, they are saying because they mean it, and not because they are mean. Sixth, they have credibility, they actually try to live up to the values they believe in. You can — to a very high degree — see a consensus conference as a negotiation between values.

We have had five biotechnology-related consensus conferences in Denmark: gene technology in industry and agriculture, mapping the human genome, technological animals, childlessness, and gene therapy. There are not many conclusions about patenting in those, because, as I said, everything depends very much on what the lay people focus on. Particularly the one on technological animals examined 'patenting'. And I think it is important that the general advantages of patenting were accepted by the panel. The panellists said that it ensures public access to knowledge, that it becomes easier to finance research, that research is stimulated, that it safeguards competitiveness for a while, that patenting gives a uniform international policy on copyright questions. Those are advantages they accepted. However, they found patenting regarding animals problematic. It tends to make us consider animals as things, objects. Furthermore patenting may create monopolies which could result in higher prices. Some of you may say, 'well, that is the meaning of patenting'. But in the eyes of the lay people, this is not necessarily good. Farmers may have to pay in the future to have licences on their own breeds. The farmers' privilege and things like that were very important to the lay panel. Another thing they said was that there is a lacking logic regarding biotechnology and patenting. I quote: 'We find it difficult to comprehend the logic of the patenting area. In our opinion, one cannot apply for a patent on a gene since this is a discovery, and not an invention. With synthetic production of a gene, which is identical to the original gene, it is possible to patent the synthetically produced gene. Thus, a discovery becomes an invention.' That may be logical to some of you, but it is not at all logical to many other people. The conclusion on patenting at the conference on technological animals was: 'We in the lay panel do not find patenting of life acceptable and we call on the Danish Parliament to keep to its decision prohibiting the patenting of animals. Also, we ought to use the possibility of influencing the development in the EC, thereby influencing international patent corporations.'

At the gene therapy conference in September 1995, the lay panel was told by the experts that medical treatments cannot be covered by patents, referring to Danish patenting practice, and that genes as such cannot be subject to patenting either; in connection with the production of a pharmaceutical product made through knowledge related to gene technology, patents are feasible. And finally, of course, the lay panel was told that discoveries cannot be patented, only inventions. They did not have comments in the final document about that. But — and now you have to believe me — in the discussion afterwards, they stated that one of the reasons why they had not made a conclusion was because they did not understand the logic. One of them remarked: 'How can a discovery, a gene, turn into an invention? The same gene through a research process?' And so, what you can learn from this is that the logic creates a lot of confusion. Personally, I could translate that into: 'If you want to make discoveries patentable, then make discoveries patentable; do not try to use word games, or ways of changing the meaning of words in order to make it possible.' That is the problem with the public, they simply do not understand the logic.

If consensus is defined as the highest common denominator among the participants — which is how we define it in a consensus conference — although my 'empirics' are few, I would say that the lack of logic may play an important role. It may be impossible to agree on the functionality of genes as inventions. There seems to be a widespread value-conservatism regarding the ethical limits for patenting. For example, the patenting of animals is not generally accepted. The general idea of patenting is accepted, but limits are drawn regarding biotech innovations.

That leads me to two conclusions. I think that the only way consensus — defined as the highest common denominator — can be formalised seems to be through regulation, as I see it. I do know that this is controversial. The reason is that I do not think anybody will go into a debate, seriously, constructively, and try to seek consensus, if they know that it will be a 'yes' anyway. So what I am saying is that if you want a substantial debate and actually want everybody to take the debate seriously, then the debate has to have a weapon. It has to be possible for the debate to end up with a 'no'. Trying to create consensus case by case on top of regulation, people cannot accept. It is a dead-end project. If processes of consensus are going to go on, then it has to be possible for the participants to take it seriously.

As a concluding remark, I would say that I have read the project description in the framework of which this seminar, this conference, is organised. And in that description, two definitions are used: one is 'substantial consensus' and another is 'procedural consensus'. I think the experience we have with consensus conferences and how they can be taken seriously shows that you can get a procedural consensus, but if there is no room for consequences around the substantial consensus which is reached in that process, it is ridiculous to start off with the procedures.

Part Three

ETHICAL, SOCIAL AND CULTURAL ASPECTS OF THE BIOTECHNOLOGY DEBATE

8 Ethical aspects of genetic engineering

Etienne Vermeersch

By way of introduction I would like to draw your attention to the following caveat: this is not an anti-scientific or anti-biotechnological speech. I have always been a rational philosopher, deeply influenced by the rather optimistic and pro-scientific conceptions of the Enlightenment. But I am also a human being of the end of the twentieth century, who has come to the awareness that, although progress of science and technology continues to be an essentially positive aspect of our culture, this can only continue to be the case if you look at its development with some caution, even with some concern. Science is still very valuable, but there are some dangers inherent in it and these dangers can only effectively be dealt with by those who are aware of them. Moreover, when you take into account that there is a growing concern among the public, especially about biotechnology, it would be naive to wipe all criticism from the table as coming from uninformed or irrational people. You will have more chances to be accepted as a valuable discussion partner if you show some awareness of the fact that there are indeed some problems around.

With your permission, as a philosopher, I will start from a rather general approach and progressively try to come nearer to the specific social and ethical problems pertaining to the patent issue. It is indeed my main point that all the issues concerning patenting of genetically modified organisms, should be treated in a kind of comprehensive methodological and ethical theory.

Technology consists in the imposition of form on a material or energetic substrate by human beings. We can thus make a distinction between natural objects or energy states — whose form is determined by non-human natural processes — and technological objects or energy states — which derive their form partially or completely from human intervention. Since imposing form, or 'shaping', is a rather general characteristic of human activity, also manifested in language, ritual, play and so on, it seems convenient to restrict the use of the term 'technology' to those forms which are economic goods. Furthermore, information is form 'par excellence'. Hence all types of detection and organisation of knowledge for the achievement of practical purposes pertain to the general notion of technology. Traditional technology begins in the Palaeolithic era with the production of tools, including weapons, utensils and fire. In the Neolithic period, the first types of biotechnology are to be seen in the domestication and artificial selection of plants and animals, and in the unconscious use of bacteria and yeast for the production of bread, cheese, wine and so on. We also witness the introduction of new material substrates, especially metals, and the use of animal muscle power as an energy resource. In the so-called 'high cultures', earlier types of pictorial representations evolve into an information technology, viz. 'scripture'.[1] Modern technology, from the 18th century on, had essentially the following characteristics:

- a progressive replacement of muscle power by other energy sources, especially fuels,

- the use of scientific research for the design of new forms and transformation procedures,

- the imposition of form on energy, especially electricity and electromagnetic waves, which resulted in a new information technology.

Contemporary technology, since about 1950, can be characterised as follows. Firstly, we witness an enormous expansion in scale and diversity of the production capacity. This led, in interaction with science and economy, to a rather coherent system which we call the 'scientific technological capitalist complex'. This so-called 'STC complex' has a thoroughgoing impact on needs, consumption patterns and social life in the industrialised societies, and has spread gradually over the whole world. Secondly, the information technology has achieved a degree of complexity which makes the simulation of evermore typically human cognition processes possible. Artificial intelligence is on its way to replace human intelligence. Thirdly, the mastery of recombinant-DNA techniques and other types of interference in the reproduction have led to a breakthrough in biotechnology. Instead of being limited to procedures like artificial selection and cross-breeding, based on the choice of phenotypes, the imposition of form can now be affected at the level of the embryo and the genome. Finally, technology is increasingly producing objects and energy processes which are completely new, as far as their nature or scale are concerned. They constitute a type of foreign elements, which will never more disappear from the environment, or will only do so very slowly (for instance: radioactivity, new isotopes, new organic and inorganic compounds and genetically modified organisms).

These profound changes mark a radical break between contemporary technology and all the foregoing stages, because they imply a blurring of the distinction between producer and product, between creator and creation. In the three characteristics mentioned above, I suggested that man brings such objects and processes into being which have the capacity to fundamentally modify man himself in his social, mental and biological functioning. Regarding the latter innovation, one could say that nature — through one of its components, mankind — can undergo a disturbance in its equilibrium which could last for millions of years. When a product can exert a drastic influence on its maker, when man-made technology can — in interaction with science and economy — modify human beings themselves, there is a danger of the emergence of an autonomously evolving system that slips away from the rational control of man and generates its own rationality and finality, to which human values and needs could become subservient.

In what follows, I will restrict myself to the study of only one of the four characteristics of contemporary technology mentioned above, namely the interference in the genome of organisms. I hope that this elementary analysis may serve as an example of the research we need and of the type of problems we will face with regard to other characteristics.

This topic exhibits a paradigmatic character in the sense that it did give rise to a new type of technology criticism. Critics from the nineteenth century onwards were mostly referring to the loss of the natural, the individual and the spontaneous, due to mass production and consumption. Those from the middle of the twentieth century drew attention to the risk of a gradual dehumanising of the social structure in which technology is evolving from a means in the service of human needs to an autonomous system that strives after its own aims. The criticism concerning biotechnology has an even more radical character. The step taken here is considered by some as the fateful crossing of a sacred threshold, by others as the violation of a taboo or the incursion into forbidden territory. The expressions 'playing God' or 'better than God' suggest that the domain of living organisms is considered so sacrosanct as to make it a sacrilege that it be subject to the dominion of man. This conception constitutes a radical break from the view on culture and technology that has been prevalent for more than 2500 years. Indeed, Hippocrates was expressing a generally accepted sentiment when he said in the fifth century B.C. 'ton men bion he phusis edoke, to de kalos zen he techne': 'biological life has been given to us by nature, but the good life is given to us through technique'. In the Greek, the Roman and later the Christian civilisation, it was the task of man to transform nature in such a way that it would become 'hemeros', 'excultus', that means, domesticated, civilised, polished and adapted to the well-being of man. If technology has always been a way of changing nature, of 'playing God', why did it suddenly acquire such an ominous character?

Before we focus our attention on this and other ethical problems generated by biotechnology, let us try to look at its development from a more general or, if you like, more abstract point of view. When you consult books on philosophy of science or technology, you will find a number of definitions implicitly or explicitly suggesting that there is such a thing as 'science', 'pure science', 'basic science' — which is a method of investigating nature to satisfy our need to know. Alongside there is 'applied science' or 'technology' — which is the organisation of knowledge for the achievement of practical goals. Though subtler distinctions have been made by some authors, in general we get the impression that the striving after knowledge, the pursuit of a rather passive state of insight into the structure of the universe, is the essential characteristic of science and that the more active part, the production of objects, is the proper domain of technology. Although this view has always been a simplification of the real state of affairs, it was gratefully accepted by a number of scientists after the second World War, when the development of nuclear weapons led to growing criticism of science. Pure science, they said, was not to blame and should never be subjected to control by society; only its application in the field of technology could be misdirected by governments or by industrial organisations. Knowledge of truth, with its first passive stage, was an unquestionable good in its own right. The progress of science was the progress of mankind itself; knowledge for the sake of knowledge could not do wrong. Yet, in our culture there has always been the curious suggestion — made in the Bible, Ecclesiastes, Chapter I, verse 18 — 'For in much wisdom is much grief and he that increases knowledge increases sorrow'. Of course it is not my intention to invite you to embrace wholeheartedly the total mistrust of science which we find in 'Ecclesiastes', a book which, according to Schopenhauer, 'contains the quintessence of pessimism'. I do submit, however, that the recent development of science and technology in general and the rise of biotechnology in particular, force us to a position where an unqualified optimism does not have a place any longer. There are several reasons for this stance. First of all, since the twentieth century science and technology are so interrelated that they could no longer lead a separate existence. Technology cannot fully expand without the theoretical basis provided by science and science itself cannot pursue advanced research without the enormous technological arsenal needed for the detection and measurement of new phenomena. Purely theoretical science has itself acquired an intrinsically technological character. It follows from this that a sharp distinction between inventors and users, and even between discoverers and inventors, is no longer tenable. You know that this distinction between discoverers and inventors is still an important distinction in the patent world. This obstacle to a rigid distinction between science and technology manifests itself in many domains, but it is exemplified in a privileged way in bioscience and biotechnology. Let me give two examples which are paradigmatic as far as the relations between science and technology are concerned. The first one forces us to cast doubt on the simple dichotomy of a rather passive, receptive, pure science versus an active, productive, applied science or technology. When you open a rather recent book on molecular biology, for instance the work 'Genes and genomes' by Maxine Singer and Paul Berg, published in 1991, you are immediately impressed by the fact that a major part of this purely theoretical textbook for graduate students is devoted to the recombinant DNA breakthrough. This means that molecular biology is no longer restricted to the study of objects or phenomena existing in nature, but also constantly creating part of its own subject-matter. One tries to change macromolecules and to make new ones, the recombination of DNA-strings leads to the existence of new viruses, new bacteria, new plants and animals. I would like to emphasise that this creation of new organisms did develop within the context of pure basic research. To prove this, let me cite a few sentences from the book I just mentioned: 'As more and more was learned about how genetic information is organised and expressed in prokaryotes, the ignorance about eukaryotes became increasingly frustrating. What was needed was a general methodology that would facilitate the molecular analysis of eukaryotic cellular genomes. Ideally, such a breakthrough would permit the isolation of discrete genes and the determination of their molecular structures. That objective finally became a reality during the first half of the seventies, when the recombinant DNA techniques were developed.' Let me formulate this in a more abstract way. One could say that scientific theories have an intrinsic tendency to generality. Scientific claims apply not only to the objects we know, but to all entities that could exist in a given field. To test the generality of the laws and to find new ones, we have to create new situations and finally new objects. This is not applied science or technology but an inevitable characteristic of an advanced theoretical science. Having ascertained this, we should now realise that these products of science can begin to lead their own lives as well. The bare fact of their existence may change the conditions of man's interaction with his environment. This insight seems to suggest the following conclusions. First, the simple distinction between pure science on the one hand, which is essentially innocent, and applied science or technology on the other hand, which alone could be misused, cannot be defended any more. The fiend could be hiding within the walls of science itself. The second conclusion is a more general one. It pertains to both science and technology. The example of the recombinant DNA technique should alert us to the fact that negative aspects of science and technology are not necessarily due to their abuse by wicked or thoughtless people. The very creation of new entities, especially when capable of reproduction, may cause disruption in the delicate balances and may even eventuate in disaster.

After these rather abstract remarks about the properties of biotechnology and their relevance for our understanding of the nature of technology and science in general, I would now like to proceed to an analysis directly related to the patent issue. I would like to do so because it is my impression, after reading a lot about the problems of patenting, that the issue has not been tackled from a general theoretical point of view, but rather as a kind of 'piecemeal engineering': a problem is brought to the fore, an *ad hoc* solution is sought until the next problem comes up, etc. This remark is not a criticism directed to specific persons or institutions: this approach is a consequence of the fact that the way of tackling these problems was seen as a simple extension of the previously existing approach of the patent issue in Europe as well as in the US. It seems to me however, that the new technologies, including the biotechnological one, have some rather original characteristics.

Firstly, one has to realise that the old dichotomy between the natural 'nomothetic', and the human 'idiographic' sciences (W. Windelband) is no longer tenable. It implied that the natural sciences were characterised by laws and the human sciences by description of structure. It is clear, however, that in sciences like chemistry and especially biology where millions of compounds or species are studied, the description of particular structures is an essential part of the theory itself. For instance, one wants to know the details of the human genome or of the genome of a particular nematode exactly. One of the consequences is the following. In physics, there was a clear distinction between a 'theory' which was 'discovered', say Maxwell's equations, and a 'technological system' which was 'invented', say the electromotor. A theory contained essentially laws; a technological product was a particular man-made object which had to be invented and extensively described. In molecular biology on the other hand, the detailed description of say a DNA-sequence or a polypeptide is an essential aspect of the purely scientific research; the technological product may consist in a way of isolating it and translating it to other organisms. Once the methods to do this are known, the difference between scientific description and technological application is a rather delicate matter. The non-obvious character of a so-called invention, compared to the equally non-obvious discovery, is not very obvious if I may say so. One can always say of course, that in every case the patent office — or in the US even the courts — should decide in this matter, but that is a way of avoiding the problem rather than solving it. Courts can use different criteria and have no general view on the subject-matter. So it seems to me that only a rather detailed theory on this aspect of the patent problem can lead to a just and decent solution.

Secondly, another fundamental novelty has comparable consequences. I already referred to the blurring of the distinction of what is natural and what is artificial. Until recently, patented procedures or products were essentially artificial, that is, man-made. Now, the new organisms are not really new but a modification of an existing organism. Once it has come into existence it leads its own life and reproduces itself as natural organisms do. A lot of problems follow from these considerations. For instance, did the original organism belong to the public domain or was it the property of a group of human beings? For instance, does the famous 'Neem tree' belong to the whole world or to some people in India or to India as a whole? If it was public domain, why aren't the oil and coal deposits in Arabia or China public domain too? The answer to questions of this type is not obvious. In the colonial era, e.g., islands were considered to be public domain until someone planted a flag on it. The autochthones, of course, were not taken into account: Captain Cook planted his flag on a shore and the island became British territory. But even then there were problems. Did his flag give him dominion over one square meter, one square kilometre, the whole island or perhaps a whole continent? Analogous problems arise in the patenting of organisms when you put your flag on a sequence of the genome. Did Cohen and Boyer put their flag on the whole continent of rDNA or only on a part? Or perhaps Watson and Crick were there already before them. I cannot expatiate on all these problems of course. I again submit that they should be treated in a comprehensive way, with all the aspects that could come into consideration.

A third characteristic is that genetically modified organisms do not come from a standardised assembly line. In the course of reproduction they may undergo mutations or phenotypic changes. How far may these go, or can these go before the organism is no longer the patented one? And here is an important aspect relevant to the so-called 'morality clause' in European patent regulation. When artificial products prove to be dangerous, the production can be stopped. But when a strain of organisms becomes dangerous, in a lot of cases it may not be possible to stop it. Does the patent holder still receive his money or does he have to pay for the damage? In other words, the patent holder, in general the producer, seems to have a more complicated responsibility. Does this need to be reflected in the patent regulations? In the European approach it seems to be so. However, I think it is necessary that a general decision throughout the world should be taken on this issue, and that we should have a general appreciation of what the notion of moral responsibility is, in all aspects of patenting.

A fourth problem, when plants are concerned, is that they may be introduced in existing ecosystems, in which they will develop a specific relationship with the other organisms after some time. So a new biotope comes into existence; the other organisms are conditioned by it. In which way is the new biotope related to aspects of the organism and the patent? Could the patent holder say that when in the biotope something interesting appears in another organism, that this is also due to his patented organism? Similarly, when it is a dangerous thing, could he say: 'I have nothing to do with it'? Take as an example the work of an author — a book or an article. In the spread of information there is some parallelism between the fact that an author has his author's rights, but on the other hand, the author is responsible for what he is saying; when it is slander he may be called responsible. There is some parallelism between 'rights' and 'responsibility'. Should this also be the case in patenting? It is a general question, again, that should be approached within a general theory.

When we refer to the modification of animals, there is inevitably the link with the possible well-being or suffering of the animals. Here also the European approach tends to link these moral aspects to the clause of 'ordre public'. I think this is a positive step forward, but a more detailed analysis is needed to weigh up the pro's for mankind and the con's for the animals in several cases. You cannot do that without a general theory, for instance, about animal suffering, animal well-being, and so on.

Time is running out, but to conclude I would like to say a few words about the patentability of the human organism and the human genome. I think that the human organism as a whole, including the gametes and the genome, should not be open to patenting. That is a statement, but we need a theory about that.

Note

1   We use the term 'information' for entities whose form aspects are so dominant that the nature of the substratum has become almost irrelevant.

113

9 Biotechnology policy and ethics

Johan De Tavernier

It is commonly acknowledged that biotechnology is intellectually exciting and also promising in the economic sense of the word. What is special about it from the moral point of view is the main question for my introduction here.

We all can see that, during latter years, biotechnology has become the object of intensive social and ethical reflection, both from within and outside the scientific community. The question is why that is the case. To know the answer, I propose to look at the work of one of the first ethicists who has written about the ethical questions involved in the genetic manipulation of animals. His name is Bernard Rollin. He is a professor in philosophy, but also in physiology and biophysics at Colorado State University in the US. By the way, Professor Rollin has introduced the first course offered in the US on ethical issues in intensive agriculture, for animal science students. He has written an article in 1986 with the famous title 'The Frankenstein thing', an article which has been re-published many times. It is still very illustrative for everybody who is interested in the relationship between genetic engineering of animals and ethics. So my main focus will be on the genetic engineering of animals, 'animal biotechnology'. Here and there I will refer to plant biotechnology. I am not going to speak about patenting and ethics, because I know that speaking about patenting is a minefield for an ethicist.

To begin, I want to summarise the article of Bernard Rollin. Secondly, I will take up some presuppositions of his speech and give some comments, finally I want to give a short presentation on the opposition to agricultural biotechnology, an overview of all the ethical considerations that have been made by different authors.

Rollin's article 'The Frankenstein thing'

Rollin is of the opinion that in the case of genetic engineering both the scientific community and also the general public often miss the mark in their attention to ethical issues. The public, because of its ignorance, is influenced by the media in distinguishing actually existing social concerns on the one hand and specific moral concerns on the other hand. The scientists are missing the point because of their

114

failure to see the moral issues in science and their belief that moral issues are nothing more than emotional issues, emotive responses, individual preferences or individual tastes. Rollin is convinced that science is not value-free. This is not new, it is an old idea. All science includes ethical values. Also our new technology is permeated with value presuppositions. For instance, the use of animals in science is not only a scientific discussion but also a moral discussion. At this point he arrives at the 'Frankenstein metaphor'; I will give a summary in three points.

The first element is that some people are thinking as follows: there are certain things we simply ought not to do, and species-modification by genetic engineering is one of them. This is the idea, Rollin says, that certain scientific knowledge or activity or applications of scientific knowledge in itself — irrespective of the consequences — is taboo, for instance crossing species lines, introduction of genetic material derived from humans into animals, and so on. The first element of the Frankenstein metaphor is not acceptable to Rollin. He is against that idea. It does not represent, he says, a defensible moral claim, even when perceived — and this is the case — by public opinion. It is very necessary, he says, to respond to this idea and he asks the research community to make a great deal of effort through public education in order to prevent the confusion between public ignorance about science on the one hand and moral concerns on the other hand. Until now, I have not spoken about applications.

A second element of the Frankenstein metaphor is the possible danger to humans that grows out of scientific curiosity, mentioned as follows by Rollin: 'There are certain things that are wrong to do because they must or will lead to great harm to human beings'. This element says that, despite the noble intentions of the scientist — and you know that Dr Frankenstein's purpose/intent in the novel was to help humanity — his activity was morally wrong. Not merely because of 'hubris', as told in the novel, but because of his failure to foresee the dangerous consequences of his actions. If we can do it, should we do it? That is the first question, and on the other side of the problem, what are the potential dangers inherent in genetic manipulation of animals? What could be the effects of one's narrow selections for isolated characteristics? Are there, for instance, unsuspected harmful consequences to humans, who consume beef and so on. What about the narrowing of the gene pool, the tendency towards genetic uniformity, although I do know that genetic engineering can also have the opposite effect. For Rollin, it is evident that fear of further consequences of scientific research is rather a prudential reason. It is not a moral reason. Sometimes, he says, a question of human self-interest, rational self-interest, prudential reasons for instance, that is more prudence than a moral way of thinking. And we all know that, following Kant, prudential reasons are not moral reasons. Therefore, the second element of the Frankenstein metaphor — according to Rollin; I would discuss that — cannot be seen as a purely moral question.

I come to the third element of the Frankenstein metaphor. The third dimension concerns the plight of a creature engendered by abuse of science. Translated into the area of genetic engineering of agricultural animals, it is the question of the moral status of animals and their potential suffering. Only the third element, according to

Rollin, is acceptable as a full moral reason. He refers to his famous book of 1981, 'Animal Rights and Human Morality', where he extends the notion of individual moral rights from humans to animals. According to him, animals with some form of consciousness have basic interests, not only secondary interests, which have to be protected. This is one of his statements in the debate. Precisely this third point in the Frankenstein metaphor has been defended by many authors, and I refer to the inspirators: Feinberg (1974), Tom Reagan (1975, father of the Animal Rights movement), Peter Singer (1975, Animal Liberation movement; slight difference), Stephen Clark (1975, the moral status of animals). Thus, the last twenty years a lot of authors are defending this viewpoint. The people mentioned are defending the idea that animals have a 'telos'. By a telos is meant that animals have a set of needs, a set of interests, psychological and physical, genetically encoded, which make up the animal's nature. Let us say it is the 'dogness' of the dog, the 'pigness' of the pig, and so on. Thus, opposed to the old obligation to 'avoid cruelty to animals' — the central idea in the traditional animal welfare movement, typical for England during the 19th century — they ask 'equal moral consideration', which cannot be understood as equal treatment. There is a difference between equal treatment and equal moral consideration. They ask equal moral consideration for non-human animals. Contrary to the people mentioned — Singer, Reagan, Feinberg, Clark, and so on — and in a sense very surprising, Rollin says that, with respect to genetic engineering, in his view, the telos of the animal is not inviolable. I quote: 'The genetic engineering of animals in and of itself is morally neutral, very much like the traditional breeding of animals. Obviously, therefore the considerations of the animal's welfare, independent from the effect on humans, should be weighed and considered before a piece of genetic engineering is undertaken.' What Rollin is arguing is the following: given an animal's telos, certain interests, which are part of that telos, ought to be inviolable. Given, for instance, a burrowing animal, it is wrong to cage such a burrowing animal, like a rabbit, so that it cannot burrow. However, Rollin has never asserted that there is anything wrong with changing that telos of a burrowing animal, so that burrowing no longer matters to it. That is the position of Rollin, and that is different from the position of defenders of animal rights, or of the Animal Liberation movement. He adds that the main moral challenge involved in the genetic manipulation of agricultural animals is to avoid modifying the animals for the sake of efficiency and productivity, at the expense of the animal's happiness or satisfaction of its nature. It is no problem to change the telos of the animal, but you may not do that for reasons of efficiency or productivity. That is in essence more or less a contradiction in the viewpoint of Rollin.

To conclude: it is clear that for Rollin, genetic engineering of animals in and of itself is considered as morally neutral, very much like traditional breeding. It only becomes problematic if it has consequences for the individual animal's welfare, independent from the effect on humans. Thus, in that sense he is defending animal rights, but not in line with Feinberg or Reagan.

116

Rollin's presuppositions

First presupposition: the more self-interest is involved — understood is: the more 'human' self-interest — the less pure the moral question becomes and vice versa. The more one is concerned about the well-being of the other individual, the more pure morality becomes. In environmental organisations, animal organisations and so on, that is more or less a typical traditional way of thinking. But is this true? I refer to Henk Verhoog from the Netherlands. He admits that there is a certain element of truth in this statement. I can agree with that. Personally I can accept genetic engineering of animals, as far as it can be recognised as therapeutic treatment of animals; a somatic gene therapy must be acceptable under certain conditions, that is clear. That, in my view, is not under discussion, but further questions could be 'what is a therapeutic treatment?', 'what do we see as a therapeutic treatment?' A second remark about this presupposition: there is more at stake than just saying 'the more human self-interest is involved, the less pure the moral question becomes'. In a certain sense it is true, and I refer — without going further into that — to the discussions about patents in biotechnology, the self-interest of the companies. It is a question of self-interest, of course. But human self-interest in itself could also be seen, according to me, as a highly moral question. For instance, who benefits from agricultural biotechnology? who will suffer from it? poorer farmers? What about the consequences for the Third World, and so on. As far as I can see it, it further develops the agro-business. It promotes the unequal access, as far as I can see it. A second way of reasoning: many people fear that the experience will be applied to humans at some stage. Let us say the human self-interest involved can lead also to huge moral questions. For instance, look at the experiences concerning fertility methods in animal breeding, which have been largely used on humans, with positive and negative effects. Well, the same could happen with genetic engineering, also with positive and negative effects. The conclusion concerning the first presupposition is: if the unhappiness of animals counts as a pure moral issue, why not the unhappiness of some humans, those humans who, just as the animals, have not asked for this technology, but are still affected by it. That is my first conclusion.

The second and third presupposition — I will take both together: 'the concern about the others should not be limited to other human beings, but to sentient animals as well' and 'the concern in relation to genetic engineering of sentient animals should be directed to the pleasure and satisfaction of the individual animal'. This particular presupposition is the result of a discussion within environmental ethics. We all know the anthropocentric way of thinking: strong anthropocentrism, moderate anthropocentrism, and so on — what we call the dominating traditional view according to which animals have no value of their own, have no intrinsic value, have no inherent value. The only value they have, in an anthropocentric way of thinking, is an instrumental value, traditionally called the 'instrumental value of use'. In that view, the differences between humans and non-humans, between humans and animals, are stressed. Moreover, those differences are believed to be both fundamental and also morally relevant. That leads us to a legitimation of the use of animals for almost any human purpose. Rollin, and many others, disagree with this point. In my view they are giving strong arguments why at least sentient animals, with some form of consciousness, should have moral status and intrinsic value as well. It is what is called in environmental ethics the 'pathocentric way of thinking', referring to Jeremy Bentham. He lived at the end of the 18th century, the period of the French Revolution, of the first declaration on human rights. In 1789 he has written in his 'Introduction to morals': 'The question is not: can they reason? The question is not: can they talk?, but the real question is: can they suffer?'. Thus, who must have rights? 'Suffering', 'consciousness', 'having pleasure', 'the capacity to feel what is wrong, what is good' and so on, that is the inseparable line, according to Bentham, for giving rights, for respecting life or for not respecting life. So, also higher animals do have the right to life. Thinking further, if you recognise that some animals, conscious animals, have intrinsic value — and I do that — then their use by man is no longer evident. It has to be legitimated by ethical arguments. In the case of conflict, for instance, between human interests and animals interests, good reasons must be given why we should give priority to human interests. At this point, you have two opinions. Some will say — referring to the Animal Rights movement — 'equal moral considerations' means 'humans may not suffer, animals may not suffer'. That is a strong way of pathocentric thinking. Fundamental interests of humans could never have priority over fundamental interests of animals. That is the way GAIA (Global Action in the Interest of Animals), for instance, an animal movement here in Belgium, is thinking. A second line of thinking — more my line — is the line of 'moderate anthropocentrism'. Moderate anthropocentrists recognise that animals have an intrinsic value, but at the same time I am a 'moderate' anthropocentrist. I use a proportionalistic way of reasoning. For instance, the suffering involved must be proportional to the importance of the goal for which the animals are used. So you have to discuss the interests involved: are they basic, are they serious, are they non-basic, are they peripheral, and so on. I could accept, I think, proportional reasons, but only proportional reasons are accepted. For instance, for me, with respect to transgenic pigs as so-called organ factories, if there are no alternatives and under certain conditions and if, let us say, the interest of humans are serious or basic interests, e.g. organ transplantations, then I could accept genetic engineering of pigs under certain conditions. That is my position. So I can accept that proportional reasons override the fundamental interests of animals. But if there are no proportional reasons, or if the interests of humans are not serious, are not basic, I cannot accept it. It is a bit complicated, but that is my way of thinking. In conclusion, if human interests are not fundamental, then we have to respect the telos of the animal and genetic engineering is out of the question. That is my reaction to the second and the third presupposition of Rollin.

The fourth presupposition is about changing the nature of animals, the so-called 'telos' or the 'genetic make-up' of animals, by means of genetic engineering. For Rollin that only becomes a moral issue if it affects the individual animal's happiness. The last presupposition of Rollin makes clear that it is not the nature of the animal itself that matters, but the realisation of certain interests which are part of the telos.

According to Rollin, if genetic engineering is used to better adapt the animal to its environment it is not morally problematic. Think of the housing conditions in the intensive husbandry, for instance. If maybe we can alter the telos of chickens, the physical or the psychological needs of chickens to adapt them to battery housing systems and so on, then for Rollin it is not morally problematic. You see the difference. Rollin would say that he certainly would not be comfortable with this solution, but not on moral grounds, probably on aesthetic grounds. And aesthetic grounds, that is a human argument, that is human interest, not the interest of animals. Well, precisely this argument of Rollin which is sometimes accepted in scientific communities, should never be accepted in the animal rights movement. What is the argument? The argument is that you have to take the concept of the telos of an animal in relation to the concept of the intrinsic value. You have to take that seriously and this implies the concept of a species-specific behaviour in a particular habitat. That is the reason why Reagan, for instance, or many other animal liberationists are against species-modifications in the case of animal biotechnology. They are referring to the moral status of animals. Rollin is not referring to the moral status of animals. He is referring to happiness and unhappiness. And in that way, he is not a defender of a deontological way of reasoning, he is defending a teleological way of reasoning.

The opposition to agricultural biotechnology

The potential impacts of biotechnology are great. Clearly, the practical applications of agricultural biotechnology hold many promises for a better world. I am not blind to it. For instance, take plant biotechnology. The existing limitations on crop production because of diseases, pests, droughts, length of growing seasons, because of soil, nutrients and so on, may be greatly reduced as a result of advances in agricultural biotechnology. Such advances have major implications to help resolve problems of world hunger, to help resolve problems of cleaning up environmental pollution or problems with soil degradation and so on. That is clear and I also see it positively. But taking all the ethical considerations concerning agricultural biotechnology into account, there are at least five more considerations. Firstly, human health concerns, health risk perception. You all know that perceived risks are not always based on expert judgements or empirical verifications. A system of risk-assessment starting from consumer demands, that is what we need. Agriculture is part of a food system that must be responsive to consumers. Concerning this point, for instance, there is the ethical discussion on labelling. Is labelling necessary for biotechnological products, yes or no? That is a question of informed consent. I think it is, but I am not 100 per cent sure of it. This first question is a question of safety.

The second question is a question of some people here in the audience — a 'common good' question. The common good-issue is a question of justice, equity, fair competition. We all know that the commercial introduction of biotechnology will change the structure of all national and international agriculture. Questions like 'who will have access to biotechnology?', 'can access be equalised?', 'is it fair?', 'to what extent should it be fair?', are ethical questions. Another question is 'will biotechnology contribute to further vertical integration in the food system, and what are the consequences for the various size and types of producers?'. A third question: biotechnology will change the definition of the food system to include a broader set of firms, companies, pharmaceutical companies, and it will also change the aims of agriculture. Pigs as insulin factories and so on. Is this a positive evolution? Is the further industrialisation of the food system a good thing? Will it not speed up food system industrialisation? What do we think about that? And a last point concerning the 'common good'-issue: how do we encourage democratic citizen participation in biotechnology? Those are questions concerning the 'common good' issue. That is the not the field of environmental ethics, but the field of social ethics. That is another view, one of the particular views on biotechnology.

The third question, or the third set of ethical considerations, has to do with the concern for human character development. I refer to Wendell Berry, for instance, in his opposition to mechanisation and specialisation. He says that traditional family farming embodies a set of values, knowledge, specific roles, and so on. A competent farmer is his own boss in traditional family farming. That system will be changed — I also am referring to Jeremy Rifkin, the classic opponent of agricultural biotechnology — because it attacks the human character development and classic values in traditional agriculture.

The fourth set of concerns has to do with animal welfare, animal rights. I have already talked about that. The last set concerns the effects on natural environments. The criteria used here are the safe production, the safety-issue — is it safe for the environment? — and the guarantee that biodiversity will not be diminished. It must be accepted as a contribution to a more sustainable agriculture. What is sustainable agriculture? It must lead to better integration of agriculture into environmental concerns. Is that the case? Yes or no? The quality of the biotechnological products must be as high as the present quality of agricultural products, and so on.

As a conclusion, I want to say that the variety of approaches to agricultural biotechnology, and also to environmental ethics, indicate the complexity of the current debate. It is important that new environmental problems have drawn attention to the insight that ethical questions are raised by human behaviour. Once you have understood the difference between environmental ethics, the so-called 'human oriented control of the environment', and on the other hand ecological ethics — the direct responsibility with respect to the non-human life — you are entering, in my view, a new understanding of life, a new understanding of the relationship between humans and nature, humans and animals. I think we need this insight in order to take new debates about biotechnological applications seriously.

10 Cultural background of the ethical and social debate about biotechnology

Tom Claes

This seems to be a sort of philosophers' afternoon. I am going to surprise you: I am not a philosopher. I am even worse, I am an ethical scientist. And this is a very strange breed of scientist indeed. My work is mainly related to ethics, the relation between morality and ethics, and the relation between culture and ethics. It is a sort of intellectual history, combined with present-day ethical concerns. I have been asked to speak about the cultural backgrounds of the biotechnology debate. The thoughts I put to you today are tentative.

Two different issues

We need to distinguish between two, although related, different issues. Behind the issue of patentability of biotechnological inventions lies the larger issue of the moral status of biotechnology. To whatever degree both issues may in practice be connected, for clarity's sake they have to be distinguished. The first one is basically a problem of ownership. The second one is related to the eu- or immorality of man-induced modifications of natural organisms. Sometimes patentability-related problems and questions are mobilised as arguments against biotechnology in general. This is not, I think, what one should do. One could argue against the patentability of inventions as such, but this is not what is at stake in the present discussion. What stirs up all the commotion is the question of the patentability of a specific kind of invention, viz. biotechnological inventions. What then is so special about biotechnology?

The common basis

While preparing this speech I read through a lot of discussion papers on biotechnology filled with both pro and contra arguments. Not only the triumphalistic cries of those who are pro startled me. What startled me most was the type of arguments the opponents to biotechnology deploy against biotechnology. To me they seem to converge towards some common basis. I would like, therefore, to focus on this basis.

This background I will call the idea of natural integrity. Succinctly put, it is the idea that both the living and non-living world posses a natural destiny or finality, that animals posses an intrinsic worth or value and that their worth or value should be graded according to their proximity to man.

The idea of 'natural integrity'

My basic hypothesis is the following: biotechnology's problematic status is (at least) partly the result of its discord with some of the most basic and widespread assumptions held in Western culture, concerning nature and natural beings. Man-induced genetic modifications, seen as changes in the fundamental 'structure' of nature, are, in this view, considered by many as violations of the 'natural integrity' (*cf* supra) of these living organisms. This idea of 'natural integrity' is itself a container holding a cluster of the following assumptions: all living organisms (a) are characterised by a sort of natural destiny or finality, a 'telos' perhaps, (b) they possess an intrinsic worth or value, and (c) their value should be graded according to their proximity to man. An implication of this view, this cluster of ideas, is that some types of changes to nature are to be left to nature itself — be it by way of biological evolution or, for those who believe, by means of divine intervention. When man intervenes — so the story continues— he somehow is transgressing/violating his legitimate domain of action. In the next sections I will briefly touch on the parts that make up the cluster.

Natural destiny / finality

What is this natural destiny or finality? The basic idea is the following: all creatures, and perhaps even everything in nature, is among other things characterised by that to which it naturally strives. Everything strives towards expressing its 'fundamental' nature. This finality has to be respected. This idea already has a long history. In the older version, you could hear something like the following : 'Horses have those inviting backs, so that we can go and sit on them'. This is, e.g., a clear case in which the finality of an animal is thought to lie in its usefulness for man. In the newer version the arguments and examples are somewhat more sophisticated. For example, what to think about the following: the natural finality of cows is having offspring and enjoying themselves being well and having a sort of natural life.

Ethics comes in when you go against the finality, the telos or the destiny of an organism. Whenever the natural potentialities — whatever 'natural' may mean in this context — are blocked, diverted or distorted, one does something wrong. Changing the genetic structure of a cow, e.g., so that the finality of the cow — enjoying itself in a pasture, for instance — could no longer be respected or attained, is (doing something) wrong. This is not my point of view, but it is part of the background of some people who argue against biotechnology.

122

This way of thinking goes back a very long way and is usually labelled under the general heading of some sort of natural law theory. One of the crucial assumptions within this theory is that that what an organism strives after constitutes a 'good' for that organism and hence constitutes a value. Of course, a general idea like this has to be refined before it can be put to use in the natural law theory. One has to select which potentialities/characteristics will be singled out as constituting the telos or destiny of a living creature. And here things become difficult. Let me illustrate this as follows. I often use the example of a mouse I once saw in a short documentary. This mouse had some electrodes implanted in its brain. In its cage there was a switch and whenever it pushed this switch, the mouse experienced an intense feeling of pleasure — sexual pleasure or otherwise, I do not know. What did this little mouse do? It pushed itself to death. This story illustrates that not all potentialities or not all things an organism strives after can be included as something that should be promoted.

I sometimes ask the following question to students of mine: 'What would you do if you were given the possibility of experiencing 24 hours of very intense pleasure, sexual pleasure, and then die?' Choosing for the pleasure would earn you probably one of the most fun-days of your life, but a very short one at that, and it will probably be a very exhausting, potentially even deadly experience! This leaves us with the problem of finding criteria for the selection of potentialities that have to be promoted. Which criteria should be used for this selection? For instance, what is essential to man? Is it enjoying sex? Raising children? Making friends? Gaining knowledge? Loving God? It is no surprise that, in the early days, they chose the love of God as the highest potentiality and the enjoying of sex as the lowest. Modern day discussions sometimes tend to reverse this order a bit. But the basic problem of course is, which potentialities are to be preferred and why? You have to assess the criteria to establish the finality or the destiny, or the telos, or whatever you may call it, of a creature. That is not so easily done. Postulating a finality does not exempt one from defending the selection of what goes into this finality and what not. One just shifts the problem.

Intrinsic value

A second element which makes up the cluster of natural integrity is 'intrinsic worth' or 'intrinsic value'. This is a very big fish! I keep hearing it over and over: 'natural beings posses an intrinsic value and this value has to be respected'. Then I start wondering what this 'intrinsic value' could be. It clearly is the idea that an organism — or a thing perhaps, I don't know, a landscape for instance, but surely an organism — has some kind of 'value in itself.' This is normally linked to what is called the 'essential quality' of it (cf the 'finality' in the previous section). This kind of value constitutes a big problem. A question will make this clear: would there be value in the world or in the cosmos if no valuing agent existed? To put it bluntly, if the totality of humanity would be destroyed, would there still be value in the world or even in the cosmos? One could of course answer by saying that not only human beings have a valuing capacity. Since other animals have the capacity to value, the answer could be 'yes, there would still be value in the world'. But what if you wipe out more than humankind, and include other natural beings capable of valuing? The problem remains. Those who defend the existence of 'intrinsic value' have to answer that the existence of this kind of value is not dependent on there being any natural beings that value.

One should make very clearly the important distinction between value as a psychological category and whatever is called intrinsic value. If an organism has needs, desires, wants, hopes, etc., that can be satisfied by obtaining an object, going into a relationship, making friends, etc., then some kind of value does exist. I like my cat, therefore my cat has value for me. When I am hungry, I value food, etc. Value in this sense is a relational concept. You cannot have value without a valuing instance, without a valuing agent. Intrinsic value — and this is often overlooked — is a non-relational concept in the sense of non-relational to man. Without valuing organisms, there still would be value of this type in the cosmos and this type of value is called 'intrinsic value'. The basic idea of intrinsic value is that things (perhaps) and living organisms (certainly) possess a value, a worthiness independent from whatever contingent desires we or other valuing organisms could have. They have to be respected — and this is a sort of implication of the foregoing — not only because they, e.g., serve our needs, but mainly because of what they are.

In this view, certain changes, for instance changes brought about by genetic engineering, disrupt this intrinsic value, because this kind of value is connected to the essential qualities of this organism, whatever these may be, and whatever their relation to other natural organisms. For instance, an organism, such as a pig, cannot, in this view, be used for spare parts. Animals cannot be used as a factory, because it is not the basic telos or destiny of a pig to be used as some sort of spare part factory. Using pigs this way means going against whatever intrinsic value the pig, or a cow or my cat or man, etc., possesses. Why? Because it reduces these organisms to objects, things, which are entities without intrinsic value or whose value depends solely on their usefulness to us.

Proximity to man

The third element in the cluster of 'natural integrity' is proximity to man. Take for instance the case that everything in this cosmos has intrinsic value — which is highly problematic in the sense that we could not eat a plant, a potato, we could not eat parts of sheep, etc., because they have intrinsic value. Intrinsic value is by no means a sort of absolute 'no-eating sign'. Intrinsic value does not imply, in the basic framework which I am describing, that one cannot eat something. Some people who strongly defend the idea of intrinsic value, for instance of chickens, do themselves eat chickens. How is this possible? Either one says something like 'well, perhaps it is (part of) the finality of the chicken, of its telos, to be tasty meat'. Or one can construct some kind of 'hierarchy': you begin with inanimate objects, move a bit

124

higher to bacteria and then to worms, insects, etc., small furry creatures and others, and then larger creatures who look us in the eye, who have their eyes in front of their faces, and on the top of the pyramid, of course, man. In this scheme man is justified, up to a certain point, to use whatever is situated in a lower position. So, the idea of intrinsic value does not have to exclude the fact that man can eat animals, that man can act upon nature, that man can modify nature.

However, modifying or acting upon nature has certain limits which cannot be transgressed, because otherwise the 'natural integrity' (*cf* supra) of this organism is violated. Man can act upon nature, can use nature in a way, can transform nature, but of course man can never (ab)use another man. Another human being should always be an end and not merely a means towards our own ends. Something similar was for instance once said by Kant, a rather important philosopher.

These three elements — finality, intrinsic value and proximity to man — make up the cluster of 'natural integrity'. They function as the background of many arguments against biotechnology. The sum effect of this is that — when one starts out from these three premises — biotechnology and genetic engineering are either bad or highly problematic, because genetic engineering destroys precisely this type of natural integrity. Changing the genetic structure of man constitutes the highest possible breach of natural integrity. It not only goes against the finality of man, it also constitutes a disrespect towards the intrinsic worth of man, and — perhaps most importantly — it touches on the creature that is situated on the highest level of worthiness.

Problems with these assumptions

I want to point out that am not referring to *risk arguments*. Risk arguments are very important arguments. If the salmon which people were referring to earlier today would escape, this would be a very big problem of course. I am also not referring to arguments against biotechnology starting from the issue of animal suffering. Whenever a pig suffers, it hurts me too up to a certain degree. Both 'risk arguments' and 'suffering arguments' are important arguments. However, the arguments that follow from either a teleology or finality argument, or an intrinsic value argument are a lot more problematic.

In this paper I am interested in the assumptions that lie behind the aversion towards biotechnology. These are widespread assumptions, explicitly cherished by some and quietly taken for granted by others. I think that the cluster I identified and labelled as the idea of 'natural integrity' is a cluster of basic assumptions of our culture. However, — and this is the main problem I would like to put to you — what if some of these assumptions, so central to our culture, prove to be wrong or at least problematic, as I think they are?

Take for instance the assumption that all living organisms are characterised by a natural destiny or a finality. The 'finality'-concept is proven wrong by evolutionary theory and is highly problematic with respect to selecting which potentialities are

125

the most worthwhile. Selecting these potentialities/characteristics does not solely depend on factual arguments.

The idea of intrinsic value, a value that can never be granted or given by any valuing agent, but is rather a value inherent in the organism itself, a value that can only be discovered but never created (so never patented?) is highly dubious and depends, I think, heavily on a religious context.

Religious context

What is this religious background? I will try to keep it short. The background of this natural integrity cluster is the assumption that the natural world is not arbitrary and in itself not meaningless. The cosmos expresses a purposeful and meaningful order. This is a cosmological perspective which has a lot of biological and philosophical implications for everything that grows and runs and quacks and squeals on earth.

This cosmological perspective with its biological implications is intimately and ultimately linked with its religious origins. What scheme is this? We have a creator who creates a world and man in this world. This creator is therefore both inside and outside of its creation. This creator is the first and ultimate source of all value. It is not man who creates value. Of course man creates psychological value, but not this intrinsic kind of value. The intrinsic kind of value is given by the creator who is both not human and (in part; I do not want to go into a theological discussion here) external to creation. For biology, this means that all creatures great and small are created and have to be evaluated according to what they express in creation. Man, of course, is the king of creation in this old cosmological idea. He may not be the Prince of this world, but he surely is the king of creation. In my view this image is completely out of tune with modern science in particular, and reliable knowledge in general.

This poses some kind of a dilemma for us. Arguing against biotechnological innovations and genetic modifications by employing the elements of the afore-mentioned cluster — intrinsic value, telos and proximity to man — places one in line with one of the sources of our cultural tradition, but at the same time alienates us from the modern secularised world of science and technology. Those who want to liberate themselves from ethics based on religion, as I would prefer, still have a long way to go, since a large part of our moral concepts and vocabulary is itself still deeply rooted in this persistent religious cultural tradition.

The alternative: secular ethics

I think the only viable option is some sort of secular, scientifically based ethics and hence, of course, the idea of an 'ethical science.' We are only just beginning to develop this secular ethics. One of the ideas I would like to put to you — and then I am going to stop — is that we should take into account the scientific findings (cosmological, astronomical, etc.) and also the experiential world of man and other animals. Take for instance the idea of Descartes: if you pulled the muscles of an animal or pinched an animal and it would move its legs, it would be the same as playing guitar. This view is no longer valid. Animals can suffer. Furthermore, we have to use a sort of consequentialist analysis including risk assessment. We also have to develop an empirical theory about value, not a metaphysical one, not one about intrinsic value. We have to ensure democracy and democratic control, but at the same time ensure some sort of workability.

The secular perspective on man and its place in nature, is the only possible and scientifically viable solution, and the only possible and very much needed basis from which we can debate about biotechnology. This would mean of course that we should increase scientific literacy as well as cultural literacy. The basic problem I have is that a lot of people who have very good ideas, and who stand for things I could very well stand for, use the wrong kinds of arguments, like for instance intrinsic value or a telos or natural destiny of an animal, etc. Get rid of these notions, try to base your opinion on an empirical basis combined with a secular outlook and start talking about what animals do feel, what they could feel, what man feels, what he should take care of, etc. Make a rational analysis, make a risk assessment, etc., but please do not go back to the Middle Ages.

11 The virtual reality of the biotechnology debate

Dani De Waele

Do I have something to say about biotechnology? Do I have something to tell about biotechnology in the company of philosophers and ethical scientists? In the company of social scientists? In what follows, this question reveals itself to be more than rhetorical, to say more than it asks.

For what has been my experience? I am not a philosopher, but I work with them or, at least, among them. I was an experimental scientist, a young geneticist before the term biotechnology was introduced. And as a natural scientist, I was used to focusing on 'what is really going on' in nature, to focus on concrete observable entities. Moving towards research that offered me more space to reflect upon science, I still have this predilection of 'letting speak' what I have under my eyes. What did this imply when I did research concerning different aspects of biotechnology, concerning social-economical aspects, concerning environmental-ethical aspects? I still wanted to let the biotechnology scene speak for itself, to unravel 'what was really going on'. What are biotechnologists doing? What do they publish? What are they producing or making? And what are they saying? What are the facts and what is the interpretation? What is unsaid, but done? Or not done, although required within their own discourse? How are they financed? What are their affiliations, their allies? And for agricultural biotechnology: in which agro-industrial world are they operating? In which world of production and consumption? In which world of whose interests, whose needs, whose dreams and desires?

So I tried to start from this material, from these questions, from this reality. I did not want to theorise at this level, I did not refer to the Bible nor to ethical theories or environmental philosophies. I followed an analogous research path for studying the sayings, writings and actions of the diverse parties involved — organised or not — in the so-called 'biotechnology debate'. These are not the primary actors in bio-technology but they feel concerned by it. They try to force their agenda upon those of scientists, industry and policy makers. I am talking about trade-unions, professional associations, ecologists, patients, consumers, groups of diverse religious and philosophical convictions, ethicists, animal rights groups, third-world movements, and so on.

I heard a lot, read a lot, gained a lot of information/knowledge about products, processes, financing, and about lobbying, of course. I noticed actions, motives, convictions and I touched on blanks, omissions, distortions and on the limits of transfer of information. All this was very interesting. I was able to structure my findings by grouping them, for example, in clusters of shared interests. I found differences in attitudes towards specific biotechnological developments between senior and younger scientists, between top and base, between engineers, chemists or molecular biologists and ecologists, between scientists at universities or in R&D firms and scientists in publicly funded institutions for applied research. Risk assessors gave information about specific risks of unwanted gene transfer in the field, different from that of the new gene combiners themselves. Almost all interest groups intermingled different levels of analysis, intertwined facts with not-yet-known-things, with beliefs or disbeliefs, with interpretations and so on and did this with sometimes remarkable rhetoric talent. Of course, almost everyone prefers biotechnological developments in human medicine to developments with animals or plants for food or non-food consumption. Human health is what counts. Food and food supply is at first sight disregarded as but a problem of the third world. And then we have the environment, Nature! For biotechnologists, nature comes in as a comforting reference when arguing that biotechnology is 'natural' (because genes are 'natural things'), that biotechnology is 'just the same as what happens in nature or in natural evolution' (when gene transfer is questioned). In the same discourse, nature serves the opposite argumentation camp because of being 'imperfect', because 'we won't find in nature adequate solutions for current problems'. Nature serves many causes, as does agriculture; an agriculture which is often ignored, not known or idealised — I mean contemporary as well as 19th century or 15th century or whenever and where-ever concrete factual agricultural practices, including tools, methods, social conditions, trade, etc. In front of concrete, specific, localised, historicised or contextualised aspects of biotechnology, I cannot make much use of those moral theories or environmental philosophies that remain abstract, theoretical, idealistic, immaterial and the like.

This was but a glimpse of all the intricate analyses I could make and did make after several years of studying different aspects of biotechnology. You can read the reports and each day you can find articles, congress proceedings and books in more or less detail on this subject. The question is: 'what can I do with all this?'. I have a complex view on biotechnology, a more 'balanced view' as one is used to say. I can introduce relevant distinctions in the complex of facts, interpretations, pep-talk and beliefs or ideologies. I can render more explicit what remains implicit, I can point out blanks and so on. Nice work, nice analyses. Now what? I really thought that these analyses can help differentiate the biotechnology debate, can help differentiate discussions about biotechnology. What debate? Which discussions? Policy discussions? In Flanders, in Belgium? Wasn't I paid by them, didn't they ask for this kind of 'policy-supporting research'? They remained very silent. Hadn't they just pumped almost one billion Belgian francs in a new Flemish Institute of Biotechnology, composed of leading Flemish biotechnology research groups, this same amount per year for 5 years? Yes, after some contestation from, among others, trade unions, it was decided that 20 million Belgian francs per year, or around two per cent, will be spent on studying 'social aspects of biotechnology', on studying...; thank you! Together with the politicians, the so-called 'social partners' were mainly interested in maintaining employment. However, even employment incentives, less than two months ago, were considered by, for example, the Walloon socialist party as not having priority in this Christian Democrat/Socialist coalition, 'because of Maastricht and the national debt' ('Of course, we misunderstood this', they said the day after in a press conference). Anyway, besides Maastricht, we are being served the 'strong international competition', the 'necessary high technology innovation', 'the information highway' and so on. Policy makers have other things to do than 'differentiate the biotechnology debate'. Discussions with biotechnology researchers revealed to be 'very interesting'. They do recognise different problems in the biotech debate, but... they are carrying out Fundamental Research (after all they are 'free'), they acquire Knowledge, they cannot help what is going on in industry and politics, and they like to keep their jobs too.

So 'to differentiate the biotechnology debate' is very interesting on paper, as 'discussion-material' at congresses — even called 'workshops' like this one. But has it any real interest? Is the biotechnology debate operating in a real context? Has it a context except for an academic audience? Does it work, does it have influence, is it embedded in concrete processes of policy, of steering, of control? And is that even possible? One does not like to hear these questions. 'That's not new!', they say. But on the other hand they have a mouthful to say on democracy, on participation, on pluralism, on public attitudes and tuning them, on bridging the gap between politics and the citizen, on reaching a consensus. Who is really interested in the whole complex of different aspects in the biotechnology debate? Who is acting on account of the totality of these aspects? Young researchers in a biotechnology lab? Science policy makers? Economic policy makers? Third world policy makers? Large farm owners in an American state? The board of directors of a seed company? The shareholders of an agrochemical multinational? Who has real impact on the biotechnology process? Not one person, that I am sure of. Are there any non-biotechnologists who feel concerned by biotechnology, who have a balanced view, who even differentiate this from that development? Do we have real impact on what is going on? Can science and technology be monitored by more than the actors themselves? Even the actors are afraid of loosing control, of being bought up, of loosing markets, of not-getting-what-others-can-get. They don't like too much public control, they don't like media interest except the one they organise themselves, they don't like the green boys and the alternative girls who will never grasp that biotechnology is necessary for human health and for the environment, and that they, scientists, entrepreneurs, co-operative politicians are on the good side, the side which knows, which holds intelligence, which cares, which propagates the Western humanistic tradition, yes Culture indeed.

And then there is an 'Oncomouse' and a patent application for this Oncomouse. There is the mouse, there are the applicants, there is the application file, there are patent laws, there is the European Patenting Office and its different officers, there is the European Parliament and its different elected politicians, there is a legal possibility to contest a patent application and indeed there are opponents and there is quite a public platform for discussing all this. What does this mean? Only, and I underline, only on occasions such as this — a patent application for a biotechnological product, *in casu* an Oncomouse — a real biotechnology debate is taking place. It concerns a realised, concrete, specific biotechnological development. The makers are known, the product is described. The so-called 'general public' has a legal opportunity, yes a right, to respond, to give comments, to vent opinions. And the whole debate has a reason and a goal, has meaning, has sense, has to produce an answer, a 'yes' or a 'no' at a concrete step in the commercialisation process of a biotechnological product. At stake of course is the approval or the dismissal of the applicants' right to exploit exclusively — though limited in time and after having made public the invention — the commercialisation on the market of their new biotechnological product. In both cases, approval or dismissal of the patent, the research is done, the product is made and the mouse, plant or gene or whatever can in principle be brought on the market under the system of free competition. But that will be debated in detail tomorrow. At the moment, what is important is to highlight that the patent debate functions as a unique (in the sense of rare) outlet for public opinion, as a unique opportunity for meaningful discussion, as the unique space where one, in principle you and I, can interact concretely with biotechnology. For some of the parties involved, that is slightly annoying. Being questioned in public by greens, by non-experts... yes, by fundamentalists! Having to hear all their nonsense! Of course, lawyers are paid for anticipating all this stuff, but... time is money and lobbying requires both. No, democracy must have some limits. Are they going to feed the world? Or cure us from cancer, AIDS and Multiple Sclerosis?

'That's nothing new!' I hear you thinking. Indeed, that is not new. 'That is pessimistic', some say. Indeed, that is pessimistic, for some. But meanwhile a lot of people behave as if we, all of us, have something to say... that counts. As if we, all together, determine the course of science, of technology, of industry, of trade. Have we not democratically elected our representatives? Are we not allowed to debate freely here? To give critique?

What in fact is bothering me? That there is a platform but no anchorage. That there are words and reports and theories and debates but no concrete participation. That there is little part-icipation, part-having at all. That we are particularised, specialised. That hierarchy exists. That power is used. That we are not equal or that we are not treated equally, in spite of that famous fraternity. That elites consider themselves as such and that they are active as such. That there are experts and non-experts. That knowledge means power. That nobody knows everything, but that some know more than others (I have not only scientific knowledge in mind). We have to recognise this. This should not be ignored. We must open our eyes on our human world as it really is at the moment, even when we see but blankets of fog. We must face what is, even if it is hidden. At least we must face that something is hidden. How can we propose to change something in our world if we are entangled in curtains and blinded by images? If we are living in a virtual world? I would like real participation, not virtual participation. I would like an anchored biotechnology debate. But is that possible?

I could propose little things, on specific levels, in concrete situations. Imagine, for example, a real kind of biotechnology debate at the level of a specific biotech lab. Could it work? How would it work? Biotech researchers already participate actively in biotechnology, don't they? Can they influence the research strategy of the lab? Do they want to influence it? Do they share the same interest? Or are there dissidents? How critical are they? Would they accept interference from outsiders? About anything? At each step of their research? Would they consider all the different aspects of biotechnology which we, social scientists, philosophers, ecologists and so on have so nicely analysed in detail? And if they want to change something, can they? Would it have a real impact? Impact on the functioning of the lab itself? Impact on national science policy? Impact on the agro-industrial world in Europe, on GATT-agreements?

No, it is not self-evident to anchor a biotechnology debate, it is not self-evident to participate really, little parts — more and less little parts — as we are. It is not self-evident to determine the course of a technology. We feel rather determined, for the worst as for the best. How? Why?

In an interesting Dutch television 'story' — as Wim Kayzer, the maker, calls his series — a physicist tried to analyse how and why he could not intervene, being a RAF officer during World War II, in the course of the decision process to bomb Dresden, a military operation that he himself and others estimated — even strategically — useless, senseless. According to him, there was no way for him and his nearest colleagues to influence this decision. There was no way back. It was the result of a hierarchical structure that showed but one flow: downwards, from top to bottom, from an idea downwards to its execution. This structure subsisted on the splitting up of a global task into little parts which were arranged along a downstream hierarchical line and on the strict compartmentalisation of each part. It existed on behalf of a top that had installed itself, that had declared itself, that considered itself untouchable, unapproachable. And each lower part behaved similarly downwards. The true face of bureaucracy. Using as excuses: 'Wir haben es nicht gewusst', 'Befehl ist Befehl', 'We are but the executors'.

What about technology, what about biotechnology? How are they structured? How is society, our society structured? And why? What is the real face of current democracy? To start with one thing: biotechnology reveals, more than any other technology, that there is a crisis or at least that there is a knot, a Gordian knot. Biotechnologists at universities or in industry do not like this phenomenon but they know it, they feel it. It is underlying their strategies, it accompanies their public relations, it explains some of their agitation. For what is the case? The case is that biotechnology and our so-called 'biotechnology debate', put their finger on the sore spot. Biotechnology, a man-made technology, born out of fundamental scientific research, born out of knowledge of the ancestral fundaments, of the fundamental elements of life, can recombine these elements (the genes) in a novel way, selective, goal-directed, precise, from species to species, by-passing mutation and selection in agriculture, by-passing the non-interfertility of different natural species. No matter if they say that it is a technique as old as agriculture or that naturally occurring gene transfer in and even between species makes out the dynamics of biological evolution, in essence biotechnology is new, it is not 'natural', it is a knowledge-based product of human culture. The whole patentability affair is based after all on the innovative character of biotechnology.

Biotechnology, in being novel — and in what sense — arouses commotion in a lot of people. Let us focus e.g. on the genes. Until now, genes were considered as being the ultimate elements in living beings that were 'natural', invariable, unchangeable by goal-directed, purposeful human intervention. Biotechnology puts an end to this naturalness and opens up the way for human cultural interventions, transformations, recombinations, manipulations, engineering. It brings us further from or farther than agriculture and animal husbandry brought us or than what human culture ever imagined possible for itself, with itself, within its own genes.

Let us go a little deeper into this matter, for I want to get at the very centre of biotechnology as well as the biotechnology debate. In my view, or the way I experienced biotechnology — throughout its processes and products, its persons involved, its publications and public relations, its pro's and con's — biotechnology is indeed an intertwining of nature and culture, but an intertwining which for us, humans, is not self-evident. We succeeded technologically in a process that we cannot/ cannot yet/can never (?) assume. And the closer and more frankly biotech developments are reaching us, the more vividly we feel disturbed. We feel uneasy. We do not like it. We did not ask for it. 'What the devil are they doing?' 'OK, for curing cancer it's all right.' For overcoming an illness that can strike anybody, for which no personal causes can be found, biotechnology seems OK to most people. 'These are clearly mistakes of nature, failures. Away with them!' We will swallow that pill. We will undergo that genetic therapy. But... hmm... developing an Oncomouse for that, that's a bit annoying. The mouse had no cancer, right? The mouse received cancer from us, didn't it? The mouse suffered from cancer in our place, no?... But they say the Oncomouse can save our lives... 'oh well, it is but a mouse'. And that goes on and on. Wait until — and if — we can choose between a tomato labelled 'genetically engineered' and a usual one. Imagine the test: you may choose between, let us say, two, usually chemically treated, tomatoes, one from Spain, one from Belgium; a genetically engineered tomato, all of them neatly packed under plastic film; a biologically cultivated tomato bought in a specialised store and a late summer tomato out of the vegetable garden of your neighbour, there under your eyes. I know it's a slightly tricky test, an imaginary test because you cannot see, cannot smell, cannot touch, cannot taste the tomatoes, you don't have to store them, I did not inform you of the tomato prices, of the tomato market, of European tomato subsidies and the like. It is but an imaginary test, a purposeful imaginary test of, indeed, your inner images of 'tomato', of how you symbolise 'tomato', of your associations with 'tomato'. Well... which tomato do you have in mind in the first instance? Don't tell me you spontaneously choose the genetically engineered one. At least, that's not what my willing victims told me. And that is why Campbell delayed the use of genetically engineered tomatoes for its tomato soup. And that is also why biotechnologists plead against labelling. Biotechnology may not come too close, may not be that openly introduced. Who bothers in fact about biotech stuff which is far from us, remaining invisible in industrial processes and processing or grown in third world countries?

What about those images we carry along? How important are they? And why? First, images are vivid things. Human beings have acquired the capacity to symbolise, to conceptualise, to think in abstractions, and they use this capacity. Furthermore, aren't we nowadays living in a society that certainly is called materialistic, but where the sense of matter, the experience of matter is more and more replaced by a symbolisation of matter. You think more of tomatoes than you really see them. You rather have an image of tomatoes, of how they grow and ripen, of soil and sun and humidity and the like, than you really bodily experience. We are living with images, images 're-deconstructed' from memory, from publicity, from dreams and desires, ideas and ideals. We are living as if. We already are living virtually, not bodily. Task division helps, growing complexity helps, extreme specialisation helps, mondialisation helps, automatisation helps, robotisation helps, artificialisation helps, Virtual Reality helps, civilisation helps.

But I am going too fast, although I must come to an end. We were at the centre of biotechnology, of a biotechnology that I circumscribed as the intertwining of nature and culture, an intertwining that we accomplished technologically but that we experience as not that self-evident. Two components. First of all: 'Nature'. We are more and more distant from nature, we don't know it anymore, we cherish it, we created an image of it. We live with an image of nature that is past, gone by. Nature, wilderness is shrinking and we are reconstructing for ourselves an image of nature that is in fact re-deconstructed out of a past reality and out of a never-been-like-that reality. Therefore, not so deeply inside ourselves, we want 'natural' tomatoes, the tomatoes we are imagining, the tomatoes portrayed in advertising, tomatoes made of nostalgia we constructed. We cannot even live anymore in a more or less natural environment without ten survival manuals or without following a survival training. And the so called 'survival trainings' are not conceived for training us in agriculture, they sort of imitate gathering cultures modernised with the help of a survival kit and war & disaster experience. They do not teach us mechanisation either. That brings us to 'culture': we are also more and more distant from culture, especially from techno-culture. We don't know it, we don't understand it, we just use it and consume its end products. We buy it. We seem to be proud of it. We brag about it, we feel 'flashy' with it. 'Look at my car' or 'Look at me in my car with my GSM', 'I have a video!' But we don't make it. We cannot make it. We cannot even repair it. We are but using it, for a while, for a short while. Or we are only possessing it, showing it. A majority is not 'making culture', is not producing it, is not creating it, fabricating it. There are creators, makers, fabricators, producers. But they are few. There are specialists, more and more highly specialised specialists. But nobody is a specialist in everything. We are distant from our cultural environment, from culture, especially from a techno-culture that is indeed growing and growing in complexity. We don't understand it anymore. We cannot keep up with it anymore. We are coping reluctantly with its development. Culture has already passed us by.

To summarise: we are no more 'ape-man' and no longer 'homo faber'. And we feel uneasy with that. In essence we are two of us. And the two of us live at the same time in another time. We are disconnected from two realities. Alienated, but still alive. An age-old ambiguity. We are too modern for a nature we almost pass by and we are 'antiquated' in front of a culture, that already is passing us by. Coming back to our biotechnology debate: please, do not have it at a purely abstract, theoretical, idealised, symbolical level. For it is the only opportunity, the only platform left to face a difficult-to-cope-with intertwining of nature and culture, indeed, to face our nowadays human living condition.

Part Four
LEGAL FRAMEWORK:
PATENTABILITY IN EUROPE AND THE US

12 Biotechnology patents in Europe: from law to ethics

Geertrui Van Overwalle

For the last 25 years we have seen a flood of new developments and new applications in the field of biotechnology and genetic engineering, a lot of which have been discussed already at this workshop. The question facing researchers, patent practitioners, patent authorities, courts and legislators is whether or not, and if yes, to which extent, protection for this technology can be accommodated under patent law.

In the first section, I will offer some background on patent law in general and on patent law and case law concerning biotechnology patents, as I was asked by the organisers, for the benefit of the people in the audience who are not familiar with patents and the like. In the second section, I would like to present to you an overview of the conceptual and technical objections which were brought to bear over the years to deny patent protection for micro-organisms, plants and animals, and I would like to explain to you how these objections can be dealt with nowadays. In the third and final part, I would like to focus on the ethical objections which are raised.

The present situation in legislation and case law[1]

Let us start with the first part. Genetically engineered organisms can be protected and exclusivity can be achieved by the use of patents. A patent, as you probably all know, is a legal title granting its holder the exclusive right to exploit his invention within a particular territory and for a given period. The essence of all patent systems is that the owner of the invention receives the exclusive right to control commercial exploitation of the invention for a limited number of years, in return for disclosing details of the invention in a written published document.

For the time being, it is possible to obtain patent protection by separate application to each of the national Patent Offices within Europe, the so-called 'National Route', for example a Belgian patent, a German patent, a Dutch patent, or a British patent. As an alternative, one can also apply for an European patent at the European Patent Office (EPO). This is the so-called 'European Route', where on the basis of one single application and examination procedure, one can protect an invention in up to

17 European countries, all contracting states which have ratified the European Patent Convention of 1973 (EPC). Once issued, a European patent comprises a bundle of separate national patents, one for each country requested, which become then subject to national law.

The 1973 European Patent Convention

Let us now focus on the 1973 European Patent Convention and the patentability of biotechnological inventions.

The EPC contains one provision which is highly relevant to the question of whether or not living matter can be protected under patent law, namely article 53, as I will show you. Art. 53 has two parts. I will start with the second part, Art. 53 (b), which stipulates that 'European patents shall not be granted for plant or animal varieties or essentially biological processes for the production of plants or animals', which provision does not apply to 'microbiological processes or the products thereof'. There is, though, a second provision which is relevant, namely, the first part of Art. 53, Art. 53 (a), which says that 'European patents shall not be granted in respect of inventions, the publication or exploitation of which would be contrary to the public order or morality'. In what follows, I would like to examine the effect of this exclusionary provision on different types of biotechnological inventions.

I would like to start with subcellular fragments. What is the effect of this article on subcellular fragments? As you have read in the article, this article does not give a clear-cut answer to the question of whether or not subcellular fragments like DNA-sequences, genes, plasmids and vectors, can be subject of patent protection. But it is generally accepted that subcellular fragments are patentable.

Second type of inventions: micro-organisms. Art. 53 (b) speaks of microbiological processes and the products thereof, and makes an exception to the exclusion of plant and animal varieties not being patentable. Although it does not speak clearly about micro-organisms per se, it is now generally accepted that patent protection can be granted for microbiological processes and for micro-organisms per se. As to the effect of Art. 53 (a) on micro-organisms, there are generally considered not to be any implications of Art. 53 (a) on the patentability of micro-organisms. As to the day-by-day-granting-policy by the European Patent Office, I would like to leave that to Mrs Gruszow, who is going to tell you something more about that in the next paper.

Let us then turn to the effect of this article on plants, and now it becomes a little bit more complicated. As you can read in Art. 53 (b), there is a double exclusion: an exclusion of plant varieties and an exclusion of essentially biological processes for the production of plants. One reason for the exclusion of plant varieties from patent protection was that several European countries at that time had developed special legal protection for plant breeding, which had been created by the International Convention for the Protection of New Varieties of Plants of 1961, the UPOV Convention, which was the basis for the coming about of national laws on breeders' rights.

140

As to the effect and as to the scope of Art. 53 (b), I would like to say that the EPC itself does not make the problem more simple, since it does not provide any clear guidance with respect to the term 'plant variety'. Since the EPC itself is less than totally unambiguous as to the exact scope of the exclusion of plant varieties, one has to rely on the interpretation of the Technical Boards of Appeal of the European Patent Office, in order to get some insight into the limits of this exclusion. The Technical Boards of Appeal made four decisions in the field of plant genetics: Ciba Geigy (26 July 1983), Lubrizol (Hybrid Plants; 10 November 1988), Lubrizol (Transgene Expression; 31 March 1992) and Plant Genetic Systems (21 February 1995).

As to Art. 53 (b) EPC and the extent of 'plant varieties', it was clear from the decisions of the Technical Boards of Appeal since 1983 — especially in the Ciba Geigy and the Lubrizol Hybrid Plants case — that the excluded area was to be equated with what is protectable under UPOV and corresponding national laws of plant variety protection. Art. 53 (b) prohibited only the patenting of plants or their propagating material in the genetically fixed form of the plant *variety*. The exclusion, however, did not apply to plants which do not meet the profile of a variety and which belong to a classification unit taxonomically higher than that of a variety, in other words, for the plant *per se*. In concrete terms, according to the EPO case law from 1983 till 1995, patent protection is not possible for, for example, the potato variety *Charlotte*, but protection is possible for the potato as such *(Solanum tuberosum)*. This line of reasoning is acceptable from the assessment that traits inserted by modern modification techniques cannot only be introduced in one specific variety but in a great number of plants. Things have changed, however. Since the famous Plant Genetic Systems Decision of 1995, claims on plants per se are no longer considered acceptable, whereas claims on plant cells are regarded as patentable. But as we will see later on, the restriction introduced by this decision should be put in perspective. So far on the exclusion of plant varieties.

Let us turn to the second part of the sentence of the exclusion of Art. 53 (b), namely the exclusion of 'essentially biological processes for the production of plants'. The Guidelines have defined this wording and stipulate that the question of whether or not a process is essentially biological is one of degree, depending on the extent to which there is technical intervention by man in the process. Essentially *non*-biological processes for the production of plants, therefore would not be excluded from patent protection. As to Art. 53 (b) and the wording 'essentially biological processes for the production of plants', the Technical Boards of Appeal of the EPO have formally confirmed the interpretation expressed in the Guidelines, thus giving greater authority to the latter.

With respect to Art. 53 (a) and plants, the question of public morality first came up in the Lubrizol Transgene Expression case, where several non-governmental organisations filed notice of opposition in which they claimed that patents on plants are immoral under 53 (a) and prohibited by 53 (b). The Opposition Division decided that the exclusion of patentability in Art. 53 (a) for inventions which are contrary to public order and morality only concern extreme cases which are universally regarded

141

as abhorrent. In view of the consideration that the actual patent related to an invention which might be used for creating new plants, the nutritive value of which exceeds that of conventionally obtained plants, and the consideration that the plants covered by that patent might give rise to a better management of food shortage in the world, the Opposition Division ruled that the exploitation of such an invention cannot consequently be considered immoral or against public order and decided that a violation of art 53 (a) was not evident. In another case, Greenpeace UK versus Plant Genetic Systems, the Opposition Division reached a similar decision.

As to the granting policy of the European Patent Office, I would also refer to Mrs Gruszow who will tell you something more about that, but I would like to tell you something about it already. On 30 March 1992, about 500 patent applications in the field of plant genetic engineering were submitted world-wide and 16 European patents were delivered. An analysis of the description and the claims of these 16 European patents shows that, as far as the kind of *technology* is concerned, patent protection has been granted for new realisations in the field of in-vitro culture as well as for recombinant DNA technology or protoplast fusion. As far as the type of *claims* is concerned, the analysis has shown that, in the case of in-vitro culture and protoplast fusion, process claims occur almost exclusively, whereas in the case of recombinant DNA technology, product claims which relate to the end product, the modified plant cells, the plants and seeds, or to intermediate products, vectors and plasmids etc., occur as well. So far on Art. 53 and plants.

Let us now turn to the animals. As to section 53 (b) and 'animal varieties', the Board of Appeal decided in the Harvard Oncomouse case that mammals and rodents constitute a taxonomic classification unit higher than 'animal variety' and are therefore *not* excluded under this provision of patentability.

As to section 53 (a) and animals, the Technical Board of Appeals decided in the well-known Harvard Oncomouse case that for each individual invention the question of morality has to be examined, and possible detrimental effects and risks have to be weighed and balanced against the merits and advantages aimed at. As to the granting policy of the European Patent Office, I would like to refer to Mrs Gruszow.

Let us finally have a look at the effect of Art. 53 on human beings. I can be very brief. The European Patent Convention does not contain a specific provision concerning the admissibility of patents on human beings, but it is generally accepted in juridical and scientific circles that human beings cannot be the subject of patent protection. But what about the human genome? What about human cell lines, gene therapy, etc.? The old proposal for a directive on the legal protection of bio-technological inventions introduced several limits to the patentability of human material and I hope that Mr Vandergheynst can tell us what the new proposal looks like concerning this subject.

This is what I had to tell you about the problem of patentable subject matter. Supposing living matter constitutes eligible subject matter; then patents can be granted if the biotechnological invention satisfies the substantial patentability requirements, being novelty, inventive step and susceptibility of industrial application (Art. 52 1° EPC). As to novelty, the invention claimed in the patent application must be new, which means that the invention shall be considered to be new if it does not form part of the state of the art. The novelty requirement constitutes no special impediment for the patenting of biotechnological inventions at first sight. As to the second requirement, inventive step, this would mean that, having regard to the state of the art, the invention is not obvious to a person skilled in the art. The inventive step requirement is intended to prevent exclusive rights forming barriers to normal and routine development. To my knowledge, the inventive step requirement constitutes no bigger obstacle to patenting for biotechnological inventions than for other inventions. As to the last requirement, industrial application, the invention shall be considered as susceptible to industrial application if it can be made or used in any kind of industry, including agriculture. The industrial application requirement constitutes to my knowledge no bigger obstacle to patenting for biotechnological inventions than for other inventions — maybe with the exception of DNA sequences, because most often only the nucleotide sequence is known without knowing the function, so that a meaningful use cannot be demonstrated.

When all the substantial patentability requirements have been fulfilled there are also some formal requirements which have to be taken into consideration. European biotechnology patents may be filed with the European Patent Office itself or with the National Patent Offices in the contracting states.

Once filed, the patent application follows three phases: the first phase being *filing and search* in The Hague. The search is carried out to determine whether or not the invention is new. This is done by the examiners in The Hague, using a unique and vast collection of some 24 million documents (patents, books, periodicals, articles, etc.). The results are set out in a search report, citing documents likely to anticipate the application or to cast doubt on its patentability on grounds of obviousness. If the applicant decides to pursue the application, a *substantive examination* takes place in Munich, where the application is examined by an Examining Division comprising three examiners who are experts in the field concerned. They verify whether the criteria for patentability have been met and maintain a constant dialogue with the applicant or his representative. If the Examining Division considers that the invention is patentable, it decides to grant a European patent. If it concludes that the invention is not patentable, in particular because it is not new or it is obvious, it refuses the application.

Within nine months from the date a patent is granted, one can *oppose* the patent. Any third party can file notice of opposition if it believes that the patent should not have been granted. There is also a right of *appeal*, which is mostly used by the applicant after rejection of the patent application, or by the opponent after rejection of the opposition. Appeals, as you all know, are handled by the Boards of Appeal.

I would like to say one more thing about these formal requirements. When filing a European patent application, two special regulations should be borne in mind. I would like to focus on the second one. Applications relating to microbiological processes or the products thereof should be deposited in a depository institution recognised by the European Patent Office.

143

That is what the scene looks like for the European Patent Convention and the European Patent Office. But what is the effect of Art. 53 on the national patent laws? What do we have to expect in our national countries? As to the eligible subject matter, several countries in Europe — including Belgium, Germany, the Netherlands, France, Great Britain and even countries which are not a member of the European Patent Convention like Norway — have adopted the exclusionary provisions of Art. 53 EPC in their national patent law.

As to national case law, there is no national case law in Belgium yet, but there is case law in Germany and The Netherlands. Courts have already ruled on biotechnology inventions. Mostly however, they did not rule on the question of whether or not patents on living matter are valid, but on the question of whether or not the substantial patentability requirements were met in the particular case.

Dual argumentation for the extension of patent law to living matter [2]

The textual argument

This is what legislation and case law on the European level and the national level look like at present. The situation seems rather comfortable, since the European Patent Office overcame the exclusion of Art. 53 (b) by giving a restrictive interpretation of the terms 'plant and animal varieties'. The problem, though, is that this approach is not always very convincing and that this textual or semantic approach has not put an end to the ongoing discussion in legal, industrial and societal circles of the acceptability of biotechnological inventions.

The intrinsic argument

In an attempt to put an end to this discussion and finally get an answer to the question of whether or not micro-organisms, plants and animals are patentable, I carried out a comparative analysis of the conceptual and technical objections which were brought to bear, in doctrine and jurisprudence, in denying micro-organisms, plants and animals patent protection in the period prior to the establishment of the European Patent Convention (1850 until 1970). Next, I evaluated these objections in the light of recent technological developments, which made me conclude that there is an additional, intrinsic argument for the extension of patent law to micro-organisms and plant inventions.

Let me show you the objections which came about in the last 100 years. Close reading of the Belgian, German and Dutch doctrine and jurisprudence, prior to the establishment of the European Patent Convention, shows that the following objections were raised to deny patent protection to living matter. First of all, the product of nature objection: a first objection that was put forward by the doctrine was that micro-organisms, plants and animals were not the result of a creative process, and hence were not inventions as such. Micro-organisms, plants, animals were products of nature and were not inventions or 'nicht Erfindungen'.

Another equally widespread objection was the living organism objection: patent law is tailored to inanimate techniques and biotechnological inventions should be excluded from patent protection because the subjects of such inventions are living organisms. In this option, micro-organisms, plants and animals were not excluded because they would be creations which lack a creative idea, but because of the special nature of the subject of the invention, out of an inveterate distrust of techniques which affect living nature. Another series of objections arose out of a non-compliance with a number of substantial patentability requirements, more specifically with the lack of novelty, the lack of inventiveness and especially the lack of industrial applicability. The lack of industrial applicability objection was at the core of a heated dispute in Belgium and the Netherlands, especially with regard to plant inventions. The majority point of the doctrine on the scope of the term 'industry' was not simple at that time. Agricultural inventions were not to be excluded in principle from patent protection. Agricultural products which were clearly definable and which were manufactured in industry — such as fertilisers or agricultural machines — could be subject to patent protection, but agricultural methods — such as fertilisation methods, preparation of fertilisers and sowing techniques — were not patentable, because of the lack of an industrial character. Breeder's inventions, products and breeding methods could not be the subject of the concept 'industry'.

At last, there are objections which relate to a non-compliance with a number of as formal patentability requirements substantiated requirements, in particular the impossibility of describing the method of making the end-product and the impossibility of repeating the original process of making. As to the non-reproducibility objection, this objection equally aroused emotions, especially in Belgium and Germany. The invention had to be reproducible. This reproducibility requirement was not literally present in the national patent laws, but was derived by the doctrine from the requirement of industrial applicability of the invention: a method was not industrially applicable, and hence not patentable, when it could not be reproduced. One tendency in the doctrine — the followers of a rigid interpretation of the reproducibility requirement — felt that the process of making, which led to the first specimen of a micro-organism or a new plant variety, should always be repeatable. So the breeding method should always be repeatable. As such a repetition was not possible in practice, patent protection should be excluded. Another school of thought — the followers of a more flexible interpretation of the reproducibility requirement — felt that it was sufficient that additional specimens of the first specimen of the micro-organism or the new plant variety could be obtained by another process, namely a *multiplication process* of sexual or asexual reproduction. In most cases this requirement could be met, hence patent protection should not be excluded.

This is what the scene looked like during the last 100-150 years. My point of view is that nowadays, a lot of these objections can be overcome. I would like to explain to you the reasons why, in my opinion, these objections are not relevant anymore. As to the product of nature objection, a lot of authors refuted this objection, arguing that the criteria for patentability should not be the natural or unnatural character of the product, but the degree of human intervention necessary to obtain such product. The German Bundesgerichtshof also formally refuted this argument once and for all in what is known as the Rote Taube case of 1969, by arguing that a technical invention can also exist in the systematic application of biological forces of nature.

As to the living organism objection, it was mainly an objection that was raised in the United States. The majority of the doctrine in Europe mostly dismissed this argument.

As to lack of novelty and lack of inventiveness objections, the molecular developments in biotechnology of the past years made it possible to overcome these impediments in a lot of cases. But as the gene transfer technique becomes more and more established, it could very well be that the specific gene transfer will have to be more difficult or uncommon to meet the requirement of inventiveness of patent law.

As to the lack of industrial application objection, critics of this objection argued that any invention which is of practical use to mankind is of an industrial character. This objection was rendered largely unfounded when the national legislators formally confirmed — when adapting their national legislation to the Paris Convention and to the EPC — that the term 'industry' should be understood in its broadest sense, i.e. including agriculture.

As to the impossibility of description objection, this objection should be put in perspective in the light of the modern identification methods.

Finally, as to the non-reproducibility objection, I want to say that that was the most difficult one. The problem of reproducibility of biological inventions has, like a Sisyphean rock, rolled back and forth over the years. But the German Bundesgerichtshof finally decided that the strict reproducibility requirement as such — by which I mean the repetition of the process of making — would only apply to *process* protection for the end product, but that in order for *product* protection to be granted for a new organism as such, the repetition of the process of making was not necessary, and the deposit of the new micro-organism together with the description of the multiplication method would suffice. Transferring the jurisprudence of the German Supreme Court to plants would signify that *product* protection for plants is always possible, because the most important dogmatic impediment to patenting plants, namely the repetition of the process of making, is removed.

Conclusion: the conceptual and technical objections raised in jurisprudence and doctrine to deny patent protection to micro-organisms and plants may at present be regarded as dated, which made me conclude that there is an additional, intrinsic argument for the extension of patent law to living matter. This brings me to plead for the revocation of Art. 53 (b) of the European Patent Convention.

Role of meta-juridical arguments in the debate on the patenting of living matter

But... it is not all that simple! Lately, a number of voices — both in non-juridical and juridical circles — have been heard to deny patent protection to living organisms, not on juridical grounds, not on objections of a patent law nature, but on what I would like to call meta-juridical impediments: ethical, societal, ecological and safety grounds. And this brings us to the theme of Art. 53 (a) of the European Patent Convention.

From a historical perspective, I can see a dual development. First, there is a shift in the character of the objections. Over a long period of time, the objections raised against patent protection for living matter were of a *patent law nature*: non-compliance of the invention concept, lack of industrial applicability, lack of novelty, lack of inventiveness, impossibility of description and impossibility of repetition. For the last 10 years, the objections raised to deny patent protection are of a *meta-juridical nature*. Secondly, there is a shift in the forum where the discussion is being held. Initially, the discussion took place in *doctrine and jurisprudence*, during let us say a 100 years. Nowadays, the major debate is held in *extra-juridical circles*: in non-governmental organisations, consumer associations, environmental groups, animal rights' groups, Third World organisations etc.

What is my attitude towards these new impediments? My central thesis is that patent protection for living matter, especially micro-organisms and plants and probably also for animals, can be justified from a juridical point of view. The revocation of Art. 53 (b) is justified on legal grounds. That is manifestly clear to me.

A different question, however, is whether or not patent protection for plants and animals is also *desirable*. My opinion on this is the following: the ambivalence of the new technology has rightly made the necessity of ethical monitoring even more urgent. The confrontation of ethics and patent law is difficult, to say the least, and usually ethics are seen in this context as a disturbance, as 'die grosse Störung'. It is in my view, however — with all my respect to Mr Schatz and Mrs Gruszow — *not* the role of the European Patent Office to engage in this ethical debate. Although the Examining Divisions, the Opposition Divisions and the Technical Boards of Appeal of the European Patent Office are staffed with highly competent technical experts, their make-up is ill-suited to this wider role.

I feel that the European Patent Office should consciously part with this question. It should model itself on the American Supreme Court, which stated explicitly — in the Chakrabarty case — that it is not competent to rule on ecological and ethical matters and that such issues should be addressed by the political branches of government, i.e. the Congress and the Executive, and not by the courts.

The answer to the question of where this discussion should take place, is by no means obvious. Should the debate take place in the national parliaments? In the European parliament? In an EPC revision conference with all the members states? Or would it be preferable for the legislator to postpone entering the debate himself, leaving public opinion and ethicists first to reflect on the ethical and social consequences of patenting living creatures in general?

Considering the fact that hardly any theoretical reflection and research has been done with regard to the interaction between patent law, biotechnology and ethics, I believe that it would be premature to give an answer to this particular question. This brings me to my final conclusion: although patent protection for micro-organisms, plants and probably also animals is justified from a juridical point of view, the final decision to revoke Art. 53 (b) of the European Patent Convention will probably be a decision where not only juridical, but also economic, social and ethical motives will play a decisive role.

Notes

1   *See* Van Overwalle, G. (1996), 'The Legal Protection of Biotechnological Inventions in Europe', in *Intellectual Property Rights and Strategic Alliances in the European Union*, in *Leuven Law Series*, Universitaire Pers: Leuven (forthcoming).
2   *See* Van Overwalle, G. (1996), *Octrooieerbaarheid van plantenbiotechnologische uitvindingen. Een rechtsvergelijkend onderzoek naar een rechtvaardiging van een uitbreiding van het octrooirecht tot planten - Patentability of Plant Biotechnological inventions. A Comparative Study towards a Justification of Extending Patent Law to Plants (Extensive Summary)*, Bruylant: Brussel.

13 Types of invention in the field of genetic engineering, arising in the practice of the European Patent Office

Larissa Gruszow

The patenting of biotechnological inventions by the major patent offices in the world has prompted a lot of controversy in various sections of society. But it is in Europe that the controversy has been, and is, most pronounced. The European Patent Office therefore welcomes this workshop because it offers a good platform for presenting the different approaches at issue and for open discussion.

My task initially was, firstly to remind you of the legal framework which is binding on the different EPO bodies involved in the procedures for granting European patents and, secondly, to show you a selection of European patent applications and of European patents illustrating the types of invention dealt with in the EPO. Now I can renounce presenting you the legal framework, so as not to repeat the very interesting lecture of my predecessor, Mrs Van Overwalle. This will of course shorten the presentation and I think it will be good, because it will leave us more time for discussion.

Before I go into the examples, I will come briefly to the point of statistics, already presented by Mrs Van Overwalle. Our statistics of the EPO show that until now, which is during the 20 years of activity of the European Patent Office, the EPO has granted approximately 320,000 European patents, distributed among approximately 30 main technical fields. Approximately 12,500 of these patents relate to biotechnological inventions in general, and about 2,400 to inventions that involve genetic engineering. If we look into the 5,000 or so European patent applications filed in that field but which did not go through the whole granting procedure, we find, classified in the genetic engineering unit, approximately 500 relating to transgenic plants and 200 to transgenic animals.

I chose to present to you the following types of invention and to illustrate each type with examples of European patent applications or granted patents:

- animals

- gene therapy (germ line and somatic gene therapy)

- isolation of human genes to produce medical substances

149

- plants

- micro-organisms

Of course, you will recognise in some of my examples well known cases and I selected expressly known cases, as far as possible, to make things easier. I should say that some of those examples could be classified under more than one of these categories, depending on which claims and which subject-matter of claims are considered. I could have chosen a completely different order, but I have selected this order because I was impressed by Professor Vermeersch' introduction to the workshop, that this could be best illustrated by two examples, the germ-line gene therapy and the Oncomouse.

I will begin with animals and of course with the Oncomouse. The examples — the sheets you have in front of you — are in fact each a simplified facsimile of the corresponding European publication, either the European patent application as published or the European patent. Next you have some of the bibliographical data with the number of the publication and also a kind of abstract and some illustrations, which in general is not the illustration on the very European publication.

Oncomouse

It is well known that the Oncomouse-invention (*cf* Figure 13.1) provides a method for producing a transgenic non-human mammalian animal having an increased probability of developing cancer. The method comprises injecting into the embryo, at a very early stage in its development, an activated oncogene sequence. As a result, the genes and somatic cells of the mammal contain an activated oncogene sequence. The European patent claims the method and the resulting non-human mammal *as such* — e.g. a mouse — to be used as a test animal in cancer research.

After the European Patent Office's Examining Division refused the application in July 1989, the applicant filed an appeal. This prompted the Board to lay down the following principles in its decision T 19/90: first, under Art. 53(a), EPO Examining Divisions are not permitted to avoid the evaluation of ethical provisions. This is a very important statement in view of what Mrs Van Overwalle said, namely that the patent offices are not the right place to decide on that matter. Within the current legal provisions, the Board of Appeal said it is not possible for the Office to escape this duty. Secondly, the evaluation of the morality criteria should involve a careful weighing up of the suffering of the animals concerned and possible risks to the environment on the one hand, and the utility of the invention for mankind on the other.

The patent on the Oncomouse was eventually granted, but 17 opponents objected to the patent so that an opposition procedure took place before the Opposition Division. Recently, the Opposition Division called an oral proceeding but could not reach a decision, so that this procedure is going on the first instance. It is not to be excluded that, when the first instance will have decided, subsequently one of the parties will appeal before the Appeal Boards of the EPO. That is all about the Oncomouse.

Europaïsches Patentamt
European Patent Office
Office européen des brevets

EP 0 169 672 B1

EUROPEAN PATENT SPECIFICATION

Application Number: 85304490.7

Date of filing: 24.06.85

Proprietor: The President and Fellows of Harvard College (US)

A method for producing a transgenic non human mammal animal having an increased probability of developing neoplasms, said method comprising introducing into an animal embryo an activated oncogene sequence. The animal may be used in testing a material suspected of being carcinogenic or of conferring protection against carcinogens.

This is a simplified presentation of the original publication

---

Figure 13.1   Method for producing transgenic animals / 'Oncomouse'

Sheep Tracy

The second example concerns the so-called sheep 'Tracy' (cf Figure 13.2) and relates to protein production by a mammal. Protein production, as you know, is certainly a useful source for producing some medicaments. As from the late 1980s, the EPO received a relatively high number of applications for this kind of invention, in which non-human mammals are used as bioreactors. In this particular application, a sheep (or any other non-human mammal) is genetically engineered by incorporating into its germ line a genetic construct comprising a DNA sequence coding for a useful protein. The genetic construct provides for expression of the protein in the sheep's milk. The protein is subsequently recovered from the milk. As an example of useful substances that may be recovered, the patent suggests human insulin, tissue Plasminogen Activator and Alpha-antitrypsin. This last protein is of primary importance in the treatment of Mucoviscidosis, an illness which affects the lungs of many children. In this case, the patent application does not report that the sheep would suffer because of its transgenity.

PCT

WO 88/00239 • EP-87 904286.9 • A-0274449

INTERNATIONAL APPLICATION PUBLISHED UNDER THE PCT

International Application Number: PCT/GB87/00458
International Filing Date: 30.06.87
Applicant: Pharmaceutical Proteins Ltd. (GB)

A method of producing a substance comprising a peptide,
involves incorporating a DNA sequence coding for the
peptide into a gene of a mammal animal (such as a sheep)
coding for a milk whey protein in such a way that the DNA
sequence is expressed in the mammary gland of the adult
female mammal. The substance may be a protein such as a
blood coagulation factor and AAT (alpha antitrypsin).

This is a simplified presentation of the original publication

Figure 13.2   Peptide production / 'Sheep Tracy'

Gene therapy

Now I will come to gene therapy. Gene therapy on human beings, although still in
its initial stage of development in practice, has given rise to fundamental questions
in society. Two essentially different categories of gene therapy have emerged. First,
the human germ line gene therapy, which aims at correcting the genome of germinal
cells, thereby eradicating a genetic illness in a human being before birth, and in
future offspring. May I show you the only existing example of a European patent
application relating to this kind of invention (cf Figure 13.3). It is a rather recent
application, of 1993. Since it is a Patent Co-operation Treaty application, the
procedure is somehow postponed, which may be the reason why the substantive
examination has not yet begun in the EPO. In this case, the current claims are
intended to protect a method for destroying the endogenous germ cell population in
the testes of a male mammal and repopulating them with primitive cells which are
not native to the said animal. Transgenic animals and the offspring resulting from
that method are also claimed in this application. It is suggested that the method can
be applied to domestic mammals to confer on them inherited positive traits. In
addition, the patent application claims that the method can be applied in germ line
gene therapy to a human patient with genetic traits which would cause serious illness.
As a representative of the EPO, Mr Gugerell has already declared in the *New Scientist*

152

of April 1994 that at least the claims directed to human germ line gene therapy will be considered as referring to methods of treatment of the human body, which are clearly excluded from patentability under Art. 52(4) EPC. Such claims would have to be scrutinised further in view of the exclusion on morality grounds.

PCT

WO 93/11288 • EP-93 900799.3 • A-0619838

INTERNATIONAL APPLICATION PUBLISHED UNDER THE PCT

International Application Number: PCT/US92/10368
International Filing Date: 07.12.92
Applicant: The Trustees of the University of Pennsylvania (US)

An animal harboring a non-native germ cell, its corresponding line, and the corresponding germ cells, are obtained by colonising the testis (or testes) of a host animal with primitive cells followed by raising and/or breeding the host.

This is a simplified presentation of the original publication

---

Figure 13.3   Repopulation of testicular seminiferous tubules with foreign cells / 'Germ line Gene Treatment'

The second example in this field relates to somatic gene therapy (cf Figure 13.4). This may be defined as the insertion of foreign genetic material into the somatic cells of a human being, with a view to achieving a therapeutic effect. The application relates to gene therapy for AIDS. It is known that the AIDS virus infects and destroys human cells by sticking to a specific so-called 'CD4 receptor' by which the virus penetrates into the cell and destroys the cell. The invention provides for a vector capable of expressing *in vivo* a foreign protein that interacts with the native CD receptors on cells. Subsequently it causes the virus to stick onto this foreign protein and to block the effect of the destructive work of the virus in the body. The Examining Division has recently issued a positive notification to the applicant, saying that it is prepared to allow the claims directed to the vector designed to be transfected e.g. in the bone marrow of the patient to express the blocking protein. However, the Examining Division is opposed to the claims directed at the therapeutic method based on the vector as not being allowable under Art. 52(4) of the EPC. This article excludes from patentability therapeutic, diagnostic and surgical methods applied to the human and animal body.

PCT

WO 90/01870 • EP-89 909945.1 • A-431045

INTERNATIONAL APPLICATION PUBLISHED UNDER THE PCT

International Application Number: PCT/US89/03541
International Filing Date: 21.08.89
Applicant: The United States of America (US)

The invention relates to an expression vector containing a promoter sequence and a DNA (RNA) sequence insert which, when transfected into a mammalian cell, will cause said cell to express soluble CD4 receptor protein. Such a protein blocks interaction between the CD4 receptors present on cells and viruses which would otherwise bind to the CD4 receptors.

This is a simplified presentation of the original publication

Figure 13.4   Retroviral vectors expressing soluble CD4 / 'Gene Therapy for Aids'

Isolation of human genes

Now I come to the isolation of human genes for medical purposes. The first example is the 'Relaxin-case' (cf Figure 13.5). This invention relates to a DNA fragment able to encode a specific human relaxin. Relaxin is a hormone capable of influencing smooth muscle contraction. It is used in many medical situations such as childbirth in the case of women lacking sufficient endogenous production of the hormone. Human sources for obtaining medicines by genetic engineering are preferred to animal sources because of their greater immunological tolerance in the human body. The DNA sequence coding for relaxin is obtained from human ovarian tissue taken from a pregnant woman at a necessary gynaecological operation. The DNA is integrated into a bacterial plasmid acting as vector for transforming that micro-organism which is then used for industrial production of the relaxin. The patent was granted in April 1991 and was then opposed. The opponents maintained that the claimed DNA sequence was open to objection under Art. 53(a) of the European Patent Convention (morality and 'ordre public'). In an extensively reasoned decision the Opposition Division rejected that objection from the opponents. This decision has been published, as most of the decisions of the Board of Appeal and of the Enlarged Board of Appeal, in the Official Journal of the European Patent Office. I do not want to go into the details of the reasoning because it would bring us too far, but I must say that this granting prompted one of the opponents to lodge an appeal which is currently pending before the EPO Board of Appeal.

Europaïsches Patentamt

European Patent Office

Office européen des brevets

EP 0 112 149 B1

EUROPEAN PATENT SPECIFICATION

Application Number: 83307553.4
International Filing Date: 12.12.83
Proprietor: Howard Florey Institute of Experimental
Physiology and Medicine (AU)

├─Signal Peptide ──→

```
                                        -20                           -10
     Met Pro Arg Leu Phe Phe Phe His Leu Leu Gly Val Cys Leu Leu Leu Asn
H2   AUG CCU CGC CUG UUU UUU UUC CAC CUC CUA GCA GUC UGU UUA CUA CUC AAC
     *** *** *** *** **  **  *** *** *** *** *** **  *** *** *** *** ***
H1   AUG CCU CGC CUG UUC UUG UUC CAC CUC CUA GGU UUC UGC UUA CUA CUC AAC
     Met Pro Arg Leu Phe Leu Phe His Leu Leu Glu Phe Cys Leu Leu Leu Asn
```

Genes and DNA transfer vectors for the expression of human relaxin; sub-
units thereof, including genes and transfer vectors for expression of human
relaxin and the individual peptide chains thereof. Methods for synthesis of
the peptides involving recombinant DNA techniques.

This is a simplified presentation of the original publication

Figure 13.5 **Molecular cloning and characterisation of a further gene sequence
coding for human relaxin / 'Relaxin'**

Europaïsches Patentamt

European Patent Office

Office européen des brevets

EP 0 093 619 B1

EUROPEAN PATENT SPECIFICATION

Application Number:
83302501.8
International Filing Date:
04.05.83
Proprietor:
Genentech, Inc. (US)

This is a simplified presentation of the original publication

Figure 13.6 **Human tissue plasminogen activator, pharmaceutical compositions
containing it, processes for making it, and DNA and transformed
cell intermediates therefor / 't-PA'**

A second example of the isolation of human genes is the Human Tissue Plasminogen Activator (*cf* Figure 13.6). Since the Human Tissue Plasminogen Activator was presented yesterday extensively by Professor Collen, I will not repeat it, but as you know this protein is needed in relatively large quantities for resorption of blood clots in thrombosis situations. Before genetic engineering was developed, the previously known activators were directly isolated from various human tissues, giving rise to small quantities of the product with insufficient purity. The Genentech patent claims specific recombinant DNA expression vehicles and various host cells useful in recombinant production of t-PA in sufficient quantities.

Plants

Now I will come to plants. The first example is the Plant Genetic System (PGS) 'Basta'-case (*cf* Figure 13.7). As we have already heard yesterday, it is a patent which has been granted and then opposed and appealed. The decision of the Board of Appeal was given in February 1995. The invention in this case is based on the 'discovery', I would say, that a certain enzyme called Glutamine Synthetase (GS) is of primary importance for the growing process of plants. The activity of the 'Basta' herbicide is based on inhibiting the GS enzyme of the plant. The PGS invention involves transforming plant cells with a vector containing a foreign DNA that codes for an enzyme which inactivates these GS enzyme inhibitors contained in the herbicide. The PGS patent contained claims aiming to protect the method for genetically engineering plant cells with the foreign DNA, the vectors used for that purpose and the transformed plants. The PGS patent was opposed by Greenpeace on two grounds. Firstly, it contravened Art. 53(a) — morality and 'ordre public'— and secondly, it contravened Art. 53(b) EPC — the exception from patentability of plant varieties. In the decision in appeal, which was already referred to this morning, the Board of Appeal confirmed the opinion of the first instance, the Opposition Division, that the PGS invention is not in contradiction with morality and 'ordre public'. On Art. 53(b) however, the Board of Appeal said that the claim directed to the plants as such in fact embraces a certain number of varieties and therefore falls under the exception of Art. 53(b).

The second example is another PGS patent application, directed to genetically modifying plants to acquire a higher nutritional value (*cf* Figure 13.8). As in the previous case, a DNA sequence is inserted into the cells with a vector, a plasmid, so that the yield of tryptophan, which is an important storage protein, will be enhanced. The economic importance of that invention is very clear, first of all for countries having hunger problems. This application is still under examination.

EP 0 242 236 B1

EUROPEAN PATENT SPECIFICATION

Herbicide containing
inhibitors of GS

Application Number: 87400141.5
Date of filing: 21.01.87
Proprietor:
Plant Genetic Systems N.V. (BE)
Biogen, Inc. (US)

vector

Plants contain glutamine
synthetase (GS), a natural
enzyme indispensable
for growth

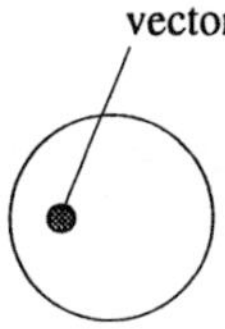

Plant cell transformed
with a vector containing
a foreign DNA that codes
for a protein causing
inactivation of the
herbicide inhibitors

This is a simplified presentation of the original publication

Figure 13.7   Plant cells resistant to glutamine synthetase inhibitors, made by genetic engineering / 'PGS-BASTA'

EP 0 318 341 A1

EUROPEAN PATENT SPECIFICATION

Application Number: 88402650.1
Date of filing: 20.10.88
Proprietor: Plant Genetic Systems N.V. (BE)

The genetic patrimony of plants has been engineered to comprise a precursor-coding DNA encoding the precursor of an albumin storage protein.
The precursor-coding DNA is enriched with a DNA insert capable of coding an heterologous appropriate polypeptide such as tryptophane, one of the amino-acids of major value for the human organism.

This is a simplified presentation of the original publication

Figure 13.8   Transgenic plants with increased nutritional value / 'Improved Nutritional Plant'

Micro-organisms

The last example is an example about micro-organisms which have been genetically engineered with a view to increase the souring capacity of the micro-organisms in the process of making milk, butter and cheese products (*cf* Figure 13.9).

I think my duty is now finished and of course Mr Schatz will give a substantial lecture on Art. 53(a) from the point of view of the EPO.

Europaïsches Patentamt

European Patent Office

Office européen des brevets

EP 0 251 064 B1

EUROPEAN PATENT SPECIFICATION

Application Number: 87108790.4
Date of filing: 19.06.87
Proprietor: Valio Meijerien
                        Keskusosuusliike (FI)

In view to optimising the characteristics of souring micro-organisms in the dairy industry - milk, butter, cheese production - the invention develops a new cloning vector that is inserted into E. coli, B. subtilis or other micro-organisms.

This is a simplified presentation of the original publication

---

Figure 13.9   A vector plasmid suited for souring agents, dairy souring agents in particular

14 Patents and morality

Ulrich Schatz

I am very glad to be able to assist to this conference, not because I have an opportunity to speak but yesterday it was particularly interesting to listen to the speakers on moral aspects and moral permissibility of biotechnology. My subject is totally different. And you will see that it is different: I am not concerned with moral permissibility of technology, more specifically biotechnology. I am concerned with a quite different question, namely whether it is permissible to patent under moral aspects.

On the first question, the moral permissibility of biotechnological inventions, I will give you an idea of where I stand. First, I love my dog. Secondly, I hate people who do harm to animals, whether this is by kicking them or by using genetic engineering and to bring out animals who suffer. Thirdly, I am among those who think that what Stanislas Leavis predicts as the probable end of the human kind, which is ecocide, is not an absurd idea. It may well happen, but not as a result of biotechnology. It will be the result of the demographic development of the population, overuse of energy of whatever source. That is about what I, as a citizen, think about these kinds of problems. I am not specifically perturbed by the idea that we can have genetic engineering.

But now I have to come to my subject. I will deal with it in two times. As a lawyer, I am used to looking first at what the law says, before speaking out what it should say. So let us first have a look at Art. 53(a) EPC.

The bar on patenting inventions, the publication or exploitation of which would be contrary to public order or morality, is as old as patent law itself and *yet* the number of cases in which the grant of patent has been refused or in which patents have been revoked by Courts is minimal. There is no such case in the 18 years of practice of the EPO, albeit — as Mrs Gruszow already said — 320,000 patents have been granted in all fields of technology by the EPO. The picture of practice of national patent offices and courts is not much different. The very few cases which are reported in literature are old cases and concern almost exclusively things like contraceptives, abortives, or the domain of the bizarre in human sexuality. Poisons, explosives, extremely dangerous chemical substances, devices used in nuclear power stations, agro-chemicals, pesticides and many other things which can threaten human life or damage the environment have been patented, despite the existence of the public order and morality bar. I shall look at the reasons for this and, as a lawyer, once again, we must always start from the text of the provision. So let me repeat the text of Art. 53(a):

European patents shall not be granted in respect of inventions, the publication or exploitation of which would be contrary to "ordre public" or morality, provided that the exploitation shall not be deemed to be so contrary merely because it is prohibited by law or regulation in some or all of the Contracting States.

So a mere prohibition by law of the use of the invention does not matter, does not hinder the grant of the patent.

The first thing we must note in this provision is that it is not the permissibility of any kind of technology or of any kind of invention *as such* which is at stake, but the *exploitation* of it. The question to be answered is whether the making or the use of the invented product or process '*would be*' contrary to public order or morality. The assessment to be made is thus purely hypothetical, and in actual fact, almost all processes and products can be used for different purposes. Of course, when you make a judgement on an act, such as exploitation, it is the purpose of that act which gives the moral quality to the act. Explosives can be used in terrorist action or otherwise in road building or in the mining industry. A given chemical substance can become an ingredient of a harmless fertiliser or can be used for the production of a chemical weapon and so forth. Given this variety of possible uses, Art. 53(a) EPC can only apply when the non-permissible use of the invention can be deduced from the very nature of the invention, as is the case with the famous letter bomb, quoted as an example in the *Guidelines for examination in the EPO*. I can assure you that it takes considerable imaginative effort to find the example of the letter bomb, and perhaps to find one or two other examples of this kind, because you will always find that an invention which is filed is not filed for a use which would be contrary to morality or 'ordre public'. It is always filed for another purpose. There is a second reason why Art. 53(a) EPC, in practice, has little chance to apply. An inventor of an item which can only be used for purposes contrary to 'ordre public' or morality has an obvious interest in hiding his invention, rather than seeing it published in a patent application. He, moreover, has no interest in patent protection since, for such an invention which cannot be exploited without being contrary to criminal law and so forth, there can be no market. Patents cost a lot of money and nobody would file a patent for a thing which cannot possibly be admitted to the market.

Another aspect, which I wish to underline in the context of the discussion on patentability of biotechnological inventions, is the fact that the moral permissibility of the act of patenting as such is not at issue at all in Art. 53(a) EPC. It is not patenting the invention but exploiting the invention which has to be contrary to morality or 'ordre public' in order to justify the rejection of the patent application. I am sorry to make this rather trivial statement, but I feel bound to do so by the Relaxin case[1], in which the opponent (the Green party) has presented a lengthy argument that the very act of patenting human genes amounts to a form of modern slavery, infringes the human right to self-determination and is thereby contrary to Art. 53(a) EPC. In its reasons for decision, the Opposition Division has gone through the pain of refuting this argument in substance. This is certainly laudable, but instead, the Opposition Division could have made the simple statement that the morality of the act of patenting as such is not at issue in Art. 53(a) EPC. For a lawyer, that would be a completely valid argument. By the same token and more generally, the claim 'no patents on life' — to the extent that it implies that patenting living matter is immoral per se — cannot possibly be satisfied under Art. 53(a) EPC. Those subscribing to this claim — and I do not say whether they are wrong or right — must therefore appeal to the legislator, rather than to patent offices who have to apply the law as it stands.

Let me now turn to the question what the law means by 'ordre public' and morality. Here, I have to say that these two concepts are interpreted in different ways in the context of national patent law on the one side and European patent law on the other. As far as national patent law is concerned, doctrine and courts agree that 'ordre public', as opposed to morality, is a body of positive law, the only question being what *kind* of legal provisions qualify as pertaining to this body called 'ordre public', rather than being simple laws or regulations, the contravention of which is no obstacle to the grant of a European patent in accordance with the second phrase of Art. 53(a) EPC. I shall not go into the detail of this distinction, but I suggest that the distinction between 'moral reasons' and 'prudential reasons', made yesterday by Prof. De Tavernier in the context of the admissibility of biotechnology, perfectly fits the distinction made in Art. 53(a) EPC between 'ordre public' and other kinds of laws and regulations. I think Art. 53(a) points to the moral reasons, rather than to the prudential reasons. In any event, I wish to underline that, as a consequence of the concept of 'ordre public' being a body of certain laws, the exploitation of an invention can *only* offend 'ordre public' if it is *prohibited by law*. That law could be general criminal law, which would apply to the case of the letter bomb. There is no prohibition for the making of letter bombs, but there is a prohibition of murdering and of using explosives for other purposes than those that are precisely indicated and permitted. That law could also be constitutional law or special laws protecting human life and dignity and other basic values of society. I fully agree that this 'ordre public' legislation also now comprises provisions concerning the protection of the natural environment.

Still in the context of national patent law, 'morality' is seen as a body of ethical norms which are generally accepted by those concerned with these norms, for example the deontology observed by all honourable members of the medical profession, ethical principles generally recognised in the different branches of scientific research, codes of conduct observed in industry and business, and the like. These are ethical norms which exist, which are accepted and which have a range of identifiable applications.

As already indicated, the Boards of Appeal of the EPO have taken a different line of interpretation in interpreting both the concepts of 'ordre public' and 'morality'. In the famous PGS - 'Basta' decision (T 356/93), dated 21 February 1995,[2] the

Technical Board of Appeal, by referring to the legal history of Art. 53(a) EPC, starts from the assumption that there is no (uniform) European concept of either 'ordre public' or morality and that the interpretation of these concepts is therefore a matter for European institutions, specifically the European Patent Office. Here we have the starting point of what you [Dr Geertrui Van Overwalle] do not wish and we will see whether you are right. It then goes on to say that the concept of 'ordre public' covers the protection of public security, the physical integrity of individuals and also encompasses the protection of the environment. Accordingly, says the Board, inventions, the exploitation of which would be 'likely to seriously prejudice the environment', are to be excluded from patentability as being contrary to 'ordre public'.

The concept of 'morality' is, according to the Board, related to the belief 'that some behaviour is right and acceptable whereas other behaviour is wrong, this belief being founded on the totality of the accepted norms which are deeply rooted in a particular culture.' The Board continues by saying that for the purposes of the EPC, the culture in question is the culture inherent in European society and civilisation. It then concludes that, accordingly, under Art. 53(a) EPC, inventions the exploitation of which is not in conformity with the 'conventionally accepted standards of conduct pertaining to European culture' are to be excluded from patentability as being contrary to morality.

You can see that there is a tremendous difference in approach between the traditional interpretation of the concepts of 'ordre public' and morality in the context of national laws on the one hand and the Board's interpretation in the context of European patent law on the other. In the context of national patent law, the only question national patent offices and Courts have to answer under the national provision corresponding to Art. 53(a), is whether the exploitation of the invention would fall under a law pertaining to 'ordre public' or violate an ethical norm prevailing in the relevant profession or social group (doctors, scientists, or what you have). If this is not the case, they do not have to care about possible hazards involved with the use of the invention, it being understood that the essential control and prevention of technological hazards is the exclusive task of special authorities created to this effect, which are not patent offices or patent courts.

According to decision T 356/93, this is different for the EPO. Since the concepts of 'ordre public' and 'morality' are no longer a reference to a body of positive legal and positive ethical norms, which you can identify and which have a real existence outside patent law, the EPO has to define *by itself* the normative contents of these concepts and then apply them. The decision expressly states that the assessment of whether or not a particular subject matter is to be considered contrary to either 'ordre public' or morality is *not* dependent upon any national laws or regulations and that, by the same token, particular subject-matter shall *not* be regarded as complying with the requirements of Art. 53(a) EPC merely because its exploitation is *permitted* in some or even in all Contracting States.

This statement clearly implies that in the field of 'ordre public' and morality, the EPO combines the roles of both legislator and judge. Since 'ordre public' is defined as covering the protection of public security and physical integrity of individuals

162

and the protection of the environment, and since the concept of morality is not understood as a reference to any specific ethical norm, the question whether the exploitation of an invention would be contrary to 'ordre public' or morality boils down to a question of technological hazard or undefined moral acceptability.

Clearly — as long as this is the leading case in case law — Art. 53(a) EPC is no longer a 'sleeping law', and the EPO, as the Board expressly states in this decision, finds itself side-by-side with the increasing number of other regulatory authorities and bodies having the duty to assess the hazards stemming from the exploitation of certain technologies, in the field of biotechnology as much as in any other field of technology.

These *other* authorities to which the Board is referring — public health authorities, institutions responsible for biotechnological security (field research and field dissemination) and the like — act on the basis of detailed laws and along specific procedures, precisely those laws and regulations which are irrelevant for the EPO in the Boards' view. They have extensive powers of investigation and sanctions at hand, which the examiners of the EPO do not have. How can examiners of the EPO possibly live up to the new responsibility ascribed to them by the Technical Board of Appeal and what is the precise standard, on the basis of which they have to make their decisions?

The first question which arises is whether the examiners of the EPO have a duty of investigation concerning technological hazards. The Board's answer to this question is not absolutely clear. It recognises, however, that in most cases, potential risks in relation to the exploitation of a given invention cannot be anticipated merely on the basis of the disclosure of the invention in the patent specification. This is typically the case in the pharmaceutical sector and also in the field of biotechnology where, for instance, regulatory field testing may be performed only after the time of the grant of the patent. The Board then goes on to say that Art. 53(a) EPC shall only apply when the threat to the environment is 'sufficiently substantiated' at the time when the decision (on grant or revocation) is made. This seems to imply that Examining divisions of the EPO do not have any particular duty of investigation and that they are allowed to base their assessment of technological risks on what is disclosed in the patent application as filed, except perhaps in the case that evidence has been filed by a third party under Art. 115 EPC. This article provides that third parties may file written observations without thereby becoming a party to the procedure. If the assumption is correct that, apart from information contained in the file, examiners have *no particular duty of investigation*, then it is very likely that Opposition Divisions and Boards of Appeal will arrive at an assessment of risks which is different from that of Examining Divisions, since, as a rule, relevant arguments and evidence will be presented by the opponent, especially in cases in which the opponent is a non-industrial entity.

The second question concerns the yardstick for making the decision that a given risk or undesirable effect of the exploitation of the invention is of such magnitude that it is unacceptable for the purposes of Art. 53(a) EPC. In the Harvard Oncomouse case (T 19/90), the Board of Appeal has sent back the case to the first instance with the order to carefully weigh up the suffering of the animals and the possible risks to the environment on the one hand and the invention's usefulness to mankind on the other. The case is now before the Opposition Division and we will see whether they are able to perform convincingly the 'balancing exercise' to which they have been invited.

In the PGS case (T 356/93), the Board of Appeal itself recognises that the balancing exercise ordered in the Harvard case is only applicable when *actual damage* (in the case of the Harvard mouse: the suffering of animals) is at issue, but not in the context of mere technological risk. When the question of *risk* is at issue, the Board draws the dividing line between 'possible hazards' on the one hand and 'conclusively documented hazards' on the other. Hazards which are only 'possible' escape the ambit of Art. 53(a) EPC, hazards which are 'conclusively documented' fall under it. The Board, however, explains this dividing line by reiterating that a patent does not *per se* amount to an authorisation to exploit the invention claimed in the patent and that in the field in question, namely herbicide-resistant transgenic plants, regulatory approval must be obtained. It then says that, should 'the competent authorities', after having definitively assessed the risks involved, prohibit the exploitation of the invention, the patented subject matter could not be exploited anyhow. This means that, whatever we do, grant or reject, it makes no difference, because the competent authority will prohibit it if it is a danger. If, however, regulatory approval is given, *then*, says the Board, patent protection should be available. There is no clearer way of saying 'it is not our business to assess technological hazards'. If the thing is admitted to the market, patent protection should be available. If not, patenting it does no harm, because it cannot be made.

If this means that the exception under Art. 53(a) EPC *only* comes into play when, on the basis of convincing documentary evidence, it is clear from the outset that regulatory approval for whatever form or use of the invention *will never be obtained*, then the Board would have made an important step back behind its starting point, namely that laws and regulatory provisions in all or some of the Contracting States are irrelevant, and would have come closer to the traditional way of applying the public order and morality bar on patenting, namely as a reference to a body of existing, generally accepted legal and ethical norms of fundamental importance for the cohesion of society. The future will show whether this is the case.

I shall now conclude this first part of my presentation with just two points. First, the claim that *patenting* living matter ('no patents on life') is morally not permissible and should therefore be excluded, cannot possibly be satisfied under Art. 53(a) EPC and equivalent provisions of national patent law. The act, the moral permissibility of which has to be assessed under these provisions is *not patenting but exploitation* of the invention. Secondly, my personal view is that patent law, and specifically Art. 53(a) EPC, does not provide for any legal basis in relation to the control and the prevention of technological hazards in biotechnology, any more than in any other field of technology. And since rejecting a patent application on the basis of Art. 53(a) EPC has no other effect than abandoning the invention to the public domain, where anybody can pick it up and use it, Patent Offices, including the EPO, cannot possibly play, and in my view should not even try to play, any role in the prevention of technological hazards.

Does this mean that Art. 53(a) EPC must be looked at as being a tree in the forest of dead law? By no means! There are two situations in which this provision clearly applies.

The first situation is that the 'exploitation' of a certain category of technology is generally prohibited by laws of 'ordre public' or is incompatible with the moral standards generally accepted in the relevant professional circles. There is no doubt, for instance, that the production and use of bacterial weapons is contrary to both public order and morality. It is indeed prohibited by international treaties and, as far as production is concerned, by criminal law. In the particular field of genetic engineering, any kind of modification of the human genome, even if it is for the purpose of eradicating severe hereditary diseases, through germ line therapy, is prohibited by law in most of our Contracting States as being contrary to human dignity and for the same reason, it is also incompatible with the deontology of the medical profession and is in fact not practised. We can discuss whether this is a definitive situation. I remember we have had a symposium on patents and ethics two or three years ago in Munich, and Baroness Warnock has said 'what is wrong with a germ line gene therapy, the only objective of which is to eradicate a terrible disease like haemophilia or other severe diseases of this kind?' The arguments are twofold. The one is to say 'if we do that, who will prevent us to do the next thing and create a better man?' This, we do not accept morally. Let us say, if we only repair a defective gene by introducing into the genome a gene which is in order, which does not cause disease, then what is wrong with that? The second objection is of a more technical nature. The state of the art today is that introducing such a repair gene in the human germ line involves that hundreds of failures will arrive and perhaps only one will be a success — even about this, we cannot be sure that there will be no secondary effects. In other words, the medical requirements of applying such a therapy in accordance with medical deontology are not there. The state of the art does not allow that. Therefore, we can say that germ line gene therapy is prohibited both by public order — because the laws are there in most of the Contracting States — and it is contrary to ethical norms, so there is no problem here. It is prohibited. No European patent shall be granted in these fields under Art. 53(a) EPC, not to mention the fact that germ line therapy, in particular, also falls under the exclusion of medical treatment exempted from patentability by Art. 52(4) EPC.

The second situation in which Art. 53(a) EPC applies concerns inventions which, without belonging to a technology which is prohibited as such, have effects or lead to results which are prohibited by 'ordre public' or specific ethical norms. This is typically the case with genetic engineering applied to animals. There is no general bar on the making, the production and the use of transgenic animals. The relevant laws and regulations — such as, for instance, the Council Directive 90/220 EEC or the German law on genetic engineering ('Gentechnikgesetz') and corresponding laws which exist in all our European states — do *not* prohibit the making of genetically modified micro-organisms, plants or animals, but merely submit the exploitation of that technology to requirements and procedures necessary to the protection of public security and of the environment. These provisions are thus clearly based on 'prudential reasons' and thereby are to be looked at as 'laws' or 'regulations' within the meaning of Art. 53(a) EPC, second phrase. At the same time, however, the exploitation of an invention, the subject-matter of which is, for instance, a transgenic animal, may fall under the ambit of laws protecting the welfare of animals, e.g. the German 'Tierschutzgesetz'. The prohibition provided for in these laws of the creation, by traditional breeding or otherwise (e.g. by genetic engineering), of vertebrates where certain parts of the body are missing or who are suffering from pain or other damages, is a prohibition that does not apply to transgenic animals or bred animals required for performing experiments which are indispensable for the purposes of basic research or in the context of the prevention or the treatment of human or animal diseases. These provisions clearly belong to 'ordre public' and at the same time reflect the ethical standard of the professionals who are active in this field. It is my view that the 'Gentechnikgesetz', together with the 'Tierschutzgesetz', and the similar laws in other countries, make up the public order which should constitute the basis for making decisions. Instead of an imaginary 'balancing exercise' determined in its own right by a patent office, these provisions are therefore the appropriate basis for applying Art. 53(a) EPC in cases in which animal welfare is at stake such as, in particular, the Oncomouse case.

Up to now, I have talked about what the law says and how I think the law (Art. 53a) should be applied. Now I come to the second part of my contribution, which concerns the exceptions to the patentability of biotechnological inventions which can be lawfully made and used in industry and trade. So far we have been considering the question of patentability of inventions the *exploitation* of which would be contrary to public order and morality, which is dealt with by Art. 53(a) EPC. I shall now turn to a different question, namely whether the *patenting* of certain inventions as such and specifically the patenting of certain inventions in the field of genetic engineering may be morally impermissible, albeit the exploitation and use of these inventions may be perfectly lawful and in conformity with ethical norms.

To come to grips with this question, we must in the first place make sure what 'patenting' actually means and what is the content or the nature of the right conferred by a patent on its proprietor. As to the content and nature of the right conferred by a patent, I have been quoting the decision T 356/93 where it says that a patent does not *per se* amount to an authorisation to exploit the invention. To be even more precise, we should say that a patent gives no right whatsoever to exploit an invention. The sole right which a patent does give is to enable the inventor or his successor in title to *exclude others* from exploiting the invention.

The patent system therefore is no more and nothing other than a *regulation of competition*, based on considerations of both justice and utility. The consideration of *justice* (or morality, if you like — this is the ethics of the patent system) is that it would be unfair to the inventor to allow his competitors to exploit the invention which is the fruit of his own substantial intellectual effort and financial investment. It is simply unjust to allow people to steal from somebody else the result of his

166

creative effort. This is the basic ground of intellectual property as a whole (copyright and all the other items of intellectual property). It has been unquestioned for several hundreds of years as a basic moral principle in our society. The consideration of *utility* is that patents are an indispensable motor of technological innovation in that they afford the inventor at least a chance — a chance, not a security — to recover the costs of research and development, through sales on the market. This is the basic system of financing research and development in industry, in all countries that have a market economy system. You take money from the client, from the buyer of goods, and bring it back to research and development. This is how almost 100 per cent of industrial research is financed. If you take away this system, you will find no money because you would be a fool engaging in research when you can take it from your neighbour. Yesterday, we have heard that the patent system also finances about 40 to 60 per cent of scientific research in university organisations. They make inventions, which they license out to industry. Perhaps you can confirm this or correct me if I am wrong, but this is what I have been told on several occasions. To sum it up, 'The patent system adds the fuel of interest to the fire of genius', to quote a perhaps somewhat romantic citation of Abraham Lincoln.

Let me here open a parenthesis which I shall close very soon. Sometimes, and specifically in the debate of the European parliament on the proposed Directive on biotechnological inventions, it has been said that the patent system favours big industry rather than small and medium industries (SME's). Admittedly, big industry has more power than small industry and also holds more patents, the ratio in biotechnology being about two-thirds to one-third. But this gives us as much insight as saying that a fat man is heavier than a thin one. The fact is that many companies which are large today owe their growth to sustained technological innovation which they initiated at a time when they were small and that, if a small company is engaged in technological competition with a large one, its only chance to stay on the market, to stay in business, is patent protection. Take that away and you will wipe out the SME's from the market in Europe as well as in other countries and in biotechnology as much as in any other field. — End of parenthesis.

I shall now look at the different areas of genetic engineering in respect of which the morality of patenting has been challenged, without the underlying technology being unlawful or qualified as not being morally permissible. First, germ line gene therapy does not fall under this category, since it is as such contrary to 'ordre public' and morality and therefore, as we have seen, falls under Art. 53(a) EPC. But, even if it did not, it would still be a form of medical treatment excluded from patentability as a whole by Art. 52(4) EPC. What is the reason for this general exclusion of medical treatment? The reason is that the patent system is a regulation of competition in industry and trade, whereas the medical art has to abide by medical deontology rather than by the rules of commercial competition. We do not consider that the exercise of medical art is a business which should be governed by the laws of commercial competition. Patent law should not apply to the medical profession. What the means of medical treatment are is indifferent, be it genetic engineering, surgery, prescribing pills or whatever. It is the profession that is exempted from the reign of patent law. This is also the case with *somatic* gene therapy which also is a form of medical treatment, but this time a perfectly lawful and morally admissible one. It is nevertheless excluded from patentability under Art. 52(4) EPC to the extent that it is a form of medical treatment to which patent law does not apply because it is not concerned with commercial competition.

Secondly, the patenting of DNA sequences, in particular *genes isolated from the human body*, as well as *proteins* expressed by such genes, once they have been transferred into an appropriate host cell, has met with two kinds of objections. The first objection is the 'discovery argument'. It is said that, since genes do exist in nature (in fact not only in the human body but in any form of living matter), they may be discovered but not invented. Claiming a gene therefore amounts to claiming a discovery which, in accordance with Art. 52(2)a) EPC shall not be regarded as an invention. Confronted with that argument, one is reminded of Gertrude Steins' poem on the rose ('a rose is a rose') and is tempted to say that, in the first place, 'a gene is a gene is a gene'. So you cannot say that a gene is a discovery and you cannot say that a gene is an invention. For the purposes of patent law, however, a discovery is a teaching which we have learned from observation and interpretation of nature, but which you cannot make or use for industrial purposes. But as soon as you put that abstract knowledge or teaching to actual use, so that a new man-made product becomes available, you have made an invention. Finding out that genes exist at all was a *discovery*. The identification of the sequence of the gene and the protein for which it encodes is also a discovery because it is still in the human body and you cannot make that protein, for instance human interferon or human insulin, from a thing which is in the body of a living entity. But isolating it and putting it into a host cell gives it to you as a tool for the industrial production. What you have then is a new factory and that is the invention. This is the dividing line. This is a long-standing approach in patent law. In 1873, the USPTO granted a patent to Louis Pasteur the subject-matter of which was a 'yeast, purified of pathogene germs'. Was that a discovery? For patent law, which is about business and competition, the esoteric considerations whether a thing belongs to the sphere of fundamental science or natural science or to the sphere of applied science, is of no meaning. A thing which you can make and which can be used in trade and industry is a possible object of a patent. We are dealing with business law, with patent law, and not with scientific theories.

This brings me to the second objection to the patenting of genes isolated from the human body, one that was brought forward in the Relaxin case. The opponent submitted that the isolation of a Relaxin gene from tissue taken from a pregnant woman was to be considered immoral, in that to make use of a particular female condition (pregnancy) for a technical process oriented toward profit constituted an affront to human dignity. What is considered immoral here? The underlying technology or the patenting of it? Apparently, not the technology, since otherwise medical or pharmaceutical ethics should have prohibited it. But that was not at stake. It must be the patenting. Indeed, the argument continues by saying that *patenting* of human genes *as such* 'amounts to a modern form of slavery since it involves the dismemberment of women and their piecemeal sale to commercial enterprises'. In actual fact, this is precisely what patenting does not involve. At the very most, patenting prevents others from becoming engaged in a business which, as in the Relaxin case, the opponent has found to be shabby. But to take the argument seriously, at its heart seems to be the idea that patents do confer a kind of property right on a physical object — in the case of human genes, the property of a part of the human body — and that, as a consequence, nobody should be allowed to make money from owning somebody else's body or any part of it. But this is exactly what a patent can never do. Patents are indeed a part of what we call industrial property. But what is actually owned can never be a physical object, whether living or not. The subject-matter to which the right conferred by a patent pertains is something purely immaterial, namely the right to make use of a *technical teaching* enabling a person skilled in the art to perform a given industrial process or to manufacture a new and useful product. It is difficult to see how such a right could affect in any way the individual rights of a person. In the Relaxin case, the Opposition Division was thus quite right in making the point that patents covering DNA encoding for human Relaxin, or any other human gene, do not confer on the patent owner any rights whatever to individual human beings or to parts of them.

Quite another question is — in the Moore case, which you probably all know — the question whether somebody has the right to take a part of your body, be it a gene or a hair. I deny the right to everybody to take even a hair from my head. This is because I have a sovereign right to dispose of my body as a whole and of any of its parts. But nobody finds it scandalous when I go to the hairdresser. But how do we dispose with parts of the human body which no longer belong to the human body? You can all imagine the cases and the ways in which we dispose of it. But when we make it the basis of a useful product, then an individual is perfectly entitled to say 'wait a minute, if you use a part of my body to make a medicament, I am in the business'. Or, if he is generous, he can say 'I agree, it is good for humanity, I do not ask for any money'. But this has nothing to do with patenting. Isolating a gene from the body of an individual person without the consent of that person is unlawful, but this issue is entirely independent of the question whether such a gene, once isolated for the purposes of industrial production of a protein encoded by it, is to be regarded as patentable subject-matter. I am therefore unable to see any moral ground for creating a new exception to patentability in the field of DNA sequences which can be found in the human body.

I will conclude now, by saying three things. You will not be surprised that I conclude that there is no special feature in biotechnology which could justify any special exception of patentability in that field, except those where the general grounds of patent law not satisfied. Mrs Van Overwalle has given the explanation why we exclude plant varieties. Traditional plant breeding gives no predictable result in the sense of strict reproducibility and that is why it is and should remain excluded from patentability. But precisely genetic engineering fulfils that requirement of reproducibility and that is why we have to interpret that exclusion so that it applies only to the case the legislator wanted it to be applied to. There is no moral consideration behind it.

So there is no reason to create any further exception for patentability and we have to interpret or otherwise to modify Art. 53(b), so that it only excludes matter which is not repeatable technical teaching. Secondly, the consequence of exclusions of patentability does not mean that the excluded subject-matter — e.g. transgenic plants, animals, drugs derived from human genes — would not be invented, would not be made, would not be on the market, would not be used. I want to say this to all those who are against patenting in this field. You will, whether you patent it in Europe or not, find it on the market place. But it will not have been invented by European enterprises. It will not have been made by European workers. It will be imported from America, Japan, Korea, everywhere you can think of, because you will have admitted it to the market. So, if you are concerned with patenting these things on moral grounds, you cannot avoid to first prohibit the making and the sale of whatever you want to exclude, viz. transgenic animals, transgenic plants, medicaments made from human genes. Prohibit all that, have the courage to go to the legislators in Brussels, in Bonn, in London, in Paris, and say 'this is immoral, prohibit it!' As long as you have not done this, it is absolutely pointless to speak about patenting. The only point in that is that *others* will make these things, not European industry. That is what we are talking about and nothing else.

Notes

1 Official Journal of the European Patent Office, 1995, 388.
2 Official Journal of the European Patent Office, 1995, 545.

15 Biotechnology patents in the United States

Ronald Schapira

At the outset I want to thank the University of Ghent, Professor Vermeersch and especially Miss Sigrid Sterckx for having invited me and also for their cordiality and their efficiency. I am very impressed.

I would like to briefly speak this morning on the status and the state of US Patent Law as concerns biotechnology and morality. As you will note from my brief paper, the issues in the United States are much simpler than they are in Europe and for that reason I will try to be brief.

As far as morality goes, I want to say at the outset that I am just a simple US Patent Attorney and I will not pretend to talk about morality more than just for myself. I do not pretend to have any great visions of morality or Americans' view of morality.

I would say that in the USA, nearly all products and processes of biotechnology that are *new*, that are *unobvious* and that are *useful* can be patented. As you probably all know by now, the US Supreme Court in 1980 decided in the famous case of Chakrabarty that that which can be patented includes 'anything under the sun that is made by man'. I am sure that if the Supreme Court would reconsider its decision, it would add 'made by man or woman'. But what is clear today is that what is patentable are micro-organisms as well as new genes, proteins, peptides, DNA fragments, methods of using them, including human genes, human proteins as well as human therapies, which are not allowed under the European Convention.

In 1987, following the logic of the Supreme Court, the US Commissioner of Patents said that henceforth any 'non-naturally occurring, non-human multicellular living organism, including animals', would be patentable subject matter. Thus became patentable the famous Oncomouse, as well as in the future, transformed sheep, goats, rabbits, mice, storks, etc. This decision of the Commissioner of Patents, as you might well expect, provoked a lot of criticism among animal rights groups, farmers, etc., saying, e.g. that it would hurt family farmers, that it would have cruel effects on animals and that it would endanger many species of animals around the world. Critics also said that 'it is the final assault on the sacred meaning of life and life processes'. And we have heard echoes of this already in this conference here. The same criticisms have been made in the US.

171

In recent years, the US Congress has considered these concerns on several occasions, and there have been proposals to limit animal patents, i.e. patents covering animals. But besides that, there have been almost no serious proposals to limit biotechnological inventions, except for human beings. There has been no Court decision, but it has been said that human beings are outside the scope of patents.

More importantly for our discussion — and I wish to focus on this — the US Congress, after hearings, reported that in their view, 'the patent law is not the place to exercise moral judgements about scientific activity'. This continues to be the well accepted attitude of the USA. If biotechnology and its inventions are to be regulated in the US, it is not the function of the patent lawyers or the Patent Office to do so. Such regulation must and is in fact being done by other agencies of Government much better qualified for this task. Thus, you will note in the future that US patents will be granted, e.g. on pigs which have been transformed with human genes, so that the pig organs, such as their hearts and kidneys, will be available for transplant into humans. This will save thousands, perhaps hundreds of thousands, of lives every year in the US, as well as around the world. Also, patents will be granted in the US for transformed human germ-cell lines, thereby providing a new tool for treating diseases which have caused untold miseries in thousands or hundreds of thousands of families world-wide every year.

As far as my personal views are concerned, I believe that the American result, the American attitude, is the correct one. And that it should be adopted, as far as possible, in other countries. In my view, patent offices and patent examiners generally do not have sufficient expertise or qualifications to judge morality or even societal risks of inventions, particularly biotechnological inventions, except, one might say, where the immorality or the risks are perceived as so unacceptable by the vast majority of the members at large of that society. That might be the only case were I personally would allow patent offices or examiners to enter into this domain. I believe that in the ideal world, patent offices outside the USA should follow this example.

I wanted to respond to one point that Mr Schatz has made about germ cell-line therapies. He is correct that in Europe, of course, therapies on humans are not allowed, but I believe he underestimates the intelligence of patent attorneys who will find ways of circumventing this restriction, such as by patenting the tools of therapy, the procedures that may be used in such therapies prior to the contact with humans, such as in a centralised laboratory where human cells may be treated preparatory to applying them to humans.

To conclude, it is my view that the American view should be the world view, so to speak, that morality should practically speaking have nothing to do with patents. But that is what we are here to discuss.

16 The new proposal for a Directive on the legal protection of biotechnological inventions

Dominique Vandergheynst

On 13 December 1995, the Commission adopted its new proposal for a European Parliament and Council Directive on the legal protection of biotechnological inventions. The official text, the COM document in the eleven languages, normally will be available next week. And normally next week, the new proposal will be published in the Official Journal. If you fill out the form correctly and Sigrid gives me your address and your name, normally I will have the possibility to send you the new document.[1] Its number will be 661/95.

When the European Parliament rejected the joint text approved by the Conciliation Committee in March of last year (1995), the Commission announced immediately that it tended to re-examine the matter more closely and in the light of the reasons forwarded by the Parliament, with a view to find the best solution in this sensitive field. Why? In 1985, the Commission White Paper on completing the internal market stated that differences in intellectual property laws have a direct and negative impact on intra-community trade and on the ability of enterprises to treat the Common Market as a single environment for their economic activities. Consequently, when publishing its initial proposal in 1988, the Commission noted: 'The primary purpose of the modern patent system is to promote technical innovation as the major factor of economic growth by encouraging inventive activity through rewarding inventors for their creative efforts.' The patent system thus secures costly investments in research and development and industrial exploitation of research results.

The initial proposal highlighted a number of specific problems regarding the application of the patent system to biotechnology. These concerned the interpretation to be given to the conventional patent law concepts to be applied to biological material that is self-reproducible or reproducible within a biological system. In other words, how should animate material be treated, compared with inanimate material? The questions raised concern the definition of the term 'subject matter' of the patent: invention novelty, adequacy of description, scope of protection, etc.

For the Commission, the legal position regarding protection of biotechnological invention needs to be clarified because current European law was drafted 30 years ago, at a time when the possibilities offered by biotechnology were science-fiction. In the absence of a clear answer to the questions outlined above, uncertainty will

173

increase. All the discussions that took place regarding the initial proposal in 1988 confirmed the need for harmonisation of the Member States' law in order to avoid a proliferation of divergent legislation and case law. Thus, the initial proposal was largely technical in character. Not that the ethical dimension was ignored, but at that time, in 1988, it appeared that the exclusion from patentability of inventions contrary to the public order and morality, which was common to all Member States' legislation and to the Munich Convention, met the need to take into account the ethical dimension. However, on 1 March 1995 the European Parliament rejected the joint text.

The legal environment regarding biotechnological inventions is unchanged and today it is of course impossible to say that this vote of the Parliament creates a kind of moratorium. The Commission has therefore to acknowledge that the issues, raised by the legal protection of biotechnological inventions, have still not been resolved. The legal uncertainty that constitutes the justification for the 1988 proposal remains. Patent law now appears even more incomplete and uncertain than in 1988. Matters will not resolve themselves with time. An increasing number of patent applications are being filed and granted. There will be more and more questions to resolve. The objective of harmonising Member States' legislation in order to ensure the smooth functioning of the internal market so as to promote a more competitive economy could thus be directly called into question once again.

For my Commissioner, Mr Monti — and of course it is the position of the Commission — without common legislation compatible with the single market, European research and exploitation of its results, particularly for therapeutic purposes, will be discouraged and placed at a disadvantage, compared with competitors in third countries. Without a Directive, nor are there any legal guarantees protecting the human body when it comes to ethical questions.

After consulting widely with all the interested parties, we, the Commission, are convinced that this proposal, the new text, achieves the right balance between the two equally essential requirements of promoting research and providing ethical protection.

Further progress in the development and application of biotechnology can be made *only* if research results can be exploited commercially, on the basis of a stable legislative framework. But since the patent law system comes into play at the interface between research and product marketing, it cannot possibly provide the means of resolving ethical problems that arise at earlier stages and later stages of the process. Consequently, ethical questions relating to either humans or animals can be taken into account only as regards the requirements governing patentability, and *not* as regards authorisation to undertake research or market a given product. Art. 1 of the new proposal says clearly: 'This Directive shall be without prejudice to national and Community laws on the monitoring of the applications of research and of the use or commercialisation of its results'. Whereas (11) of the new proposal emphasises 'whereas a patent for an invention thus not authorise the holder to implement that invention, but merely entitles him to prohibit third parties from exploiting it for industrial and commercial purposes and whereas, consequently, substantive patent

174

law is not capable of calling into question national and Community laws on the monitoring of the application of research and of the use or commercialisation of its results, notably from the point of view of the requirements of public health, safety, environmental protection, animal welfare, the preservation of genetic diversity and compliance with certain ethical standards.'

The new proposal draws a clear distinction between inventions and discoveries and does not include the famous word 'as such', in relation to elements of the human body, in order to remove the difficulties of interpretation regarding the patentability. The new proposal states clearly that a discovery cannot be patentable and stipulates the conditions subject to which an invention may be patented. Art. 3 of the new proposal says: 'The human body and its elements in their natural state shall not be considered patentable inventions.' Paragraph 2: 'Notwithstanding paragraph 1, the subject-matter of an invention capable of industrial application which relates to an element isolated from the human body or otherwise produced by means of a technical process shall be patentable.' In short, an invention is something that provides a technical solution to a technical problem. An invention is artificial. The essential factor is the technological contribution, given that it constitutes the human input and that the same result cannot be possibly achieved simply through the interplay of the laws of nature. Elements isolated from the human body by means of technical processes are artificial and thus qualify as inventions, since they are technical solutions invented by man in order to solve technical problems. Nature is incapable of producing this type of elements by itself. The techniques employed in order to isolate such elements from the human body were only possible by means of human intervention. Whereas (13) says 'Whereas it should be specified that knowledge relating to the human body and to its elements in their natural state is scientific discovery and may not therefore be regarded as patentable invention; whereas it follows from this that substantive patent law is not capable of prejudicing the basic ethical principle excluding all ownership of human beings.' Whereas (15) adds 'No patent may be interpreted as covering an element of the human body in its natural environment forming the basis for the subject of the invention.'

The new proposal completely excludes from patentability methods of germline gene therapy on humans. As regards the position of principle against germline gene therapy that should have been taken when harmonising national laws of patents in respect of biotechnological inventions, the Commission can only emphasise that patent law cannot allow itself to adopt a position on principle *erga omnes*.

This exclusion from patentability is one example covering inventions the publication or exploitation of which would be contrary to public order or morality. A second example, given by Art. 9 of the new text, is the following: 'Processes for modifying the genetic identity of animals, which are likely to cause them suffering or physical handicaps without any substantial benefit to man or animal, and animals resulting from such processes, in so far as the suffering or physical handicaps inflicted on the animals concerned are out of proportion to the objective pursued.' The criterion of proportionality is justified particularly in view of Council Directive 609/86 on the approximation of laws, regulations and administrative provisions of the Member

States regarding the protection of animals used for experimental or other scientific purposes. This criterion of proportionality more specifically appears in Art. 3 of this Directive of 1986. And I have to say that this criterion of proportionality has been added in the new proposal, because the Parliament asked to introduce this criterion during the conciliation procedure.

Of course, the list of Art. 9 is indicative and not exclusive. For the moment, Art. 9 of the proposal gives two examples, two guidelines of interpretation of public order or morality. The new proposal also introduces directly into patent law a derogation for farmers in respect of breeding stock.

Finally, in order to make the proposal for the Directive clearer, it seems appropriate to alter its structure. Genetic modification of the law! Definitions are now given at the beginning of the text. They concern the notions of biological material, microbiological process and essentially biological process for the production of plants or animals. The definitions are followed by provisions on patentability. In accordance with the structure of Member States' legislation and EPC, the first description given is of what may not be regarded as a patentable invention: Art. 3, on the patentability of elements coming from the human body. The extent of and exclusions from patentability are then specified. Finally, exclusion from patentability on grounds of public order or morality is clarified. The rest of the new proposal is in fact identical to the former proposal.

If the Commission proposes, the European Parliament and the Council decide. Apparently, a lot of people forget this. The next important steps of the procedure will be the opinion of the Economical and Social Committee[2] and the opinion of the European Parliament in first reading. Afterwards, the Council will have to adopt its common position. One can imagine that the first steps of the new procedure will take one year.

Some other words to conclude my speech. Apparently, for some people in the public opinion, it is very necessary and very urgent to have today a public debate. For the Commission, the democratic public debate has the possibility to take place during the co-decision procedure. If some people in the public opinion have some concern to express, they have the possibility to try to lobby the MEPs, the Council and the Commission. For the Commission, from an institutional point of view, public debate is not demonstrations on the street.

Notes

1   OJ EC n_C 296, 8 October 1996, p. 4.
2   OJ EC n_C 295, 7 October 1996, p. 11.

Part Five

THE CASE FOR AND AGAINST THE PATENTING OF BIOTECHNOLOGICAL INVENTIONS

17 Jan Van Rompaey

Most of you who have some background in biotechnology will probably know Plant Genetic Systems. I can say that we are a leading company in agricultural biotechnology and that we can boast some important achievements in this field. We were, for instance, the first company able to make plants resistant to insect attack, to render plants resistant to herbicide and to produce male sterile plants. The last one is probably the most important and elegant result of plant genetic engineering. All these inventions are now being implemented and made into products which are expected to be marketed within the next few years.

Being at the forefront of things has of course also its more interesting aspects. PGS has been privileged to be involved in proceedings at the European Patent Office that finally led to a decision of the Technical Board of Appeal containing detailed comments on 'morality' in connection with patents on biotechnological inventions. It is therefore no surprise that this decision was mentioned so often in the preceding talk. Although some of the aspects of this decision have been highlighted, for instance by Mr Schatz, I think it would be important for you to have a complete history of this case and to see which arguments were presented and how the Technical Board of Appeal decided on the various arguments.

Perhaps first something about the invention, which was already briefly described by Mrs Gruszow. The patent involved in these proceedings was in connection with an invention on plants resistant to a specific class of non-selective herbicides. The particular herbicide that PGS used is a natural product, produced in nature by certain bacteria and acting in a plant cell by inhibiting a certain enzyme, a Glutamine Synthetase, which is essential for the survival of the plant cell. PGS and the co-applicant of the patent application, Biogen, identified the gene from a bacterium that encodes a protein capable of inactivating this herbicidal compound. We were then able to introduce this gene into plants, so that the protein was produced in the plants. We were able to show that the protein was also capable of inactivating the herbicide in a plant cell environment. Apart from the scientific importance of this achievement, PGS thought — and I think many others at that time — that this invention was indeed an important contribution to environmentally friendly weed control, which is so needed in world agriculture. Indeed, the herbicidal compound is rapidly degraded in the soil, in contrast to many of the selective herbicides that are still currently used. Nevertheless, the invention also has other uses. It provides, for instance, a tool which enables selection of transformed plants and is thus of general importance in genetic engineering of plants. PGS is, as you know, a small company and as Dr Schatz has pointed out, patents are very important to us. Indeed, they are probably more important for small companies than for very large companies. So when we made our invention, we of course applied for a European patent. As you can imagine, we very pleased to learn that the European Patent Office granted us a patent. Mrs Gruszow showed you the concise and short front page of the patent; this is the full text, which contains of course a lot of information. You can also understand that we were very surprised to learn that the granting of the patent was opposed by Greenpeace, an organisation that we otherwise quite respect, on the grounds, among others, that the disclosed invention was immoral. Fortunately, the arguments brought up by Greenpeace in connection with the morality issue were rejected first by the Opposition Division, and then also by the Technical Board of Appeal, when Greenpeace appealed. I am not going to show you again Art. 53(a), the basis for the opposition of Greenpeace on the morality ground, but perhaps it is important to show you some of the claims of the patent. They might be very difficult to read, but Claim 21, for instance, related to a plant that contained the particular gene I was talking about earlier and that provided the resistance against this Glutamine Synthetase inhibitor. In connection with the morality issue, Claim 24 was perhaps even more important for Greenpeace, because it is a process for protecting cultures of plants against weeds by applying the herbicide, given that the plants that are to be protected contain the gene and are resistant to the herbicide.

When looking at the various arguments brought forward by Greenpeace, and then later by PGS and by the Technical Board of Appeal, I think it is worthwhile to note that they can roughly be divided into three groups. A first group of arguments was concerned with general comments on Art. 53(a) and how this article should be interpreted by the European Patent Office. Another group was concerned with arguments related to the slogan 'patenting life is immoral'. A last group of arguments was concerned with the morality issues related to the use of the herbicide resistant plants in the environment. In their grounds of opposition, Greenpeace stated generally that the patent provided legislative approval for the exploitation of the invention, for instance the release of GMO's — genetically modified organisms — in the environment. They also said that the patent may have undesired international effects, mainly on the third world. On 'patenting life', the list was very extensive. I don't know whether I have covered them all, but Greenpeace asserted that the plant genetic resources are the heritage of mankind and should not be patentable as such; that plant genetic engineering and the patenting of life showed the tendency of man to control nature which is immoral, that plant genetic engineering can result in the depletion of genetic resources. They also brought a survey into the proceedings to show that the public is against herbicide resistant plants — a survey conducted among farmers in Sweden. Finally, with regard to the use of herbicide resistant plants in the environment, Greenpeace said that the EPO should balance the risks versus the benefits of the exploitation of the invention, viz. the release of the genetically modified plants into the environment, and if the risk would be greater than the benefits, the EPO should reject the patent. This of course referred to an earlier decision in the Harvard Oncomouse case, in which the Board of Appeal instructed the Examining Division to take this balancing exercise into account. Another argument was that the exploitation of this invention can result in serious irreversible environmental risk, because the treated plants could become weeds and the gene could escape and spread into other plants. Consequently, ecosystems could be damaged. Furthermore the use of herbicides would increase and biological diversity could become depleted. However, they also clearly said that it is impossible to predict accurately the consequences of a release of GMO's in the environment. Finally, they asserted that there was no regulatory framework for assessing safety.

This is all I will say about Greenpeace this afternoon. I think Greenpeace will undoubtedly comment on these aspects later on. What did PGS argue in response to these various arguments? I think a very important argument is — and this has also been repeated throughout this morning — that a patent does not provide the patentee with a right to exploit the invention. The exploitation is subject to governmental regulation and control in the various contracting states of the European Patent Convention. A patent only gives the right to exclude others from exploiting the invention and only for a limited time, only for twenty years. We also argued that there are no international effects of a European patent. The exploitation of the invention in non-EPC-countries is not affected in any way by the grant of a European patent. We also argued that many people were of the opinion that biotechnology, plant biotechnology and especially the invention of PGS and Biogen would benefit the third world. We also argued that the exclusions under Art. 53(a) should only be applied in very extreme cases, as was the case law at that time in the European Patent Office. On 'patenting life', we replied that, when we are introducing a gene into a plant, we are not depleting genetic resources. If anything, we are adding genetic resources, because we are providing to the public a new set of genes that can be useful. With regard to the survey that was cited by Greenpeace, the Swedish farmers survey, we argued that the survey was not concerned with the particular invention at stake in this patent, but related to biotechnology in general and — most importantly — that it did not represent the opinion of the European public in general. We also argued, based on a number of publications, that it was clear that the public does not consider the invention so abhorrent as to make the grant of a patent inconceivable.

On the use of herbicide resistant plants in the environment, PGS argued that the EPO is not equipped to obtain and evaluate the detailed technical data for assessing risks and benefits of the invention, and that opposition proceedings are not the adequate forum to discuss these issues. We also stated very clearly that the assessment that Greenpeace obviously desired *was* regulated by existing European directives and national legislations. We stated very clearly that Greenpeace did not provide any positive evidence of the so-called risks presented by the patent. The opponent only raised potential hazards. I must add that we *did* admit that in some instances the use of the invention should be discouraged. This was related to the agricultural practice as a whole (crop rotation schemes etc...), so in some cases we indeed agreed that it would not be advisable to use the invention. However, we also pointed out that, in other cases, the use of the invention would be of extreme benefit. Finally, we also indicated that a European patent can only be obtained when the invention is published, thus informing the public on the invention and ensuring adequate control on the exploitation of the invention. If an invention would not be published, it might become secret and could be exploited without knowledge of the public, which is clearly undesirable.

On the basis of various arguments, the Opposition Division first rejected the opposition. Greenpeace appealed. More or less the same arguments, phrased in different ways, were presented to the Board of Appeal; they issued their decision in February of last year. Dr Schatz already highlighted some aspects of this decision from a general point of view of morality standards in the EPO, but it might be useful to guide you through quotes of this decision, as far as they relate to the particular arguments which were presented by both parties. I am just going to show you — I am not going to read it because Dr Schatz already did — that it is very important that the Technical Board for the first time actually defined what they consider to be the concept of morality and of 'ordre public', public order. For the purpose of this discussion, I think the most relevant aspect is that the concept of public order also encompasses the protection of the environment, which was a major issue raised by the opponent. Finally, the Technical Board conceded that Art. 53(a) should be narrowly construed, in accordance with the general European patent law that the concept of patentability in Europe must be as wide as possible.

On 'patenting life', the Technical Board said that 'Seeds and plants per se shall not constitute an exception to patentability under Art. 53(a) merely because they represent living matter or on the ground that plant genetic resources should remain the common heritage of mankind. The patenting of wild type plant resources which may be used as starting material is not an issue in the present case. That such resources should belong to the common heritage of mankind is therefore not in jeopardy.' Thus the Technical Board obviously related to the fact that a patent provides to the public something that was not already available to the public. If a plant with certain characteristics is new, patentable, provided to the public, the old plant which is not genetically transformed is still public and can be used by the public as a whole.

With regard to the survey, the Technical Board said that a survey or an opinion poll showing that a particular group of people or the majority of the population of some or all of the contracting states opposes the grant of a patent for a specified subject-matter, cannot serve as a sufficient criterion for establishing that such subject-matter is contrary to 'ordre public' or morality.

With regard to the use of herbicide resistant plants, the Board of Appeal said that in the exploitation issue there are two aspects: there are aspects of ordre public, since protection of the environment falls under the heading ordre public, but there are also morality aspects. They commented as follows: plant biotechnology per se cannot be regarded as being more contrary to morality than traditional selective

182

breeding, because both traditional breeders and molecular biologists are guided by the same motivation, viz. to change the properties of a plant by introducing novel genetic material into it in order to obtain a new and possibly improved plant. So what they were actually alluding to was that the subject-matter of the invention was already used for several years in Europe and was considered to be acceptable. The fact that now another tool was used to achieve the same results did not change the fact that such an invention would be immoral. Like any other tool, plant genetic engineering can be used for constructive or destructive purposes. None of the claims of the patent in suit referred to subject-matter related to a misuse or destructive use of plant biotechnological techniques, because they concern activities and products which cannot be considered to be wrong in the light of conventionally accepted standards of conduct in European culture. I think this is quite clear. Then they went on with relation to the aspects of ordre public insofar as they related to the use of herbicide resistant plants. They said that the invention claimed in the patent may only be exploited within the framework defined by national laws and regulations regarding the use of the said invention. In most cases, potential risks in relation to the exploitation of a given invention for which a patent has been granted cannot be anticipated merely on the basis of the disclosure of the invention in the patent specification. They agreed that transgenic plants normally require regulatory approval in the majority of countries where biotechnological developments take place before even initial small-scale field testing can be performed, and they referred to specific European Directives. They said that no conclusive evidence has been presented by the Appellants, showing that the exploitation of the claimed subject-matter is likely to prejudice seriously the environment. In fact most of the Appellants' arguments are based on the *possibility* that some undesired destructive events might occur. They also said that the mere fact that there may be inadequacies in the existing regulatory framework does not vest the EPO with the authority to carry out tasks which should properly be the duty of a special regulatory authority or body constituted to that effect. It would be unjustified to deny a patent under Art. 53(a) merely on the basis of possible, not yet conclusively documented hazards. Since no sufficient evidence was adduced, the assessment of patentability with regard to Art. 53(a) may not be based on the so-called balancing exercise, but the balancing exercise is just one possible way of doing that, as Dr Schatz also pointed out.

So, at least on the basis of Art. 53(a), PGS found its arguments vindicated and the Technical Board of Appeal concluded that in the present case Art. 53(a) does not constitute a bar to patentability.

Thus it would seem that the decision sets a clearer framework for assessing patents in the field of biotechnology, with regard to Art. 53(a). The fact that the various divisions of the EPO are required to examine an invention within the framework of Art. 53(a) seems to be reasonable, as long as that article is in the EPC. However, the fact that a rejection under Art. 53(a) should only be invoked under very extreme circumstances seems to be equally reasonable. Society provides sufficient instruments to cope with potential problems associated with the exploitation of inventions, a point that was explicitly recognised by the Board. In this regard, I think it cannot be emphasised enough that a patent does not provide the patentee with a right to exploit the invention, and — as Dr Schatz has also pointed out — an equally important consideration is that a patent is in fact a reward, given by the government to an inventor, in return for a full disclosure of the invention to the public. European patent applications need to be published before they can be granted. Such publication serves to make the public aware that an invention has been made and could be exploited. If the public opposes the exploitation of the invention, it can address itself to authorities qualified to address public concerns, or it may request the establishment of properly qualified authorities and adequate regulatory legislations. It would thus seem that if the likelihood of a patent being rejected under Art. 53(a) would be too random, patentees might indeed refrain from filing patent applications and choose to exploit their inventions without laying them open to public inspection. This might result in a situation entirely undesirable for those interested in a society which can make informed decisions on important issues.

18 Isabelle Meister

I would like to thank the organisers for inviting me to give a brief presentation on Greenpeace's view on the patenting of life. I guess that our view does not please many of the persons who are here, because we have a rather fundamental approach when it comes to the patenting of life. We consider the patenting of living beings and genetic resources as such as immoral. Therefore, we request a ban on the patenting of life. Our concerns are actually based on several reasons. Among others, we think that allowing the patenting of life leads to the classification of life as the product of an industrial process, an artificial commodity. We believe patents should only be granted for inventions that are in the public interest and the patenting of life to some extent undermines the patenting principles and the patent system.

In relation to the first point, that we consider the patenting of life as immoral, we should look at the difference between bolts on the one hand and cows and weed plants on the other hand. The bolt is an artificial product, it is man-made, it can be produced in an industrial process. On the other hand, plants and cows are living beings. They reproduce, they differentiate and they have growth independent of human intervention. We can change them through traditional breeding methods or we can use other techniques like genetic engineering, but the basis of life — growth, reproduction, differentiation — are natural processes. When the patent system, that was designed for technical processes and products, is applied to living beings, this means that we put living beings on the same level as a copy machine, an overhead, and reduce and reclassify them as industrial products. They become commodities and they are reduced to genes that can be recombined. With the patent system this is going to be rewarded.

The fundamental problem is the way we view and value life, nature as a whole, living organisms, plants and animals and their unique capacity to reproduce. Today, nature as a whole and living beings are increasingly given their own rights and dignity, even by the public and by some policy makers; organisations like Greenpeace have fought for this. The Swiss Constitution, for instance, states that the dignity of creatures needs to be protected. We want to protect the dignity and the rights of living organisms. That does not exclude that we can use them, because if we would deny people to live from animals and plants for food we would deny the dignity and the rights of human beings. However, it is clear that living beings cannot be put on the same level as copy machines. It is the dignity of living beings that requests that they cannot be covered by such an inappropriate system like the patent system, which has been designed for technical processes. There is great public concern about this, it is a moral question and it cannot be allowed. Even the Commission's suggestion to have this balancing act does not solve these moral concerns, which we do have and which we believe are widespread among the public. From what we feel and experience in our work on the environmental issue we know that these values are becoming increasingly supported in the last decade.

As regards the patent system, a very abstract technical system which wants to protect life, we consider this as a relapse to the past century, because what is a key point in environmental protection is the value of ecosystems and living beings. On the other hand, we think that patents — since they give a monopoly over the exploitation of an invention — should only be granted when they are purely in the public interest. When we look at genetic engineering, many of its applications often do not contain a given interest. In particular, when it concerns the release of genetically modified organisms or when it concerns food, there seems to be widespread unease. When it comes to the release of genetically engineered organisms — and that was our legal opposition to the patent of Plant Genetic Systems — PGS admitted that there might be unknown effects. Taking a precautionary approach, when there could be unknown risks this cannot be allowed because it is not in the public interest. Protecting the environment is in the public interest, not the patenting of such technologies, because patenting is not only allowing something but also rewarding it. Patenting rewards the creation of 'freaks', like plants with toxic chemicals, plants with rat genes or whatever. These kinds of creations are rewarded through the patent system. On the other hand, there is the risk that through patent protection, the trend towards monopoly control over the world's food supply might be extended. As a result we might eventually end up with a couple of companies that decide which kind of genetically engineered crop is growing where; similarly they can withhold non-genetically engineered, non-patented crop varieties. We believe that no monopoly control should be allowed in the world's food supply.

In our perspective, the patenting of life undermines the patent system, because only inventions which are new and non-obvious ought to be patentable. These are the basic requirements, but when it comes to genes, plants and animals, they are not invented but discovered. What we can do is to isolate and characterise a gene, by standard methods, and transfer them into another organism. These are not inventive processes, but discoveries. Today, it is standard technology to isolate a gene and transfer it to another organism to produce a protein. This is not really inventive. Another accepted principle in European patent law is that essentially biological processes and plant varieties are excluded from patent protection. Here we come to our case at the European Patent Office. We believe that the existing patent law excludes patents on life, according to our understanding of morality and according to the exclusion of essentially biological processes and plant varieties. Therefore we decided, back in 1991, when we saw that the patent was issued to PGS, that we wanted to test out how far we could go or what we could achieve with legal arguments. We wanted to test out the system. Even in the beginning, we were not so certain to achieve any success. We were not certain which article would be the most successful as a basis for opposition. When we presented our arguments on Art. 53(a) about 'morality', we did not say that patenting allows something to be commercially used. What we said is that patenting acts as an 'enabling legislation': it rewards and promotes technology. Obviously the national legislations have to decide whether it is allowed or not, but patenting acts as a promoting and enabling legislation. We included that genetic resources are the common heritage of mankind, and now they are under the sovereignty of states, but they should not become subject to patents. We also included our environmental concerns over genetic pollution and overuse of herbicides. At the same time we submitted two surveys. The first was a survey from Sweden, asking Swedish farmers — the potential users and consumers of these particular herbicide resistant plants — what they thought about the morality and even the economic value of herbicide resistant plants. Both from a moral and an economic perspective, there was huge rejection. We submitted another survey that was carried out in Switzerland in 1992, in which the public was questioned about patenting life: were they in favour or opposed to it? 69 per cent of the public in Switzerland said they were opposed to it.

We have used this variety of arguments and we also used the case law established in the Oncomouse case to perform this balancing act between benefit and suffering, to see how far we got. The outcome was that the Technical Board considered and rejected our reasoning because we were not able to prove any actual damage.

On Art. 53(b), the article about the exclusion of varieties and essentially biological processes, our experience was that, by claiming plants to be the product of a microbiological process, plants became patentable. In the end, if you allow this, the only important event in any living being's life is when the genetic engineer intervenes and the gene is introduced. Everything else — like differentiation, reproduction and growth — is reduced to this very simple step.

On Art. 53(b), we were much more successful. The limits were defined, viz. that the product of genetic engineering goes to the plant cells but not to the whole plant. On the other hand, concerning plant varieties, there were two issues. For the first one we argued that all the examples disclosed by PGS in the patent were applied to existing plant varieties, so — although the claims were smartly rephrased — they actually claimed plant varieties. Another argument was that through the genetic engineering event, the introduction of the herbicide resistance gene, the plant becomes a plant variety based on the meaning of the UPOV Convention and according to the meaning of Art. 53(b). Indeed, through the introduction of this gene the plant becomes distinctive and the gene is stably integrated and will be in all future generations. On this argument the Technical Board of Appeal followed us. They decided that the genetically engineered plants themselves become, through the event of genetic engineering, a plant variety.

So in the end we came out with a partial victory. We failed to get the whole patent withdrawn as we aimed to. We failed on the morality issue, but we established that the plants, the seeds and the progeny could no longer be patented. We wanted to test how far we moved by believing that the patent system excludes living beings from patent protection. After this case, it is clear to us that we have to put more pressure in order to get a revision of the Patent Convention and the EU-proposal. It should be made clear that patents on life are excluded.

19 Jan Mertens

I will present a more general political argument on biotechnology, patents and morality. I will make three statements and a conclusion concerning consensus formation. I will talk about the role of politics in this domain, the fundamental ethical point of view which underlies the Green's political position in this debate and the broader political context.

First of all the role of politics. Ethics is concerned with good and bad, right and wrong, and this has a great influence on politics as well. The way I see politics, it has to do with how we should organise our society, what a good society is and how we can improve it. Therefore we have to deal with the development of concepts and instruments to improve society and organise the debate. This brings me to my first point: politicians are no scientists in the narrow sense of the word, I mean the way most people understand it. I, myself, studied literature, so can consider myself as a scientist as well, but when most people talk about scientists they mean something else. Politicians must have a broader perspective, they must look at the way society is organised. They are the ones — no matter how difficult it is these days — who should build up a global view to encompass all elements. The concept of politics itself is becoming very problematic these days, but still I think we have to look for a way in which it can function. My statement is 'the democratic process can never be replaced by a sort of *expertocracy*'. With this I want to react to a statement made yesterday, 'let the experts be the experts'. I do not agree with this, when it should come down to 'let the experts take political decisions'. The ethical discourse, which for me is of the utmost importance in politics, cannot and may never be replaced by a purely technological one. To quote one of the previous speakers: 'this is what democracy is about'. The political struggle of the Greens in the EP to block the Directive on patenting biotechnological findings has to be seen in this perspective. It was commonly acknowledged that it was a Green victory in the European Parliament that the Directive was not accepted. I have to say something about a personal frustration in this matter. I am seriously offended by the enormous amount of disrespect we experienced for the European Parliament in the media. During this conference I heard it several times as well. I must say I do not accept to be called uneducated or something else. I think it is not acceptable that democratic institutions

189

are treated in this way. My basic point in this matter is that we need regulatory mechanisms in this field. This will be difficult, as we heard yesterday. The mechanisms should be flexible and work in a sort of continuum. Maybe it is trying to put things in a box, but I think that is the way it is. In this context, I do not mind the changing of the roles, as was said yesterday. That so-called 'progressive' groups become conservative is not the point. Answering the question of Mrs Van Overwalle this morning, I think that in this political construction, this talking about regulatory mechanisms, the European Parliament should play a role.

Concerning technology and especially biotechnology, I would like to make two remarks. Too easily, Greens are put in the corner of the naive, Luddist, anti-technology freaks. I think this is a great mistake. My first remark concerns new technologies. It was also mentioned yesterday by Professor Vermeersch that the nature itself of these new technologies is at the origin of a fundamental democratic problem. As the development and application of modern biotechnology is such a complex problem, this cannot simply be solved by better informing the public. Nevertheless that is what we always hear: 'they are not informed well enough, they don't know what it is all about'. This is a typical statement by people working in the industry. I think — and this is extremely difficult for politics — that the nature of these technologies itself is responsible. It is very difficult to see how we have to cope with this. But we cannot evade it.

A second, more general statement on technology is that, when we look at the fundamental problems our Earth has to deal with in the next years, technology itself can never solve the problems. Technology can be useful, but more fundamental questions have to be asked.

Now I will move to the second part of my presentation: what is the fundamental ethical position underlying the Green's political position? It is a bit like the previous speaker said: the main point we are against or have great difficulties with is the on-going colonisation of all aspects of life by market mechanisms. We have great difficulties with the way all aspects of life are or can be made into a commodity and the way all these things are instrumentalised. Our opposition to the patents is an element of a broader criticism. So if we have to deal with these problems we need a broader ethical perspective than the traditional one. This is my point: we plead for an ecological ethics and that led to our active engagement in the Oncomouse case. To reply to a statement of Dr Schatz, I agree that we also have to cope with the problems at the root and not only fight the patents, but sometimes the patents are the only things we can fight. I do not mind trying to deal with the problems at the roots. An ecological ethics means that indeed it has to do with intrinsic value and with natural integrity. Often ecological ethicists are accused of being essentialists. It is my conviction that the reductionism which lies in the so-called 'scientific ethics' leads to an even bigger essentialism. The fundamental problem is: what is the position of man towards nature? Man is not above it, but man is *in* it. I was seriously shocked by two things I heard yesterday, and I quote: 'the nature of man is to exploit his environment' and 'we will not find the appropriate organisms in nature, we have to engineer them'. I must admit, it gave me cold shivers. What are the underlying assumptions within these statements? They are economic views. It is about market mechanisms, about competition, about business. Statements like these have been made over and over again during this conference. But that is the fundamental problem. If we accept the priority of this narrative, viz. free market economy, then patents are defensible and patents on life are defensible because everything can be turned into a commodity. Not in my view, however, because I do not accept this assumption. So, to reply to a statement made yesterday: philosophy has everything to do with economy. Economy is not a value in itself.

My third statement, and the conclusion, is that we cannot talk about patents and biotechnology separated from the context in which they function. They function in an economic context, in the concrete political organisation of this planet. The first thing is: there are limits to growth. It is difficult for us to accept, but I do not think that we can solve the fundamental problems of agriculture, for instance, by trying to expand productivity. Humanity has to cope with the difficult message that limitless growth in a limited world is impossible and we should not try to evade it with a flight forward.

A last example I would like to mention is that the discussion on patents cannot be separated from the way international politics is organised. We see the way the Indian government is put under very high pressure to change their patent laws, in the context of the GATT agreement; I cannot accept this. The case of the Neem tree has to be put in this context. We, the Greens, worked together very closely with Vandana Shiva in India because we do not accept this way of imperialism.

To conclude, if we want to go to some sort of consensus, I think two elements are important. The first element is that there is a serious need for a regulatory framework and we accept this. We, as politicians, want to talk about this, we want to negotiate, we want to look for mechanisms and instruments. This is democracy and I hope everybody accepts this. The second element is that we can only come to a serious debate and possibly to a consensus if we all accept that we also need the fundamental discussion on the aims, on the purposes. Why are things developed? In which context do they function? For my position, I think the belief in economic growth in a global free market must be questionable.

20 Steve Emmott

I am the co-ordinator of something called the 'Patent Concern Coalition' in the UK, which is an alliance of NGO's that are opposed to the extension of patent law to living material, so although I am a barrister, I am here as a campaigner.

It is becoming increasingly clear to me that it is not possible to reconcile the objections such as those we have heard from Greenpeace and the Greens — and which we support — with the existing patent system, either be it the way the EPO is interpreting the convention or indeed the new proposed Directive under EU law.

I was supposed to talk about the Oncomouse, but this morning we already heard some of the history and some of the technicalities, so I will not repeat those to you again but a little later I will deal with where the case stands at the moment.

I picked up an article on the Oncomouse in the *Bio/Technology* magazine yesterday. The last sentence says: 'Prolonging the case with imprecise morality arguments, which will doubtless evolve over time, may rejuvenate discussion about whether the EPO's role is to determine patentability or ethical acceptability'. That reminds me of some correspondence I had with the UK patent office last year, when they wrote to me to say — as they do and as patent lawyers love to say — that a patent does not give anyone the right to do something, only the right to prevent others from doing it. I wrote back, asking if they could explain why, if a patent does not give the right to do something, there is actually a morality clause in the European Patent Convention, because surely it can only be there if there are some inventions which society would not wish to endorse with a patent. The answer came back: 'Yes, it is true that there is a morality exemption,' — and this is the key phrase — 'however, patent examiners do not have the resources nor the time to carry out detailed ethical or environmental analyses'. I think we have heard it before, have we not? Ethical issues are important, but they have nothing to do with us. Please leave the patent lawyers alone.

It is true, there is not much morality if you are patenting a photocopy machine, but if you are looking at the issue of living material and the extent to which patent law should apply to them, it is inescapable. It seems clear to me that the only way of resolving the dilemma's which the EPO finds itself increasingly in, is to change the rules under which they operate, so they do not become public policy makers; that is the position they find themselves to be in — in the Oncomouse case, amongst others.

Even the GATT agreement, which is not something that I am used to speaking in praise of, does acknowledge that the system which applies in Europe may not be the only one, and it says that members may exclude from patentability inventions the prevention of which is necessary to protect human, animal or plant life or health, or to avoid serious prejudice to the environment. That is a new phrase, which does not appear in the European Patent Convention, and which the EPO has now taken to quoting as part of its set of criteria to make evaluations. So the ground on which they are being invited to make their deliberations is widening under them and they are finding it increasingly difficult to decide how to respond to that. GATT also goes on to say that members may exclude from patentability plants and animals, so there is nothing immutable about the concept that plants and animals have to be patented. The law is there to be changed if we choose to do so.

The problem, I think, is that we are dealing here with an increasing number of 'value-laden' issues. This is a phrase which I saw originated by the European Commission and it tends to be used as a category for things which are too difficult to handle. In fact these are the issues which we want to argue on and which we believe to be the fundamental ones.

We have some suggestions for help for the EPO. A respected group, the Nuffield Bioethics Council — which is an independent group in the UK — suggested that the European Patent Convention needed clarifying. What they said is that European member states should adopt a protocol to the European Patent Convention, which would set out in some detail the criteria to be used by national courts when applying the immorality exclusion to patents in the area of human and animal tissue. They had understood, as I think we are all beginning to understand, that the existing wording is completely inadequate as a guidance for anybody.

The next question would be whether the new EU Directive would help us at all in this regard. It is an opportunity which they did not take in the 1988 draft, and sadly they failed to take again in the new draft. It avoids addressing the key questions on morality, except to state the obvious, viz. that germ line therapy is to be excluded, which I think we all knew, and that there should be a balancing test applied to the animal suffering question. In fact, the debate about applying any test should be a much more sophisticated and detailed one than simply saying — as the EPO said itself and as the Commission is now following — 'you have got to make a decision'. There ought to be cases in which animal suffering is impermissible. There should be a fundamental bottom line written in there, which says that there are some things we shall not do. That is the recommendation which was made in the UK Banner report, which is an advisory committee to the Ministry of Agriculture in the UK.

Where is the Oncomouse at the moment? You may know that the oral hearing in November collapsed in chaos towards the end. It was clear that the vast majority of the oppositions were going to be on the morality issue, and the Opposition Division proved to be totally unable to cope with seventeen groups representing two hundred objectors, each seeking to demonstrate to them that in fact there were fundamental, wide-ranging and key moral objections to the granting of this patent. Opponent after opponent came to the floor to express, in terms of opinion polls, in terms of petitions, in terms of legislative resolutions in certain countries, the opposition to this patent on morality grounds. What happened at the end of the week was that there were rulings outstanding on three particular legal aspects. One was that of industrial applicability, which I would guess they will decide in favour of Harvard University. The other two were on the question of varieties and the morality question itself. It is not clear what the time scale is going to be for the Opposition Division to give its decision on this, nor whether the hearing is actually officially closed. Although the oral proceedings are closed, it was still open to opponents to make further written submissions since not all of them were heard. The one thing that does seem to be clear is that they invited the patent proprietor to make some amended claims. If it is to be upheld at all, it will be on the grounds restricted to onco-rodents, rather than onco-animals. As originally drafted, it extended to onco-giraffes or onco-whales or any other onco-mammals you could imagine. Also possibly, it will be restricted to experimental and research use. That in itself is a whole new fertile ground for patent lawyers to earn some money at, because it is completely unclear to what that restriction would apply.

In justice, the patent should be revoked, but if we insist on leaving ethical and moral judgements concerning patents on life to the patent lawyers and the patent administrators, we cannot get a humane, compassionate set of decisions, but rather reductionist and mechanistic ones. That is what patents are: they are reductionist and mechanistic. It is not the fault of patent lawyers or patent administrators, they were just born that way, that is what they are trained for. We cannot expect them to be making decisions for which they are not equipped and ill-advised. Our job as campaigners — and I suggest also as legislators — is to provide a clear, binding, unequivocal statement of the limits to patentability of living material, in language which cannot be misinterpreted.

The Patent Concern Coalition in the UK has such a statement of the law, as we think it should be. This has now also been translated into Greek, German, Dutch and Spanish and has been circulated for support in those countries as well as in Switzerland. It is already signed by hundreds of NGO's and thousands — if not tens of thousands — of individuals. I will read it to you, as it is very short:

> The undersigned organisations and individuals oppose the granting of patents on genetic material, originating or derived from humans, animals and plants. We believe that the extension of patent law to the basic genetic structure of living matter means treating life itself as a mere commodity, with adverse moral and practical consequences for humankind, the animal kingdom and the natural environment. There is presently no unequivocal bar to patenting life forms. We believe that the following should be declared to be unpatentable as being contrary to public morality:
> 1) humans, human parts, human tissue and all genetic matter originating or derived from human sources;

2) processes and techniques for genetic modification of such human matter and methods, treatments and therapies for applying such processes and techniques;
3) animals, animal parts, animal tissue and processes for the genetic modification of animals;
4) plants, seeds, plant tissue and other propagating material.

I believe that only by excluding such living material from patentability can we ever finally resolve the morality issue.

21 Daniel Alexander

I was fortunately promised by the organisers of this conference that I would not have to prepare a lengthy paper, so I have not done that. What I would like to do is pick up on some of the things that have been said this morning and this afternoon, address one of the issues which has been touched upon in brief under Art. 53(b) of the Convention and explain to you why I think that has some importance in the context of this debate. Then I will put before you some other considerations as to other ways of thinking about this issue in the context of patent law.

My role in this debate so far has been as the legal representative of first Greenpeace and then one of the opponents in the Oncomouse case. I was representing them at the respective oral hearings. Nonetheless, I propose to speak today not as a representative of anyone, but just as a lawyer with a general interest in these areas, not just in the field of genetically modified organisms but genetic engineering more widely. I will try to identify what I think the real difficulties are in this area and why it has raised so much concern.

One starts in this area with a sense of paradox. The reason for it is this: many of the proponents of doing away with discussions of morality in the context of patenting start with an argument that says 'well, it does not really matter whether a patent is granted or not'. There is of course something in that argument, in that it is right to say that the grant of a patent gives no positive right to a proprietor of a patent to do anything. It is merely, as has been said this morning and this afternoon, a right to exclude others from the particular territory claimed. That consideration points to saying 'why bother at all with any kind of morality provision in a patent convention?'. Why shouldn't, for example, a letter bomb be patentable? Surely there is no objection to patenting a letter bomb, since, by granting a patent, you do not in fact give the proprietor of the letter bomb patent any right to put letter bombs through anybody's letter slots. I think the reason why this issue *does* matter was in fact very clearly expressed by Dr Van Rompaey from Plant Genetic Systems, where he said 'a patent is a reward'; it is a reward given by a public body after examining in very general terms whether the proprietor *deserves* to have a patent. The issue of desert of a patent is addressed under a number of headings: whether it is sufficiently new, whether it is sufficiently non-obvious, whether the patentee has given the world sufficient technical details to perform the invention to entitle him to the monopoly that the patentee wishes to claim. The role of a patent office in that context is to provide a reward. Those who oppose the grant of patents in this context say, 'well, all right, we do not have a difficulty about a public body going around granting rewards in this context, but if they are to do so, they must do so in the context that those activities may be open to public challenge'; they are after all a public body and the whole foundation of the patent system is that patents are granted in the *public* interest. The purpose of patents — this is even expressed in the US Constitution — is to promote science and the useful arts. Patents are not just granted for the sake of it, in order to give a particular private benefit to an inventor, but they are granted because it is thought that they are economically valuable. It is in that context that I think one has to start with the presumption that issues of public policy can and should play a role in the granting of patents. That leads to the inevitable difficulty, very well outlined by Dr Schatz this morning, which is, 'what on earth are people in patent offices — not just the European Patent Office, but patent offices around the world — to do with an issue of this kind; how on earth are they to assess this particular issue?'. We have the provision in the European Patent Convention, there are similar provisions in the laws of member states of the Convention, and they do unquestionably place a burden upon examiners in the patent offices and Technical Boards of Appeal. It has been said time and again: 'we are not qualified to do that, why do you place the burden upon us?'. To some degree, one must have sympathy with that line of reasoning because I imagine that most people who join the staff of the European Patent Office did not do so in the expectation that they were going to have to resolve very difficult and hotly contested issues about fundamental principles of the 'commodification of nature' — as one of the most recent speakers mentioned — and whether it was right to grant patents on animals and to engage in an elaborate weighing exercise of risks and benefits, which it takes governmental committees a very long time to resolve. However, as against that, it might be said that a patent office is a body which has a very special qualification, precisely to engage in that kind of activity. The reason for this is really twofold. Firstly, it does contain people who are technically qualified. Secondly, those people are by and large decision makers as to technical issues and decision makers in relation to matters where judgement is required. Issues of obviousness have in some contexts been said to be a kind of value judgement. One is making value judgements of this kind in making decisions as to whether to grant or to refuse a patent in any event. So it might be said that they are not that well qualified to make these decisions. The question may then be raised 'who else should?'. One of the speakers this afternoon expressed a view which may be widely felt by people outside the scientific and commercial community, which is that it is only in the EPO — as the matter stands — where environmental groups and other people who are concerned about these issues really feel that they have an adequate forum to address issues of considerable importance in a particular way. That particular way is that of law — I will return to the role of the law and legal processes towards the end of what I want to say. The importance for this particular purpose is that it is only in that kind of institution that these issues may be addressed

197

on a sufficiently case-by-case basis, because a lot of these questions that one has to address are very particular questions, involving very particular facts and, for instance if one is to go into environmental risks, very detailed evidence as to what those risks and benefits are. That is one reason why one might say that an institution such as the EPO is in fact not unsuited, but might be particularly suited to making decisions of this kind and ought to do so.

The second point I want to make relates to the role of Art. 53(b) of the Convention in this whole debate. We have not heard a great deal about it today, although it was mentioned a bit this morning. What I would like to do is bring you up to date with where things stand on Art. 53(b). The reason for doing this is to explain why I think that some of the issues we are discussing in the context of Art. 53(a) may be issues which do not need to be discussed in that particular context just yet. If one looks at the law as it now stands, the position — if one looks at it fairly — is that patents for genetically modified plants and animals are going to be very difficult to obtain. The reason for that is a combination of the reasoning in the Plant Genetic Systems case and what might be called a clarification of that decision, recently delivered by the Enlarged Board of Appeal of the EPO, in late November last year (1995). Many of you will know that, after the Plant Genetic Systems case had come out, there was a degree of concern expressed, particularly — as I understand it — by those involved in the industry, that this was going to make it much harder for them to get the claims that they wanted. This is something which, no doubt, we can discuss a little bit more during question time. The upshot of that was that the President of the European Patent Office in essence referred that decision to the Enlarged Board of Appeal on the grounds that there was an alleged conflict between that decision and earlier decisions, including the Harvard Oncomouse case and a still earlier decision. Really, I think the subtext of it was to get a different view, if that was at all possible — those who are here from the EPO will correct me if I am wrong in thinking that. Anyway, the upshot of that case was that the Enlarged Board of Appeal declined to give a different view, indeed they declined to entertain the reference made by the President at all. In the decision that came out in November, which is G 03/95, they said, 'look, there is not in truth a conflict between the Plant Genetic Systems decision and earlier decisions; earlier decisions were concerned not with the genetic modification of plants, but with the issue of whether a plant or an animal per se would be excluded from patentability'. For example, one of the earlier cases concerned the application of an oxime derivative on to plant matter, in order to protect it from herbicides. An argument was advanced that said 'here you have a claim to things that are in fact plant varieties'. The European Patent Office rejected that argument on the basis that, in truth, there was no claim to a plant variety. It was just a claim to a plant with a bit of a chemical on top. The important thing they said in that case was that plant varieties are concerned with the genetic modification of plants. So, time went on and the Plant Genetic Systems case came up, which *was* a case that concerned the genetic modification of plants. In due course it was held that certain claims in that case embraced — that is to say, covered — plant varieties and were not excluded by the exclusion-to-the-exclusion in the Convention that deals with microbiological processes. The Enlarged Board made a distinction between those two kinds of cases and said 'if you have genetic modification of plants, it is going to be a plant variety, and if you do not have genetic modification of plants it is not going to be a plant variety', in very simple terms. If anybody is interested, I have got the text of the decision here and we can discuss it in more detail if you wish. The effect of that decision will be — if it continues to be followed in the EPO — to really make it rather difficult to have claims that encompass genetically modified organisms. The purpose of introducing this point in the context of this debate is that the substantial objections that have been raised on grounds of morality are precisely directed to claims for genetically modified organisms of that very kind. And if they are not allowable in any event under Art. 53(b), the whole issue under Art. 53(a) becomes one of much lesser importance. So, it was to a degree right to say this morning that the remedy for this problem is not really to direct one's fire so much at Art. 53(a), which provides a great difficulty for the EPO, but to direct one's fire as well particularly at Art. 53(b) — it was said that ought to get rid of. Unless one does that, there is no point engaging in the debate at all, because you have already lost the first battle.

That is the legislative context I think we find ourselves in at the moment and which will in due course have to be grappled with. That brings me onto some final points, points on where things stand now. A very interesting and helpful speech on the new draft Directive was made this morning. What the draft Directive has done is the traditional thing: imposing restrictions on the patentability as such of particular products. No doubt there will be very interesting legal questions raised in due course about the extent to which a Directive obliges a member state to actually seek a relevant modification to the European Patent Convention. You must bear in mind that the European Patent Convention covers territories that are not covered by the EU, but we can leave that debate for later. One thing which may be worth discussing in the context of a workshop of this kind, is to consider alternative methods for raising issues of morality and public policy in the context of patenting. So far, those issues have really only come to be raised at the stage of considering whether a patent is or is not to be granted. That is what the European Patent Convention says and that is reflected in the laws of the member states. One alternative — and I just put it to you for consideration this afternoon — is whether or not there might be some better consensus achieved by considering whether or not these issues would be either additionally or perhaps better raised at the stage when a patent comes to be enforced. Supposing, for example, that there was a provision in English law that said 'when the issue of relief for infringement of patent' — whether by way of injunction or delivery or damages or an account of profits or whatever it may be — 'comes to be assessed, account shall be taken of the extent to which the invention is beneficial to the environment' or 'in accordance with public policy' or whatever it may be, whatever little phrase one would want to insert. It may be that one would find a suitable forum for addressing issues of that kind in that context. There are institutional issues that one has to consider here, because one of the things that has generated so much trouble for the EPO is the fact that the opponents to patents of this kind are not commercial entities. From my experience, which is primarily dealing with commercial litigation, they quite often have an interest in keeping well quiet about certain of the activities of even the people they are suing for patent infringement or being sued by. The opponents to patents of this kind are people who, as it were, have no interest in the matter commercially. If one *were* to do something like that — raising issues of public policy and morality at the stage when a patent comes to be enforced — it would have to be ensured that there were some adequate role for public participation in decision making of that kind. Otherwise you would find, for example, cases like Hoechst suing Plant Genetic Systems and neither of them would want to say against one another 'our activities are to be taken into account in a detrimental way as regards any relief that should be granted'.

So, what I think one ought to be doing in this context, just to sum up, is to think not only about the issues of morality, public order and environmental concerns at the stage where a patent is granted, but also to think about those issues at the stage when a patent comes to be enforced and to ensure that, in whatever one does, one does not deprive the public of an opportunity to provide input on what, in my view, is a very important public policy issue.

22 Michel Vandenbosch

First I would like to stress why GAIA opposes in principle the patenting of, the breeding of, and the carrying out of experiments on transgenic animals. At the same time, I will try to contribute to the aim of this workshop, 'how to reach consensus on the subject'.

As an organisation defending the moral interests of sentient animals, GAIA worries a lot about the developments in the area of breeding and carrying out experiments on transgenic animals. They are a cause for particular concern from an ethical point of view, as previous speakers have pointed out already. What worries me a lot is the suffering caused to the sentient animals that are being used in the field of genetic engineering experiments. What worries me the most is the process of turning animals into inventions, into objects. The breeding of transgenic animals and the attempts to patent sentient life illustrate how, at the end of the twentieth century, the process of turning sentient animals, morally worthy of protection, into objects, is being pushed to the limit.

Transgenic animals are created to act as bio-reactors, to produce pharmaceuticals and other proteins in their milk or blood. Sentient animals are genetically engineered to develop cancer or other painful chronic diseases. Genetic engineers are attempting to develop transgenic farm animals which will be more productive, i.e. grow bigger, grow quicker, grow leaner. We have transgenic pigs, for example, that suffer from poor vision, arthritis, stomach ulcer and muscular weakness. Transgenic mice, pigs, bulls, sheep, goats and so on are now produced in large numbers and they are animals with welfare needs. They require certain living conditions. Not only are they forced to suffer in their existence on the factory farm of the intensive livestock breeding industry, but also researchers have now succeeded in transforming the bodies of the animals themselves into, as it were, biotechnological factories. Transgenic animal technology in the field of intensive livestock farming is aimed at increasing productivity and profitability. The consequence of all this is even worse quality of life for the animals. The purpose of using transgenic animals in biomedical research is said to be the 'saving of human life', but the biomedical value of for instance Oncomice for cancer research appears to be doubtful, as previous speakers have pointed out already. The suffering of sentient animals, however, is not in doubt.

It looks as if an 'animal industrial revolution' is taking place. Bioengineers have purposefully broken down the barriers between the species. Unfortunately, this has not been done to develop an ethical relationship with the animals, based on empathy, on the recognition of animals as sentient beings morally worthy of protection, and on respect for their quality of life. The breeding of transgenic animals is, in my opinion, a revolutionary step, but a step which — ethically speaking — creates a very problematic dynamic. I fear that it may lead to an absolute 'de-animalisation', an objectification and extreme instrumentalisation of animals, which are being treated as the ideal tools whereby their status as being morally worthy of protection is ignored. Creating Oncomice, genetically programmed to develop cancer, and of which the offspring too develops the disease, creates a dynamic that seems impossible to reverse.

What about the difference between what is natural and what is artificial, which is, admittedly, outside the field of transgenic animal research and genetic engineering? This difference is already not quite obvious, but now it becomes even more confusing. Whatever the answer to that question is, the language used by scientists in this field suggests that animals are viewed as scientific tools. That sort of inclination is growing. This way of looking at and considering animals is, in my opinion basically wrong, for it is contrary to the ethically sound point of view that sentient animals are morally worthy of protection. Why is that? Because they possess characteristics that are of moral significance, basically because they are capable of experiencing better or worse quality of their own life, physically and mentally. Unlike plants, they possess a nervous system or brain that enables them to experience the quality of their life in time and space, they can feel pain or enjoy pleasure, they have preferences, they have a memory, they can learn from experiences, and all this of course within the scope of their nature. For those reasons I don't see why we should not enlarge our moral horizon to non-human species that also possess these morally important and significant characteristics. Logically and rationally, we must recognise the ethical importance of animals possessing morally significant characteristics. Their moral interests should be taken into account, as our moral interests should be taken into account, and — without going too far into that debate — granting legal rights to animals would in my opinion be an option worthwhile of at least being examined when assessing conflicts of interests between human and non-human animals.

So these are sufficient reasons why non-human animals should not be treated as if they were tools, as if they were mere means to an end. The end does not justify the means, and therefore GAIA opposes the patenting of transgenic animals. In our opinion, this is precisely a step in that wrong direction. The animal becomes officially an invention. Well, it is contrary to scientific insight and knowledge about the nature of animals, it is contrary to ethical insights about assessing and considering animals in an ethical context. Moreover, there is also a risk that using transgenic animals for experimentation will threaten the reduction, refinement and replacement dynamic, which is only slowly emerging in the field of biomedical and other types of invasive research being carried out on animals.

The modern scientific ethological and ethical insights into nature and the moral importance of animals do not easily find their way into society. They will not easily become engraved in what could be called 'human collective conscienceness'. Nevertheless, there are clear indications that society develops slowly but surely into the direction of an enlargement of our moral horizon towards other animal species. In other words, I fear that the development of research on sentient animals and the patenting developments will jeopardise this tendency that is stimulated by modern science and ethics and of which modern science and ethics confirm the validity and rationality. I also fear that the development of research on transgenic animals, and maybe also the patenting of those animals as inventions, will lead to an increase, instead of a reduction, of the numbers of animals being used for experiments, which according to many scientists is one of their aims.

How many animals have already been affected by genetic engineering? Worldwide, several millions, and it would seem that their numbers are increasing rapidly. According to Home Office figures in the UK — in Belgium there are no figures available — over 73,000 genetically engineered or transgenic animals were produced in 1992 in the UK alone. This means a 65 per cent increase since 1990. In Belgium too, transgenic animals are being bred; neither the public nor the government so far appear to be concerned about it. Do I need to stress once more, as previous speakers did, that a thorough social debate is required in the light of the serious ethical problems linked to this whole issue? Indeed, I think that, until this debate is not taking place in earnest, the usefulness of stressing the necessity of such a debate again and again cannot be underestimated. No doubt information provided by NGO's can contribute to this debate, especially in Belgium, where the public appears to be more and more sensitive to animal welfare and animal rights issues, which are also increasingly appearing on the political agenda. In Belgium, recently — within the framework of the animal welfare law — a deontological committee saw the light, which has to consider animal experiments. In the interest of an open public debate, GAIA hopes that this Committee will carefully examine the ethical issues at stake and contribute to pressing the responsible authorities to agree a moratorium on the development of and experiments on transgenic animals, without which a social consensus might be very difficult to achieve. How could we reach such a moratorium, which is necessary to build up a certain sufficient level of confidence between the parties involved? Needless to say, it will not be easy because it needs unbiased and informed people, prepared to understand that the absolute truth most probably does not exist. It is necessary that, during that period, real dilemmas are detected and discussed in depth, so it would be necessary to detect the difference between real dilemmas and pseudo-dilemmas or quasi-dilemmas. What is presented as a real dilemma might not be able to resist ethical scrutiny. When profit becomes the goal, it is far from being morally and intellectually honest to present it as an ethically sound reason to allow the breeding and the patenting of transgenic animals.

In short, GAIA is opposed in principle to what is happening now, but a society where we are only humans and we are not Gods, of course, it is not an ideal society. One should be open to discussion and treat very difficult and complex problems with an open mind, but surely with an honest mind.

23 Michiel Linskens

First of all, I would like to say something about my Society. We are the biggest animal protection society in the Netherlands, with nearly 200,000 members and a staff of 50 people in The Hague, so at least you could say that we are very educated on this issue. We are a member of the World Society for the Protection of Animals, and also a member of Eurogroup for Animal Welfare, a lobby organisation in Brussels which covers all the animal welfare organisations in Europe.

I am a biologist who is also very interested in the patenting issue. Before I became a policy adviser, I wrote a book about transgenic animals and ethics. Ethics is an issue which I find very interesting.

I would like to say something about the situation in the Netherlands concerning the patent issue. Recently, in December, we held an inquiry with the Dutch public. Over 2,000 people were asked their opinion on patenting. It turned out that over 70 per cent of the people were against the patenting of animals. This figure is comparable to figures in Germany and Switzerland. Furthermore, a declaration — which has also been presented by Steve Emmott of 'The Genetics Forum' — is undersigned by over 50 major NGO's in The Netherlands, so I would say at least that there is large support for a strong opposition against the patenting of animals. Moreover, political parties in the Netherlands also oppose patenting very much. I think nearly every political party would like to have a ban on patents on animals. This resulted in an amendment, in 1994, of the Dutch patent law. Although all political parties wanted a ban on animal patents, an amendment was accepted which was not exactly what they wanted, because this amendment still makes it possible that animals are patented. However, there is one requirement: every animal that is to be patented should be ethically reviewed. This ethical review is done — or has to be done, because the legislation is not yet in force at the moment — by an independent committee, and this committee works with the principle of *'no, unless'*. This means that genetic engineering of animals is forbidden, unless you can prove the necessity of the research and the lack of alternatives and the absence of suffering of animals, at least no unnecessary suffering. So, in The Netherlands ethics and patenting are closely linked. As I said before, it was not exactly what the politicians wanted. They found out a few months ago, by asking patent lawyers and other people whether it is indeed true

204

that animals are still patentable in The Netherlands. Now they are reconsidering this issue and they will probably discuss it again within our government, probably to change the law to make sure that animals are no longer patentable. So this is the situation in The Netherlands, and the debate is still on-going. Probably that will last for a few coming years.

Our position, as an animal welfare organisation, has already been explained by Michel Vandenbosch. The only thing I will do is give a short summary of our arguments. We are also, just like all major animal welfare organisations in Europe, opposed to the patenting of animals. Animal welfare organisations, as stated in the statutes, should take care of the interests of the animals. If these interests are harmed by a certain technology, for instance by genetic engineering, then we oppose this technology.

The arguments we use against the patenting of animals are that it is morally unacceptable to consider an animal as an invention. So this is a moral argument against the principle of patenting. Furthermore, as Michel Vandenbosch said, patenting animals will lead to an increase of the use of laboratory animals. Since most of the laboratory animals are designed to suffer and die, this will lead to an increase of animal suffering.

I will not go further into detail with regard to these arguments, but I would like to discuss with you some more biological aspects of patent law. This is because I am a biologist, but also because I would like to understand the reasoning of patent lawyers. In some cases, this is very difficult. First of all, I would like to discuss the issue of reproducibility. I know an invention should be reproducible. The process of making a transgenic animal is biological, except for the first step, i.e. the micro-injection. You micro-inject a new gene construct into an embryo and afterwards the embryo divides, grows, becomes an individual. The individual breeds, gets offspring, and so on. After the first step, the engineering step, nothing can be done about the final result. It is just 'wait and see', and as you probably know, in biological processes not only genetic factors are involved but also environmental factors. A pregnant woman, for instance, should not smoke or drink too much because it could harm her embryo or foetus. So, these environmental effects will also influence the final outcome, together with the genetic factors. When we look at the genetically modified bull Herman, and we realise that this animal is produced by micro-injecting 2400 of embryos, resulting in only one transgenic animal, my question is whether you can speak about reproducibility. If I use the same gene construct, the same method and the same equipment, using 3,000 embryos, I can probably be very sure that the outcome will not be a second Herman. The micro-injection itself is specific, but they do not know how many of the gene constructs will incorporate in the genome. It could be one, it could be twenty, it could be three hundred. Even after the first seemingly precise step, you do not know what will happen afterwards. So I wonder whether you can say that, the process being not controllable, it is reproducible.

Another issue that I find very interesting concerns the 'varieties'; I am probably not the only one who worries about this. If we have a look at the animal kingdom, you know that it is divided into vertebrates and invertebrates, and the vertebrates are

205

divided into mammals, fish, birds, reptiles, amphibians. If we have a look at the mammals, a distinction is made between human and non-human mammals. So this is the first aberration of this scheme — invented by people — of the animal kingdom. If we have a look at the mammals, we have a lot of families, and one of them is the rodents. The rodents can be divided into a lot of species, one of them being mice. A fundamental characteristic of a species is that two individuals of the same species, normally a male and a female, are able to produce fertile offspring. In this scheme of the animal kingdom, varieties did not exist until man started breeding. We bred a lot of varieties: of cows, of cats, of dogs, of mice (e.g. the nude mouse). The question is: scientists produced other mice with the help of genetic engineering and they claim that they patent a new species, but if you look at the main characteristic of a species, you could wonder whether it is indeed a species that they claim, or at least a new species. A transgenic mouse is able to mate with another normal mouse. I suppose that you cannot say that new species are patented in this case. Probably, new varieties are patented. Why do they claim species? Because varieties are not patentable. From the point of view of a biologist, I have serious problems with the fact that genetically engineered mice are considered as a new species. But even if so, if you patent a species, every variety which is covered by the species is also covered by the patent. So there are two problems: transgenic mice are, in my opinion, not a species and even if so, then varieties are covered by the patent, whereas varieties are not patentable. These are difficult problems with which I have been dealing for three or four years and, until now, no patent lawyer has been able to convince me of the reason why this is possible.

Finally, I would like to focus on the new Directive of the European Commission. In Art. 9, there is a part of ethics involved in the Directive, because it says that 'processes for modifying the genetic identity of animals which are likely to cause them suffering or physical handicaps, without any substantial benefit to man or animal, and animals resulting from such processes, insofar as the suffering or physical handicaps inflicted on the animals concerned are out of proportion to the objective pursued, shall be considered unpatentable'. This is a part of ethics, and I wonder who should deal with it. As we have seen with the Oncomouse, it has been dealt with by the European Patent Office itself. I question whether this office is capable of performing such an ethical weighing up. But it is also questionable who decides about what a 'substantial benefit' is and what 'out of proportion' is. In my opinion, you need an independent committee, which is trusted and respected by all players in the field. That is probably one of the problems in getting such a committee together. But I believe that the EPO is unable and should not be enabled to do this ethical review.

To conclude, I would like to remark that, if you listen to proponents of the patenting of animals, they say that patents are necessary to stimulate research for biomedical purposes, etc. But I wonder how they can explain that, already since the beginning of the 1980s, probably over a thousand transgenic mice lines have been produced by scientists without patents at all. I suppose that they, at least scientists who are doing fundamental research, just continue with their work, with or without patent coverage.

Probably, if you look at the broadness of the patent claims in the Oncomouse case, these claims may even frustrate cancer research with transgenic mice, because after ten years, Dutch scientists who currently work with cancer mice covered by the Oncomouse patent, are faced to pay royalties to Dupont, if they would design a gene therapy based on Oncomice. So maybe the patenting of transgenic animals could even frustrate scientific research.

I see that, at least in some parts of Europe, there is a large opposition against the patenting of animals. The European Parliament, a democratic institution, is also against the patenting of animals. I wonder whether it would not be wiser for industry not to continue enforcing a new patent law, and show some respect for the democratic process in society by looking for other possibilities to protect their activities in this field.

24 Christoph Then

My presentation will be about the intrinsic value of the European Patent Convention, and especially Art. 53 of the EPC. I will try to connect Art. 53(a) and Art. 53(b) in a reasonable way.

History of patent legislation with special regard to Art. 53(b)

Special rights for genetechnology?

It was in 1980 when for the first time ever a patent was issued for a bacteria in the United States. This event also marked the beginning of a widespread international trend to extend patent law into the sector of living nature. The patent was granted in 1980 with the understanding that bacteria are 'far more similar to inanimate chemical compounds than to horses, bees or raspberries' (retranslated from German text) and therefore the granting of a patent would not be in conflict with patent law which is designed to be applied to technical inventions.

In 1988 the first mammal, the so-called Oncomouse, was patented in the United States. A parallel expansion of patenting practises occurred with only slight delay in Europe. In 1981 the European Patent Office (EPO) in Munich patented the first micro-organism, and in 1992 the Oncomouse was patented. Both in Europe and in the United States, increasing numbers of patents have been issued for plants. This has occurred in Europe without a sufficient legal basis. For that reason the industry, through massive lobbying pressure, tried to push the proposed Directive on the legal protection of biotechnological inventions through the European Parliament, by which the granting of patents for plants, animals and human genes would be, for the first time, explicitly permitted. The Directive was rejected by the Parliament on 1 March 1995. It was reintroduced by the Commission of the EU in December 1995 and will be on the agenda till 1997.

208

The basis for granting patents in Europe is the European Patent Convention (EPC). In 1996, 17 nations, among them Switzerland, belong to the EPC. The jurisdiction of the EPC thus exceeds that of the European Union. According to the EPC, patents are, in principle, issued for inventions. For that reason it is also possible to patent inventions in the field of genetic engineering, for example techniques to produce pharmaceuticals. From a viewpoint of medical research, sufficient protection by patent does therefore exist.

On the other hand, it is not allowed to grant patents for discoveries, essentially biological processes for the production of plants or animals and plant or animal varieties. As to 'discoveries' (Art. 52 EPC), this exclusion is critical concerning both the issuance of patents for human genes and for 'genetic resources' in general. As to 'essentially biological processes for the production of plants or animals' (Art. 53b EPC), this refers to procedures whose profitability stems mainly from the ability of natural reproduction or from the natural growing of plants and animals. As to the exclusion of 'plant or animal varieties' (Art. 53b EPC), we can find similar rulings on an international level: the GATT established that animals and plants can be excluded from being patented. Moreover, the NAFTA treaty between Mexico, Canada and the United States excludes species of plants and animals from being patented.

Historical aspects

Modern patent law is a compromise between inventors' interests in wide-ranging monopoly rights and the interests of society as a whole, which in return expects contributions to enhance its overall well-being.

In the case of patents which cover the ability of natural reproduction, for example the patent for the Oncomouse, the inventive achievement corresponds in no way with the extent of rights that would be granted to the inventor. Instead of giving a legal protection for an active producer, a privilege would be granted to a passive utiliser of essentially biological processes.

Overall, by issuing patents in the sector of living nature, modern patent law threatens to close ranks with the late Middle Ages privileges system. According to this system, 'indispensable, commonly known goods such as salt, beer, vinegar, glass, iron, sailcloth, dried herrings and so on, were monopolised in favour of individuals and, accordingly were more expensive for everybody.'[1]

European patent law draws a clear line concerning the granting of patents to living nature. This line is not arbitrary and should not be looked upon as being historically outmoded. The essence of modern patent law is defined by this line. Indeed, seen from a historical point of view, the patent-office develops in a backward direction. Special rights or privileges are being granted that do not correspond to what in each individual case is 'technically' being developed.

The practice of granting privileges was especially widespread in Great Britain. In 1601, the British crown was forced to dissolve the most pressing monopolies on indispensable goods and food, and in 1623, the famous 'Statute of Monopolies' was enforced. With ever-increasing industrialisation, the British Parliament passed the 'Patent and Designs Act' of 1907 in which currently still applicable principles of patent law were first formulated. In exchange for the monopoly granted in the form of a patent, an equivalent service had to be expected and the government would be entitled to make sure that this service would indeed be rendered. The monopoly had to be viewed as a valuable good for which the whole of society could expect a service in return. If this service would not be rendered, society would be authorised to cancel the monopoly. These principles were, above all, formulated concerning the sites of production. Yet these guidelines also served as principles concerning the exchange of services between the individual patent holder and society in general. Such principles are still of crucial importance to modern patent law.

There are close parallels between the above-depicted development and our current situation. We are in the midst of stormy technological changes. Then, as much as today, industry lobbyists tried to use these developments by expanding their claimed monopolies and for a new distribution of markets. Then, as much as today, society is called upon to resist this development.

Now it seems that it could become possible to get very similar privileges on the basis of patents on natural reproduction and natural growing of plants and animals. This would be the situation following the proposal for a 'Directive on the legal protection of biotechnological inventions', Art. 11 of which says that 'The protection by a patent on a product containing or consisting of genetic information shall extend to all material ... in which the product is incorporated and in which the genetic information is contained and expressed.' This means, for example, that the patentee who claims a gene has unlimited monopolies on all generations of plants or animals in which the gene will be incorporated by any technical method in the field of bio- and genetechnology and even on all plants and animals which could result from the following cross breeding procedures. If this becomes patent law, 'essentially biological processes for the production of plants and animals' will be the basis for far reaching monopolies and the patent system will get an instrument to privatise processes of life nobody could ever invent.

It is important to emphasise that the problem of granting patents to living organism has to be viewed independently from the approval or disapproval of gene technology overall. Other technologies which can lead to similar abuses of patent law, also need to be restricted by society. In the interests of society, such special privileges are not tolerable.

In general, even genetechnologists acknowledge that there is a problem. Among them is the well-known German genetechnologist Prof. Dr Winnacker, who is generally in favour of issuing patents in the field of genetic engineering. While attending a meeting of the Bavarian Agricultural Ministry on 18 May 1994, he stated his opinion concerning the Oncomouse as follows: 'If in spite of this, the patent for the Oncomouse is contrary to morality, then it is due to the fact that the patent is too far-reaching. This, in turn, will lead to a degree of monopolising which was never intended by patent law.'

Basic considerations of granting patents in view of European cultural history (Art. 53a, 'Morality')

Distinction between owner and creator

The fact that animals and plants are emphatically distinguished from man-made things can be traced backed to antiquity. This tradition can not be simply explained with technical reasoning. It has more to do with the human experience that one can only cope with one's environment and survive in it, if one accepts it as a partner. Nowadays this state of mind is more compelling than ever before. This development is also being reflected in legal procedures; accordingly, the fact that animals have an intrinsic value has been emphatically re-evaluated in a legal sense. The idea that the world is not man-made and that life is not made and not makable is a central tradition of European cultural history. On the other hand, it is historically and culturally well substantiated that a patent is based on a creative claim. Therefore, whoever bases his argument on only technical or legal grounds misses the real question at hand.

In a cultural sense, the distinction between an owner and a creator has always been clear-cut. Christian tradition in general, as well as any cases which touch upon this distinction, are proof of this. Alchemistic ideas and Frankenstein's story clearly show societal outrage at the possibility of producing or 'creating' artificial life.

The granting of patents on laboratory animals

Issuing patents for animals represents an abrupt change of course and a contradiction to European cultural history. By doing so, one does not only claim the right to experiment with animals, but animals themselves become the goal of a productive process. The animal product being protected by patent is the result of a process which is not primarily directed by medical but rather by economic motivations.

Both the necessity and the scope of animal experiments performed to develop new pharmaceuticals are topics of strong controversy. The Animal Welfare Legislation is an attempt at compromise between various interests. The law stipulates that intrusions on vertebrates are only permissible if specialists find it absolutely necessary. In Germany experimentation on animals is not allowed without 'sound reason' (Art. 1 TierschutzGesetz). The legislator thus attempts to define under which circumstances exceptions can be made to a general rule, which states that due to human responsibility for the animal as a fellow creature, we have to 'protect its life and well-being' (Art. 1 TschG). By granting patents to living things, this principle is totally reversed. The careful evaluation of scientific and ethical questions is no longer the main criterion. Animal experimentation becomes part of an economy which heats up competition and is strictly regulated by market forces, while ethical questions are undermined. The hope for a 'Superanimal-model' in which AIDS, cancer and heart attacks can be researched, is combined with gigantic economic expectations.

The patent for the Oncomouse had repercussions in Europe. By mid-1995, about 300 additional applications for patents on animals were registered at the European

211

Patent-Office. Among those are applications for patents on cows that would donate more milk and pigs and fattened turkeys that would grow even faster. This demonstrates that the Oncomouse is only a gateway to an entire zoo of genetically mutated animals, which are supposed to fulfil new profit expectations. A perspective which, under these circumstances, only keeps in mind the scientific benefit of each animal experiment avoids the real problems.

The exclusion of granting patents for inventions which are contrary to morality and public order (Art. 53a EPC) achieves a new significance at this point. This regulation can only mean that any creative claim to mammals, for example, can not be legitimate. This again indicates that Articles 53(a) and 53(b) EPC have a common perspective. The exclusion of granting patents on animals and essentially biological processes and the notions of 'morality' and 'public order' are insoluble, both ethically and legally.

Economic arguments against the granting of patents to lifeforms (Art. 53a, 'public order')

The issuing of patents for plants leads to a general redistribution in the agricultural sector and the food market

In the sector of plant breeding, developments have already advanced relatively far. Accordingly, the 1994 OECD report on 'Biotechnology, Agriculture and Food' concisely states that 'The crucial focus in this sector is on the reorganisation of the seed market resulting in a stronger integration into the agro-chemical sector.', and 'Concerning new product marketing strategies, the possibility of being a deliverer of genetechnology has become less important and a new strategy has taken its place: one tries to gain control over seed markets and, even more importantly, tries to push into the markets laying behind in order to reclaim the industrial surplus value for oneself.' (Re-translation from the German text)

These quotations illustrate that with the help of genetechnology, via a detour in patent law, the whole sector of agriculture and food market is supposed to be intentionally monopolised. In doing so, genetechnology is a primarily economic instrument. Modern patent law is abused for objectives it was not designed for.

At this time, the concentration of the seed market has already achieved precarious dimensions. By granting patents to plants, previous legal protection systems are getting needlessly destroyed. Plant breeding, agriculture and the food market could get dependant on a few international companies. The issuing of patents to plant varieties lays an important role in this development.

Agriculture utilises distinct and stable plant varieties and breeders can only sell these genetically stable kinds of plants. In the case of patents involving agricultural plants, one can therefore assume that the patent also includes plants varieties. It can be clearly shown that many granted patents essentially aim at plant varieties, because this is the only tradable form of plants on the seedmarket. Normally the word 'plant

212

variety' is strictly avoided in patent applications. But all claims directed to 'plants', 'part of plants', 'seeds' and even most of the claims directed to 'genes' (in the context of plant genetic resources) have just one common economical perspective: the marketing of plant varieties.

Compared to the previous legal protection systems for plant varieties (like the International Convention for the Protection of New Varieties of Plants, UPOV), a patent is a substantially stronger kind of protection. For this reason, both of the legal protection systems can only compete with each other if patents are clearly defined within certain borders. Otherwise conventionally bred plants and already existing varieties would be economically devalued through patent protection for plants in the context of genetic engineering.

If plant varieties are unrestrictedly included in a patent, as is currently the case at the European Patent Office, then a complete cancellation of the previous system of plant breeders' rights is foreseeable, for the following reasons:

–   It is easier to maintain patent protection than it is to maintain the normal plant breeders' protection, because the lengthy breeding to achieve a tradable variety is not necessary in the case of a patent.

–   In order to maintain protection by patent it is sufficient to pay fees to the patent-office. The maintenance of a plant breeders' right is dependent only in the case of ongoing active breeding that guarantees the availability of the variety.

–   Contractual licenses, according to the wishes of the patent holder regulating the cultivation, processing and trade of each single product, can be added to the patent that has been granted. This affects not only the seeds, but the whole food sector (cultivation, processing, selling) right up to the supermarket shelf. There is no comparable procedure in plant breeders' rights.

–   In contrast to plant breeders' rights, a patent can claim an indefinite number of varieties and species of plants at once. It is enough to manipulate a single gene in order to be granted such claims.

–   Licensing contracts enable the patent holder to profit from further breeding, while the usage of previous varieties is free of charge or stipulation (plant breeders' privilege).

Summarising this, it needs to be stated that in general the economic and temporal expenditure for genetic engineering is far less extensive than the effort it takes to breed a new variety.[2] However, patent law clearly favours genetic engineering and disadvantages normal breeding procedures. Whereas the patent is a virtual protection aiming at future options, the plant breeders' right only includes existing plant varieties. In this light, protection by patent shows qualities of a passive privilege, while a plant breeders' right presumes that plant varieties are actively produced.

The above-depicted situation will occur more or less independently, whether a farmers' privilege will be established in patent law or not. By making it possible to claim patent titles on plant varieties, every farmers' privilege will be rendered a farce. As we can see in the example of the patent being granted on the 'Flavr-Savr-tomato', the granting of patents will indeed lead to a totally new distribution of the seeds- and food market. In the marketing of the 'Flavr-Savr' tomatoes in the United States, everything from cultivation to processing and selling is regulated by respective licensing contracts.

Economic arguments can also be raised against granting patents to laboratory animals

Granting patents to animals, especially agriculturally useful animals, can endanger the public good in fields including the food industry, agriculture and animal breeding.

Issuing patents to laboratory animals also does not make sense in a larger economic context. So far, the laboratory animal market was, and still is, profitable without issuing patents. These animals are currently traded together with contracts that forbid any further breeding, often only animals of one gender are being traded and so on. In this respect, genetechnology does not essentially change anything. The actual breeding which leads to laboratory animals being traded is still in the hands of respective institutions. For example, DuPont has contracted the breeding of the Oncomouse to the firm CHARLES RIVER, because DuPont expected them to have the necessary known-how to breed animals in sufficient numbers and quality. So far, institutions such as CHARLES RIVER have been able to survive well on a market without patents. The breeding of animals in laboratories which are not specifically equipped for the purpose, is nearly impossible, because the risk of genetic deviation in the animals is too high, the results from experimenting with such animals are typically not reproducible. Thus, the economical situation between providers and consumers of these animals regulates itself automatically. We can prove this fact with a letter of the CHARLES RIVER company to our co ordination office in Munich which makes clear that the company considers the granting of patents to laboratory animals as uncalled for, even hindering.

On top of this, scientists and pharmaceutical companies fear an unnecessary rising of costs in their work. Research can easily be impeded through the granting of patents. The world's biggest laboratory that trades with test animals is the Jackson Lab in the United States. This laboratory has established as its goal the provision of as many animal test models as possible at the lowest possible price — in order to spur research and in explicit contrast to the goals of granting patents.[3]

Why then issue patents on animals for experimental use at all? It should be noted that the issuance of patents is absolutely unnecessary to secure an appropriate return for these animals. The granting of a patent to the Oncomouse clearly demonstrates that the inventory value of the laboratory animal model is not at stake. As is well illustrated, the interest of the patent holder was not the licensing of the Oncomouse, but the licensing of all pharmaceuticals developed with the help of the mouse.[4]

214

Through the monopoly on the animals, the pharmaceutical market was supposed to be skimmed. This is shown by the licensing contracts that DuPont was trying to obtain. The issuing of a patent for the mouse therefore was only a trick to profit in a completely new way from the production of pharmaceuticals without having invested a single dollar in this sector.

This incident is no accident. In general, interests in follow-up procedures that pursue the animal test model (for example, licensing contracts having to do with the pharmaceutical market) result in situations in which the animal is only an economic instrument, which through its monopoly, helps open up further markets.[5]

Overall, the desirability of granting patents to animals in light of the economic interests of the common good cannot be proven. By extending claims into the markets of pharmaceutical products, costs in health care could sky-rocket in the near future. New dependencies could therefore evolve in the pharmaceutical industry.

To summarise, Art. 53(a) and Art. 53(b) are tightly interconnected. Overall, the stipulations of Art. 53(a) are therefore to be viewed as general exclusive criteria against patenting of life, especially against the patenting of genes, plants and animals.

Notes

1   Prof. Dr Hubmann (1988), *Commercial Legal Protection*, Beck'sche Verlagsbuchhandlung: Munich.
2   This rings also true for the breeding of varieties from genetically engineered plant cells [*cf* Lange, Peter (1993) in *GewerblicherRechtsschutz und Urheberrecht*, S.801].
3   *Cf Nature*, Vol. 364, 26 August 1993, p. 755.
4   *Cf New Scientist*, 26 June 1993, p. 4.
5   *Cf Bio/Technology*, Vol. 11, June 1993.

25 Luc Vankrunkelsven

WERVEL is a Flemish working group which tries to be a point of reference in the discussion between farmers, consumers, Third World movements, peace and environmental movements. On each issue, we try to link the interests of all these parties. This is also the case for the topic of genetic engineering and patents.

We can see, in Europe and gradually in the entire world, two tracks in the development of agriculture. The first one is the dominant industrial agriculture and the second one is the sustainable family farming. In the logic of the first system, we find genetic engineering, patents, rBST hormone, herbicide resistance, etc. Life becomes a private property of industry, while WERVEL and other groups defend the ethics of respect for life. For us, a safe food production and a sustainable agriculture have the following characteristics:

- low production, linked to the soil

- low input of energy, pesticides and fertilisers

- regional and local markets

- control of the production

- a fair price for farmers and women farmers, in exchange for a quality product for the consumers

- consolidation and creation of jobs in agriculture

Twelve reasons[1] to be against the patenting of living beings

The first reason has to do with farmers. European farmers will be obliged to pay royalties on every generation of plants and livestock they buy and reproduce for production purposes. The rural community will lose its last means of control over the first link in the food chain and become totally dependent on multinational corporations.

Secondly, breeders will no longer have free access to germ plasm for developing new varieties of plants and animals. Most independent breeders will simply go out of business. As a result, the only innovation in the breeding sector will be found in the legal departments of large corporations, where patent lawyers will dictate the direction of biological research.

The third reason is that European consumers are likely to end up paying higher prices for food, medicine and other products of genetic engineering. In buying patented genetically engineered products, consumers will be unwittingly subsidising industry, as royalty charges will be passed on to the end product.

Four, public research will be undermined and effectively privatised. The public sector is paid for by all of us, but the extension of the patent system will ensure that only private industry benefits. Universities and public research institutes will be obliged to keep secret their research results, funded by the private sector, while the corporations apply for their patents. This means that the public exposure and circulation of scientific information will be restricted drastically, to the detriment of learning and innovation.

Five, European market structures will undergo a dramatic wave of increased concentration. Less firms will be able to compete on the market place and many will be bought out by the strongest multinational corporations. Stronger monopoly structures in the agro-business, pharmaceutical and chemical sectors will emerge, with their consequences on prices and quality, leaving us few choices in our needs for food, health and a cleaner environment.

Six, genetic diversity in Europe will suffer tremendous erosion as monopoly control over genetic resources severely restricts their circulation and destroys their status as the common heritage of mankind. If the resources become the exclusive property of a few corporations, genetic uniformity will increase substantially and society will have to pay the bill.

Seven, the food supply in Europe will be threatened by monopoly control over genetic resources, farmers' harvests and the processed results. Patent holders will have more power to decide what we eat. Hybridisation and patenting together make the monopolies stronger and stronger. Such excessive control over the food supply is extremely dangerous, as just a few integrated firms will dominate the sector.

Reason number eight has to do with the Third World. The Third World will increasingly lose access to scientific information and technology transfer, and will see their freely donated biological resources privatised by the North. Patenting life in Europe would also mean a total denial of farmers' rights in the South to compensation for all the work they do to provide the world economy with rich and useful genetic diversity.

Nine, the whole concept of human rights will be decimated as human beings and parts of their body can become the exclusive property of patent holders. That

217

corporations can own your organs, physical traits or intimate genetic information is a total denial of the individual right to an independent existence and to control over one's own body.

Ten, animal life will become a nostalgic notion of the past, as patenting stimulates the genetic engineering of animals to suffer. They serve industrial systems for the production of food and medicine. You already heard other speakers about this issue.

Eleven, society's relationship with nature will be reduced to a commercial enterprise, based on exploitation and profit. Patenting life means that some people can intellectually own the very foundations of living matter and life cycles, thereby undermining any respect for nature in our world. Biotechnology inventors do not create nature. They simply cut it into pieces and claim ownership over it. Such arrogance towards the world around us has already done tremendous damage and is a suicidal attitude towards the system that sustains us.

The twelfth and last reason why we should not patent living beings is that ethical and religious values, based on respect for life, creation and reproduction, will be thoroughly subverted. The patenting of genetic material forces upon us a reductionistic and materialistic concept of life as a mere collection of chemical substances that happen to be able to reproduce and can be manipulated and owned.

Note

1 Treated more indepth in the (Dutch) brochure of WERVEL 'Oranje knipperlicht voor genetisch gemodificeerde soja'.

26 R. Stephen Crespi

Introduction

Patent law exists to stimulate the advancement of technology. It is grounded in the assumption that new processes and products which provide material benefits and enhance the quality of human life in other ways are desirable and should be encouraged. But other ideas must also be given due weight if science and technology are not to lead to a sterile culture of consumerism. One of these is the ethical dimension of new technology, a topic that has captured the attention of many persons, public officials and committees, as well as the highly vocal protest groups whose activities can be frequently seen on our television screens.

Patent law is not silent on the question of morality. Morality has been brought into the patent law, at least from the time of the Strasbourg Convention (1963) onward, in terms which were later enshrined in European law through the European Patent Convention (EPC, 1973). The EPC route for European patents is the one most used by those who invest heavily in science and technology, especially biotechnology. Art. 53(a) of the EPC excludes from patentability any invention 'the publication or exploitation of which would be contrary to morality or ordre public'. The corresponding national patent laws contain essentially the same provision.

EPC Art. 53(a) has provided a convenient international forum for the Greens, animal rights campaigners and others to object to the grant of specific biotechnology patents. Most of the arguments in patent disputes can be resolved by factual enquiry and legal logic. Moral arguments are not so easily dealt with. The prevailing view in official patent circles, which is shared by the professional patent attorneys, is that it should not fall to those who administer the patent law to be the arbiter of moral disputes. Many commentators question how we are to decide the moral principles relevant for this purpose. It is difficult to answer this question in the diversity of ethical theories which vie for favour among moral philosophers today.

The morality of patenting

First it is necessary to avoid a common confusion. EPC Art. 53(a) is not concerned with intentions but only with the morality of actions. Leaving aside the publication aspect it is the act of actual exploitation of the particular invention to which the moral test must be applied. Whatever the ultimate intentions of the patent holder, the morality of the act of patenting does not come into the question. The belief that it is wrong to patent certain substances, organisms, or processes calls for an ethical judgement which is outside the patent law itself and therefore one which patent officials or judges cannot be called upon to make. Patenting an invention does not commit the patent holder to any particular course of action. Having a patent may be an encouragement to exploit an invention but exploitation can take place even without a patent, assuming other laws and regulations permit. But if one were to assert that biotechnology patenting is a morally dubious activity in itself the context in which it arises must be considered. First there is the research, which must be successful or there will be no invention to patent. Then comes the patenting step and after that the publicity and commercial exploitation. Unless we believe that it is wrong to do this kind of research, or that it is wrong to publicise and exploit the results of this research, or that both of these are wrong, it is a strange kind of ethical selectivity which focuses on patenting in isolation. Patenting, as such, is neither wrong nor right, but should be classed as ethically neutral. To refuse a patent would be a futile gesture which would not by itself stop the invention being put to practical use. On the other hand, if society were to judge that the practice of a particular invention deserved to be banned by law then no-one would bother to patent it. The various ways in which objections to biotechnology patents are being formulated are summarised below.

'Patenting Life' — the power of the slogan

There is a basic objection to the idea of 'patenting life' and thereby 'owning life'. Science cannot tell us precisely what life is but we ordinarily use the word as an abstraction from concrete living things. It follows that 'patenting life' is a meaningless notion since the law does not allow the patenting of abstractions. This unfortunately does not prevent 'patenting life' being used as an effective slogan to delude people into thinking that a sinister move is being made to monopolise the very stuff of life. This tactic has powerful emotional force. If 'life' can be patented and owned, the objectors say, it is being treated as a 'commodity' (or is being 'commodified'). Were the objection to be framed in a more honest way as one against the patenting of living things it would immediately lose most of its force since living things can indeed be commodities, such as yeast for bakers and brewers and edible plants and animals for farmers. Making this statement in no way commits one to a 'reductionist' philosophy which views living things as mere objects. Realising, of course, that farmers actually do own the plants they have grown and the animals they have bred and reared, the objectors draw a distinction between owning specific plants and animals and the ownership of 'whole species' which they claim is achieved by

220

patenting. But a species is a biological classification and not an entity in its own right. The same incidentally can be said of the plant 'variety'. These terms are also abstractions; no-one ever meets a species or a variety but only individual members of these groups. Furthermore, biological classification systems have not yet accorded species status to transgenic plants and animals.

Patenting micro-organisms, plants, and animals

Those who object to the patenting of living matter will not allow any exception to their rule. The objection to patenting micro-organisms is usually merged into the general condemnation of patenting 'life' which, it is claimed, implies a failure to respect life. This claim is entirely nebulous. Society cannot have bread, wine, antibiotics, and vaccines without the industrial use of micro-organisms. These products are wholly good and to produce them more effectively by means of improved and patentable strains of micro-organism is also meritorious. Patenting the organisms harms no-one.

Neither does the morality argument sit comfortably when applied to the patenting of plants. Usually the morality objection slides into one of challenging the use of genetically manipulated plants on grounds of public safety. Important though the question of safety undoubtedly is, it is primarily a matter of complying with other laws and regulatory procedures which have been established as matters of public policy. It would doubtless be immoral to act irresponsibly in matters affecting public health but the legal control on this has nothing to do with patents.

The morality argument appears more pertinent when it comes to the patenting of animals. Here we are dealing with a sentient being and we have to ask whether that makes a difference to the moral issue. As one American journalist put it, when in April 1988 a US patent was granted on the Harvard mouse, 'Last month the Government granted its first patent on something that can look you in the eye'. It is necessary first to address the metaphysical objection that an animal should never be reduced to the status of an 'invention' let alone a patentable one. This objection is strongly expressed in the writings of Oxford theologian Andrew Linzey.[1] The term 'invention' in patent law has to serve for a wide variety of items which are presented as new and improved over what is already known and used, i.e. over the 'prior art'. Some inventions are startlingly new in their own field but most are improvements or modifications of existing processes and products. All that concerns the patent law is whether their difference from the prior art required inventiveness to accomplish or was 'obvious' to the skilled person. In the case of a transgenic animal it is this difference from the prior art upon which the legal examination focuses. Since this difference does not occur naturally it is reasonable to describe it as a work of human ingenuity, as man's handiwork rather than that of Nature. The US Supreme Court made this point in 1980 in approving the grant of the first patent on a genetically manipulated strain of bacterium (the Chakrabarty pseudomonas) and it is equally applicable to the higher life forms. In opposing the grant of the patent in the Chakrabarty case, the US Department of Agriculture had argued that US 'utility' patents were intended only for inanimate inventions. The Supreme Court rejected this narrow view and pointed out that the distinction was 'not between living and inanimate things but between products of Nature, whether living or not, and human-made inventions'. In the case of a transgenic animal, no-one undervalues the work of God or Nature which provided the starting point, nor does the inventor claim to have 'created' the whole animal. Of course the patent has to cover the modified animal as a whole since the difference cannot be used in isolation. Animal transgenics involves the genetic modification of two main types of animal, laboratory test animals and farm animals each of which requires its own special treatment from the moral point of view.

As to laboratory animals, in the Harvard Oncomouse example the animal is genetically programmed to be more sensitive to carcinogens. In this respect the Oncomouse is superior to previously used types of laboratory mouse, which may mean that scientists can get away with using fewer mice than before. It is nevertheless an emotive case which reinforces the dilemma over the general question of sacrificing animals in the quest for cures for human diseases. The moral argument against animal experimentation is that the animals are used as mere tools for human ends and in ways that involve suffering. The animal's pain and in some experiments the indignity for the animal are distressing to many persons. But if this is a real moral issue for humans, it has nothing to do with the question of patents for improved test animals. It is the fundamental issue that society must settle rather than the side issue of patenting. Patenting is under attack because the system presents an easy target and gives the objectors a platform for much publicity. But the patent officials have to decide the question against a background in which the practice of breeding and using laboratory test animals is widely accepted as permissible and indeed necessary if medical science is to advance. If that practice is not immoral per se the objectors have the burden of convincing the authorities that it becomes immoral by the introduction of genetic manipulation or some other technique of biotechnology in the breeding process.

The second category of transgenic animals concerns farm animals. It is possible to have reverence for animals as part of Creation, as did Francis of Assisi, without necessarily being a vegetarian. It is also possible for non-vegetarians to be concerned at the conditions under which animals are industrially reared by intensive farming methods and how they are transported. These callous practices charge the emotional atmosphere but have little or nothing to do with the subject of transgenic animals and patenting. In Britain, ethical issues involved in various aspects of this technology have been addressed by two public Committees.

The first of these, the Polkinghorne Committee,[2] was asked to consider 'the moral and ethical concerns (other than those related to food safety) that may arise from the use of food products derived from production programmes involving such (transgenic) organisms'. The Committee did not address wider ethical matters but did discuss the idea of 'moral taint' attaching to foods produced with the aid of genetic modification and deriving from the 'unnaturalness' and any animal suffering that might be involved. In considering the element of unnaturalness the Committee identified, as the key factor, 'the status of the human genetic material inserted into the farm animal'. The committee distinguished beween human genes as existing in the human body and 'copy genes of human origin'. The procedure for cloning the gene results in an enormous dilution effect (10 55) so that the inserted material is overwhelmingly in the form of synthetic copies of the original. If mRNA cloning is used literally nothing remains of the original gene or message. Accordingly the essential phenotype of the recipient species is not altered. Basing itself on these essentially scientific facts the Polkinghorne committee rejected the idea of 'a moral taint that would warrant a total prohibition on genetically modified food use'.

The second committee, the Banner Committee,[3] considered the ethics of novel farm animal breeding techniques. The committee asked whether there are 'intrinsic' objections to certain types of genetic modification which cannot be overcome by consequentialist arguments. According to the consequentialist view of ethics the morality of an action is to be judged (only) by its consequences. This usually entails a weighing of the benefit against the harm resulting from the action. An intrinsic objection is one not related solely to the consequences of a particular practice or action but to the practice or action itself. In genetic modification the Banner committee considered that there is a line between what is acceptable and what is not. Whilst this cannot be drawn sharply, the committee felt that 'harms of a certain degree or kind ought under no circumstances to be inflicted on an animal'. One prominent theme in this connection is that of naturalness, contrasted with modifications which threaten the achievement of the animal's 'natural ends or good' or do not respect its 'essential nature and well-being' or its 'natural characteristics and form'. For example, the use of gene technology to increase the protein content of cow's milk or to cause poultry breeding stock to produce only female chicks could be achieved whilst still respecting the essential nature and well-being of the animal. These were therefore not considered intrinsically objectionable modifications. However, to increase the efficiency of food conversion in pigs by reducing the sentience and activity of the animal was to disregard the ends and purposes which are natural to pigs and was intrinsically unacceptable. Asking whether there might be reasons for prohibiting patents in this field, the Banner committee rejected most of the anti-patent arguments. However, having concluded that some instances of genetic modification are acceptable and some not so, it felt that the patent law should be able to discriminate between these. Having passed this heavy burden to the patent authorities the Banner committee went on to dismiss the provision in the proposed EU Directive on this subject (see later) which would have denied patents for 'processes for modifying the genetic identity of animals which are likely to cause them suffering or physical handicaps without any substantial benefit to man or animals, and animals resulting from such processes'. The committee were not convinced that this was a workable and appropriate provision since it left so much open to question. This Directive was voted down by the European Parliament in March 1995 but a revised version has been re-presented by the Commission in December 1995. In Art. 9, 2(b) of the new draft the above provision has been retained, but with the added wording: 'insofar as the suffering or physical handicaps inflicted on the animals concerned are out of proportion to the objective pursued'.

Patenting material derived from human tissue

Patents have been granted on materials isolated from human tissues and some have been challenged, including the following two examples: the Moore case and the Relaxin case.

John Moore has expressed concern over the fact that cells taken from his spleen have been 'immortalised' and patented by his physician and the University of California. John Moore's cancerous spleen had been removed by his physician who then, apparently without his patient's consent, researched its cell content and developed a useful and patentable cell-line. The patent (US 4,438,032) covers a single cell suspension of the Mo cell line and various proteins produced from it. John Moore sued both the physician and the University. On the intellectual property ownership issue Moore's claim was not upheld by the California Supreme Court. The court decided that he was not entitled to claim ownership of the intellectual property in the cell-line. The patented cell line, though developed from the spleen cells, was held to be different from the cells actually removed from the patient. Moreover the court was of the view that to grant to the 'donor' of an organ the intellectual property in anything developed from it would inhibit research of this kind. John Moore nevertheless continues to believe that others have patented and now 'own his genetic essence'.

As to the Relaxin case, the Green Party opposed European patent 112,149 granted to the Howard Florey Institute of Experimental Physiology and Medicine for a gene sequence coding for human relaxin, a hormone involved in reproduction. The gene was isolated from ovarian tissue removed in the treatment of an ectopic pregnancy. One of the Opposers' arguments was that it is 'an offence against morals to exploit the pregnancy condition of a woman by removing tissue from her ovary and using it as the basis of a profit-oriented technical process'. The Opposers also raised the bizarre objection that the use of the ovarian tissue for the purposes of the patent 'involves dismemberment of women and their piecemeal sale to commercial enterprises'.The Opposition Division roundly rejected these objections.[4] The tissue had been provided with patient consent; it was perfectly acceptable to use such material; DNA is not 'life'; a human being cannot be reconstructed from the sum total of human genes; and many other reasons were given for dismissing the arguments as thoroughly misplaced. Are there genuine reasons for concern? It is hard to see what real harm or injury to the original donor of the tissue is caused by the patenting of a cell-line or a gene derived in these circumstances or, indeed, what reason exists for any repugnance to this sort of research and development. And yet, the arguments in the John Moore case were apparently prominent in the last-minute struggle to survive of the proposed European Council Directive on the legal protection of biotechnological inventions. With the John Moore case in mind, the Conciliation

Committee, set up jointly by the European Council and Parliament, inserted various Recitals into the final text of the Directive,[5] for example Recital 12 reading as follows:

> Whereas, in the light of the general principle that the ownership of human beings is excluded, the human body or parts of the human body as such, for example a gene, protein or cell in the natural state in the human body, including germ cells and products resulting directly from conception, must be excluded fom patentability; whereas, however, an invention incorporating industrially applicable parts obtained in a technical manner from the human body in such a way that they can no longer be ascribed to a particular individual may not be unpatentable because of the human origin of such parts, even where the structure of these elements is identical to a part of the human body, it being understood that parts of the human body from which such parts are derived are excluded from patentability.

The ethical reasoning on which Recital 12 was based is strange indeed. First, what possible connection could there be between, say, the abolition of slavery and the patenting of a DNA sequence? Secondly, the practical circumstances in which anyone could conceive of patenting, or would want to patent, anything in situ in someone's body are (to say the least) difficult to imagine. Thirdly, how anything isolated from a particular individual (e.g. John Moore), and replicated, sub-cultured, and adapted for use in an industrial process, could possibly remain linked in some way to that individual stretches the imagination to breaking point. This recital is not to be found in the Commission's currently proposed text. Furthermore proposed Article 3 now reads: '1. The human body and its elements in their natural state shall not be considered patentable inventions' and '2. Notwithstanding paragraph 1, the subject matter of an invention capable of industrial application which relates to an element isolated from the human body or otherwise produced by means of a technical process shall be patentable, even if the structure of that element is identical to that of a natural element'.

Ethical views of various organisations

As to ethics in science and medicine, the 'consequentialist' view of morality does seem to be the 'establishment' view of science and medicine. For example the use of experimental animals is justified in the following exemplary quotation:[6] 'The use of living animals in scientific research can be considered justified if it is likely to produce appreciable benefit to society, if there is no other way to conduct the research in question and if all reasonable steps are taken to keep any distress or suffering to a minimum'. A very large percentage of the public would probably agree with this position. It also appeals to lawyers. When the EPO Appeal Board in the Harvard Oncomouse case instructed the Examining Division to consider the morality issue under Art. 53(a) it suggested that the decision 'would seem to depend mainly on a careful weighing up of the suffering of animals and possible risks to the environment on the one hand and the invention's usefulness to mankind on the other'. The EPO decided that the likely benefit to cancer research outweighed the other factors and granted the patent. This 'balancing principle' is criticised by the Opponents of this patent who say that an intrinsically immoral act cannot be justified by the fact that some good consequences ensue. This argument rests on the unproven first premise that it is immoral to alter the genes of animals and especially so if it results in animal suffering.

As to the official and patent professional views, the professional Institute of European patent attorneys has stated its view on the subject from a practical standpoint. EPI support the official view of Art. 53(a) as expressed in the European Patent Office Guidelines, where it is stated that 'This provision is likely to be invoked only in rare and extreme cases. A fair test to apply is to consider whether it is probable that the public in general would regard the invention as so abhorrent that the grant of patent rights would be inconceivable.' EPI also endorse the view of the EPO Opposition Division[7] that 'only in those very limited cases in which there is an overwhelming consensus that the exploitation of an invention would be immoral may an invention be excluded from patentability under Article 53(a)'. As to the test suggested by the Appeal Board, which is evidently to be used as a standard for transgenic animal cases, EPI observe that it will often be very difficult if not impossible for patent officials to assess the benefit against the harm of particular inventions, since these may only be properly revealed in the longer term and not at the early stage at which patent applications are examined. EPI much prefers the 'overwhelming consensus' test as the ultimate criterion for the purposes of Art. 53(a) and sees the balancing principle as just one factor contributing to the assessment.

The Nuffield Council on Bioethics has addressed patent issues in chapter 11 of its recent report.[8] This is a lucid summary of the issues involved in patenting inventions derived from human tissue and it discusses the following options for resolving problems. One option is to remove the morality issue from the remit of the Patent Office and to leave it to national courts. The Council notes the disadvantages of this option, especially the likelihood of different national treatment and the resulting confusion. To this commentator, the legal mechanism for this option is far from clear for patents granted under the EPC. National courts do not become competent to deal with EPC patents until these have left the jurisdiction of the EPO. It would be a most strange provision in the EPC itself which could not be raised by the EPO either during normal examination of the patent application or in Opposition proceedings after the patent has been granted.

Another option is the 'light approach', favoured by two former Comptrollers of the UK Patent Office. This is a variant of the first option in which the Patent Office invoke the objection only in extreme cases, leaving it to national courts to sort out remaining problems. This corresponds to the present attitude of the EPO in examining patent applications. How it would work in the face of a determined third party Opposition to the patent is far from clear.

A third option is to transfer the responsibility from the patent examiners to a specially constituted ethics committee which would become part of the official process either before or after grant of the patent.

The option most favoured by the Nuffield Council is for the creation of a protocol to the EPC setting criteria for national courts in applying the immorality exclusion. This might avoid the problem of differing national treatment but, to this commentator, it is open to the same difficulty as raised against the first option. Bearing in mind the predominating role of the EPC in the European patent scene, any scheme which retains the immorality exclusion in the law of the EPC but aims somehow to exclude or mitigate its rigorous application during the European phase of the patenting process may not be viable.

The Council of Europe has prepared a 'Draft Convention for the protection of Human Rights and Dignity of the Human Being with regard to the Application of Biology and Medicine: Bioethics Convention'.[9] From its very title and general character this Convention is concerned with matters affecting the human being at the personal level. Provisions relating to the disposal of human tissues, organs, and other body parts are therefore subsidiary to its main aims. Art. 21 (Prohibition of financial gain) states that 'The human body and its parts shall not, as such, give rise to financial gain.' The Explanatory Report makes clear that this article is particularly concerned with organs and tissues removed from the human body and the sale of these for 'financial gain for the person from whom they have been removed or for a third party, whether an individual or a corporate entity such as a hospital'. In a draft Opinion of the Reporting Committee (Committee on Science and Technology Document 7210), Art. 21 is considered as extending to the patenting of the human body or its parts (at section D21, p. 12, where comment is also made touching upon the distinction between discovery and invention in patent law). The extension of Art. 21 into the sphere of patent law goes beyond the proper remit of the Convention and is unwarranted. Patent law issues which arise in the field of industrial processes and products do not conflict with the protection of human rights and dignity. The opinion expressed concerning discovery and invention is over-simplified and contrary to the view prevailing in both official patent circles and private circles. According to paragraph (10) of the Explanatory Report the scope of the Convention is restricted to human medicine and biology and excludes plant and animal biology 'insofar as they do not concern human medicine and biology'. In accordance with its objectives the Convention quite properly addresses medical intervention procedures carried out on the human being. It should not venture into industrial biotechnology.

Other objections

Those who object to biotechnology patents do not restrict themselves to raising moral objections. They are also ready to deploy patent law arguments if it suits their case. This can be seen in their criticism of gene patents.

Patenting genes

Discovery or invention?

The law says that inventions are patentable but discoveries are not. The objectors argue that because genes exist in Nature they cannot be invented but only discovered. The distinction between discovery and invention is difficult to define in any of the sciences of Nature because the act of discovery so closely underpins the resultant practical application which constitutes the invention. As ethicist W. Ch. Zimmerli has put it,[10] 'every scientific discovery, if made technologically applicable, becomes an invention'. The official patent authorities are not prepared to dismiss the isolation of genes as mere discovery.

The question of novelty

Another argument against patenting genes is that, because of its pre-existence, a gene cannot fulfill the patent law test for novelty. But this test is framed in terms of availability to the public. Thus it focuses only upon what is already in the public domain through public disclosure or use prior to the filing of the patent application. Genes do not easily fit into this scheme. To be made available to the public the gene must first be isolated, preferably characterised as to its nucleotide sequence, and cloned. The contribution to the art on which gene patents are based is the making of the gene available in a form which can be utilised to produce an expression product, and to produce this in quantity, for example, as a commercial pharmaceutical product. Alternatively the cloned gene can be used to transform an organism of another species giving rise to new products, e.g. transgenic plants and animals. Genes are therefore a special case of the broad class of naturally occurring substances which in appropriate circumstances can be patented. Mere pre-existence of the substance, in association with vast quantities of other materials, is insufficient to overcome the objection. This is the official legal view.

The question of inventive step

Those who describe DNA sequences as 'mere discovery' are now latching on to the argument that however difficult gene cloning was in the early 1980s it is becoming routine in the laboratories of today. Whatever may be the state of current general skill and knowledge in relation to these techniques, this argument is only relevant to the question of inventive step i.e. as to how much ingenuity is required to isolate genes. This evaluation must always be made case by case and no broad general dismissal is justified.

In answering this question patent officials have recognised the distinction between the genes of lower organisms and those of higher life forms. They have noted that, whereas prokaryotic genes are transcribed in full, the messenger in eukaryotic gene transcription is an edited form of the primary RNA transcript. Consequently the DNA made by enzymatically copying mRNA into cDNA is not the natural gene as such. In summary, the afore-mentioned objections to the patenting of genes have so far been singularly unsuccessful when applied against European patents.

To conclude, it is undeniable that biological research throws up profound moral questions, especially in the field of human reproductive technology. But the relevant debates should begin at the right point, with the research and its potential uses, and not with the wholly subordinate question of patents. The patent issue is no shortcut to the resolution of the more difficult questions, and until these have been properly addressed, patenting should not be singled out for attack. Removal of the current morality provisions from the patent law is not a practical possibility in the foreseeable future. It is unfortunate that the legislators have put these provisions in such an inappropriate place but those responsible for administering the law should interpret the exclusions narrowly and not allow them to be used for disruptive purposes. Judging from recent experience, attempts to put specific provisions into the law as to what particular types of invention are to be unpatentable are likely to be the subject of interminable controversy. At present the debate is provoked to suit the agenda of the Greens, Animal rights campaigners, and their supporters. Once started there is no reason why it should not extend into other areas of moral dispute. For example, the question of patenting abortifacient raises for many people far more important moral questions than those of animal welfare. Pandora's box is better kept shut.

Biotechnology, patents and developing countries

The problems of the developing countries are a complex subject embracing issues far beyond the scope of the present discussion. On this broad canvass patents are of peripheral importance. Nevertheless, many of the spokespersons for developing countries insist that patents are bound up with the North/South gap in wealth and technology and therefore cannot escape from the general condemnation of this huge imbalance. Some Europeans also believe this. For example, in the discussions between the European Commission and the EC member states over the biotechnology Directive, the Danish representatives pressed the Commission to consider the possible impact of the Directive on developing countries. Considering that the Directive was addressed only to the national laws of EC countries it was never clear in what way the developing countries could be affected by it. In attempting to defend the patent system from this particular indictment, patent attorneys can only skim the surface of the subject by exposing some of the more glaring fallacies. Two types of charge can be distinguished, though they are usually intertwined in the prosecutor's case. First is the charge of exploitation by the West. Second is the claim that genetic diversity in the poorer countries has over centuries been seriously eroded and that biotechnology patenting will reinforce and accelerate this trend.

Exploitation

In one version of this charge the 'gene-prospecting' scientists are depicted as swooping down on the more biodiverse regions of the world, taking what they want without payment of any kind, and departing with their plunder. This is a fictional account designed for popular consumption. Developing countries are undoubtedly a rich source of natural medicines, insecticides and other bioactive substances which are the basis of traditional remedies and even products of local industries. But there is often an extraordinary reaction on the part of some commentators to any attempt by organisations in industrially developed countries to create and then patent technology relevant to natural products originating in the less developed countries. The furore over the Neem tree is a prime example of this reaction.

A corrosive attack on patenting in relation to the Indian Neem tree has been made by V. Shiva and R. Hollar-Bhar.[11] Azadirachtin is one of many active compounds present in bark, leaves, flowers and seeds of the Neem tree, Azadirachta indica. The remarkable properties of this compound have been utilised in India from ancient times in the form of extracts of various kinds produced by Indian farmers and small industrial firms. In recent years the US pharmaceutical company W.R. Grace and others have examined this source of potent therapeutic agents and have taken out US patents on their work. Serious objections of principle (as well as specific challenges to novelty and inventiveness) are raised against these patents by the above authors. A search reveals that these patents are essentially patents for particular extraction processes, formulated products, and synthetic derivatives. Also revealed is the fact that no corresponding applications have been filed in India or any other developing country. It is therefore impossible to see how patents of this sort can be fairly characterised by Shiva et al as 'intellectual piracy', 'expropriation of indigenous experimentation', 'controlling access to raw material', and 'making farmers' seed stock and natural pesticides into the intellectual property of multinational companies'. Even if corresponding Indian patents existed these would not interfere with the freedom of farmers and peasants to continue to use traditional methods.

'Piracy' of genetic resources

There is no doubt that the cultivated plants of the developed countries owe a great deal to traits which they have acquired by transfer from the germplasm of developing countries. This fact must of course be counterbalanced by recognition of what Western science and technology have achieved as a result of substantial investment in the development of these genetic resources. Two examples of this will now be given. Modern wheat cultivars have been made resistant to the fungal disease 'eyespot' by the transfer of a gene from the North African wild grass Aegalops ventricosa. Eyespot was a fungal disease prevalent in the cooler maritime climates of Western Europe, North America, and parts of South America but not in the North African coastal regions where the grass originated and still flourishes. This is therefore an example of genetic material derived from an underdeveloped area in which it has no special importance or use but can be utilised to improve the crops of other countries. This important achievement of French scientists took place around the time that legal systems of plant breeders' rights were coming into existence (the early 1960s) and it is probable that the resulting new varieties would have been more suited to this type of protection than to patents. In a more recent example, a gene responsible for producing a trypsin inhibitor in the cowpea (Vigna unguiculata) has been isolated by British scientists of the University of Durham for transfer to other genera of plants. The cowpea is a legume, also called 'black eyed bean', which is grown as a food crop in West Africa and in both North and South America. Resistant varieties of this plant produce a trypsin inhibitor which prevents the invading insect from digesting protein so that it dies from starvation. Transfer of the inhibitor gene to other plant genera requires the methods of plant biotechnology and cannot be achieved by traditional breeding methods. The technology is aimed at protecting cotton and cereals against bollworms which affect these crops throughout the American and African continents. It is applicable also to protect grain of wheat, maize, rice and sorghum against storage pests some of which are a particularly serious problem in Africa, India, China and Japan. In these quoted instances, which are not untypical, the valuable genetic material has been obtained from developing countries, but has left the original source material unharmed and available to fulfil its original purpose. Certain fears have nevertheless been expressed over developments of this kind. First the genetic material, if patented, is said no longer to be freely available. But until the inventor achieved the isolation of the gene, the gene was available only in the form of the original plant and in this respect nothing has changed. The new transgenic plants are of course not freely available, while the patent lasts, because the patentee will expect some share in the benefits which he has originated. But until the inventor completed this work these plants were not available either. The following possible pattern of creation and exploitation of this type of technology has been suggested previously.[12] A biotechnology research group in a scientific research institution or in an industrial research laboratory will isolate the gene from the germ plasm of the source country, and will patent the gene sequence (usually as cDNA) and the method of gene transfer to the plants targeted for protection. The patent owner will be free to develop and exploit this technology commercially on his own behalf. But it may be preferable to license the technology to commercial plant breeders in industrially developed countries and to appropriate organisations in developing countries e.g. state-run agricultural research institutes together with the know-how to transfer the gene to chosen types of plant. The plant breeders or research institutes may obtain plant breeders' rights for any resulting varieties. The new varieties will be sold to farmers who will cultivate them and, as a result of their improved pest resistance, will be able to economise in the use of chemical pesticides. The public will benefit from the advantages to the environment resulting from this technology. It is difficult to see who will not gain from this achievement. Unless the transgenic plant enables the farmer to achieve a better yield or a saving on the use of insecticides it will not be worth the higher price and it will not be purchased.

Agreements for transfer of natural resources

The claim that the rights of developing countries are ignored by the 'gene prospectors' from developed countries is simply untrue. Many efforts to regulate material transfers are underway such as the following.

Under the heading of European Community initiatives, it should be mentioned that as part of the mandate of the Concertation Action in the Biotechnology Programme (CUBE) of the European Communities, 12 March 1985, Task 6 was concerned with 'Promoting in co-operation with developing countries and relevant institutions the safe and sustainable exploitation of the renewable natural resource systems within their respective regions'. Since 1986, Biotics Ltd, a UK private company, has been supported by the European Commission to promote the phytochemical screening of developing country flora by industrial and other specialised European research organisations. In November and December 1988, Biotics Ltd reported[13] on the establishment of mutually beneficial agreements involving West African, South-East Asian, and Australasian source countries. Several thousand plant samples have been handled under this model Agreement. The participating countries will receive royalty or equivalent returns linked to the commercial development of derived products. Initial payments have been made for the supply of samples but royalty payments must await successful commercial development which may take some years to accomplish. This programme has attracted the interest of the United Nations Industrial Development Organisation (UNIDO), the US National Cancer Institute, the World Wide Fund for Nature, and the International Board for Plant Genetic Resources.

Other international initiatives, so-called Material Transfer Agreements (MTA), have also been under way for some time. For example, the United Nations Food and Agriculture Organisation (FAO) has produced an 'International Code of Conduct for Plant Germplasm Collection and Transfer'.[14] The devising of various forms of model Agreement has also been under study on behalf of the International Plant Genetic Resources Institute. These two selected examples show that the subject has been active for some years prior to the enactment of the Biodiversity Convention (to be outlined in a later section). Opponents of 'bioprospecting' have compiled a list of current industrial activities of this type[15] which make it clear that these are being carried out by agreement with local Governments or other organisations in developing countries.

In an Overseas Development Institute paper[16] the following opinions are expressed:

> ... the existence of more extensive intellectual property protection for plant genetic resources could in the long run have a substantial impact on global biological diversity ... Agriculture in less developed countries could suffer if free access to plant genetic resources ends ...

The ODI paper unfortunately does not define what is meant by 'plant genetic resources'. This term is normally taken to mean the germplasm of all existing plants. It is a basic consequence of patent law that these resources, being already in the public domain, cannot be taken from it and made the subject of intellectual property rights. The ODI paper admits that loss of plant genetic resources is:

> ... closely bound up with agricultural development itself, particularly with the introduction of new varieties ... as farmers replaced their many traditional varieties with a few introduced ones ...

The above effects are presumably a consequence of traditional plant breeding and cannot be attributed to biotechnology still less to plant patents. It is difficult therefore to see justification for the speculation that 'the granting of patents on plants involves the risk that access to a common pool of plant genetic resources essential to plant breeding is likely to become restricted.' A major investigation by the Organisation for Economic Co-operation and Development (OECD)[17] made with the help of a distinguished international team of scientific experts, concluded that:

> ... the modification of existing germplasm by valuable genes or gene combinations from whatever source is essential to achieve the necessary adaptation of existing plants to meet future targets of the developed and developing countries ... There is an urgent need to bring the cultivars of non-OECD countries up to the production standards of OECD countries in terms of yield and efficiency ...[18]

The OECD goes on to explain the many problems to be overcome to achieve this. Patents do not figure directly in this diagnosis. Against negative views of the kind expressed in the ODI paper, Western authorities stress the working of the free economy in which new technology competes on price and other factors with existing technology and also supplies competitive alternatives within its own sphere. A good example of this approach is that of J. Duesing (formerly of Ciba-Geigy).[19] Duesing notes the stimulant effect of patents in the search for diverse solutions to particular problems, e.g. at least five strategies for achieving virus resistance in plants are currently under evaluation. Duesing is writing in the context of the Biodiversity Treaty, which he sees as a positive development. The following summary and assessment of this Treaty closely relies on Duesing's own analysis.

233

The United Nations Convention on Biological Diversity, known for convenience as the Biodiversity Convention, was enacted in Rio de Janeiro in June 1992. The Convention was signed by 157 Governments and more have signed since then. At present, the Convention has been ratified by 117 States, including the European Community and its Member States, Canada, Japan, and the Russian Federation. The United States signed with reservations and has not yet ratified the Convention. The prime purposes of the Convention are to ensure conservation of biological diversity, sustainable use of its components, and the fair and equitable sharing of the benefits arising out of their utilisation (Art. 1). The Convention recognises that conservation must be balanced with development, particularly in developing countries. Genetic resources have in the past been declared 'a common heritage of mankind to be preserved, and to be freely available to all, for use for the benefit of present and future generations'.[20] This principle has now been superseded by the Convention.

The term 'genetic resources' signifies the totality of the germplasm of existing plants and animals in public use or otherwise in the public domain throughout the world e.g. in public germ banks. The question of rights deriving from the ownership of source material e.g. human cells from an individual person or plant tissue from a particular geographical location is not addressed directly by patent law. Patent law is essentially concerned with inventions which are the outcome of research carried out on or with the source material. The source material itself, in its natural state, cannot be patented, but the isolation of an active principle or a genetic component of the natural material leading to a new and useful application, e.g. medical or agricultural, can be the subject of a patent.

The Convention recognises the sovereign rights of States over their natural resources. Therefore national governments have the authority to determine access thereto (Art. 15, §1). The Convention provides that, in return for providing access to its genetic resources, a donor country should benefit through any of three mechanisms: participation in research (Art. 15, §6), sharing in the results of research and proceeds of commercial exploitation (Art. 15, §7) and access to and transfer of derived technology (Art. 16, §1). According to Art. 15, paragraphs 4, 5 and 7, access and sharing are to be dealt with 'on mutually agreed terms' and 'subject to prior informed consent'. Therefore access to genetic resources must be preceded by negotiation as to the form in which benefit to the donor country is to be achieved.

According to Art. 16, §§ 1 and 2, it is for the Contracting Parties (national governments) to provide for access to and transfer of technology which makes use of genetic resources. Art. 16, §3 requires Contracting Parties to 'take legislative, administrative, or policy measures' to achieve this objective, particularly where a developing country provides the genetic resources. Also, by Art. 16, §4, such measures must have the aim that the private sector facilitates joint development with, and technology transfer to, governmental institutions and the private sector of developing countries. However laudable these objectives may be, it is not immediately apparent how to implement them legislatively in the private enterprise systems which predominate in industrially developed countries. As noted by Duesing, 'Governments will need to consult with all sectors and interests, private and public, to determine acceptable forms of co-operation in specific industries'. As mentioned above, much is already being done toward these ends at non-governmental level. Technology transfer may be achieved by a variety of mechanisms. It will usually include the licensing of some form of proprietary right obtained either under an established statutory form of intellectual property or deriving from the possession of secret know-how and/or proprietary biological material. In Art. 16, §§ 3 and 5, the Convention recognises that the technology to be transferred may be the subject of patents and other intellectual property rights. The existence of an adequate system of intellectual property rights in the donor country will not of itself ensure that technology transfer takes place. However, as it provides an orderly method of achieving such transfer and of controlling unlicensed and unfair competitive activity, it will offer a strong inducement to the whole process of investment in research and development of the genetic resource and in the subsequent exploitation of the derived technology.

As to farmers' rights, although not formally dealt with in the Biodiversity Convention, the concept of 'Farmers' Rights' developed by the FAO[21] should be briefly mentioned under this heading. In the FAO resolution:

Farmers' Rights means rights arising from the past, present and future contributions of farmers in conserving, improving, and making available plant genetic resources, particularly those in the centres of origin/diversity ...

P.R. Mooney and H. Hobbelink, two of the most seasoned anti-patent campaigners, and others[22] continually bemoan the fact that existing systems of intellectual property make no provision for such farmers' rights. Their thesis is that commercial agriculture concentrates on the best plant varieties and thereby tends to increase genetic uniformity with loss of biodiversity. This trend, they maintain, is encouraged by biotechnology which is in turn supported by the patent and plant variety right systems. Their case assumes that agriculture would take another direction were it not for these pressures. These critics give little weight to the world-wide systems of gene banks set up to preserve germplasm, or to the Biodiversity Convention, or to any other public initiatives taken over the last three decades to address the problem. They are inconsolable because the primitive farming and rural communities of undeveloped countries are outside current ipr systems and receive no recognition for their own 'intellectual property' (sic) in preserving their diverse varieties over generations. It is not clear how farmers' rights are to be given practical expression by means of a formal legal system. Those who argue for this to be done have so far not come forward with a practical framework for such a system. It is virtually impossible to see how this concept could be grafted on to intellectual property law and it will therefore require the creation of a special kind of system (sui generis) as envisaged under the GATT Agreement.

Concluding remarks

The scientific exploration of Nature benefits mankind through discoveries concerning animate and inanimate matter. Where such discoveries are the basis of inventions which can be developed into industrial processes and products there is no reason in law or of principle to deny patent protection for such inventions. The living nature of the materials involved imposes no general restriction on what may be patented. Genetic material, isolated from a natural source or modified artificially, is likewise not excluded where it can be applied to a useful purpose. Genetic material is a resource which exists to be utilised for human welfare. Where this achievement requires the exercise of ingenuity, the award of a time-limited measure of legal protection is appropriate. However dogmatically it may be stated, it is by no means established that a causal link between patenting and genetic erosion exists now or will be created as and when transgenic plants come to the market. For a patent to be granted, there must be an invention, which means that some new and inventive product or process has been brought into existence, using existing living or genetic material as a springboard for the particular innovation. Genetic erosion stems from human choices over a wide range of factors having little or no connection whatsoever with intellectual property. Objections purportedly based on ethical principles have also been examined and found wanting. The number of 'ethical committees' being set up to deal with the wide range of issues involved in modern biological research is on the increase and many of them will want to include the patent dimension within their remit. So far, no substantial case against patenting has been made out. Most patent attorneys argue that the morality issue should be removed from patent law because patent examining authorities and courts are not specially equipped to make moral judgements binding on applicants and patentees. This argument is valid but unlikely to receive sympathy from official circles in the foreseeable future. Not all attorneys are in favour of abolition of EPC Art. 53(a). The problem is how to apply Art. 53(a) in the absence of objective criteria for the resolution of disputes over moral issues. A test based on the balancing of good versus harm, as suggested by the Appeal Board in the Oncomouse case has a strong appeal to the Anglo-Saxon mind and is one of the ethical norms of science. But it is not universally accepted and is challenged by the Opponents of the Oncomouse patent (the end does not justify the means). The burden of proof lies with the protestor. Grounds must be given in support of any claim that a particular invention offends against Art. 53(a). An appeal to popular opinion can only be conclusive if the consensus is sufficiently clear and strong. The intensity of opinion held by a dedicated minority cannot be decisive.

It is a commonly held opinion that patent law is a highly complex subject best left to specialists and other rare individuals who find it congenial. As regards procedural detail this judgement has some foundation but, as to basic principles, patent law is in fact built upon a few simple concepts. These basic ideas, or criteria of patentability, are fourfold: to be patentable an invention must be both new and inventive over what is already known; it must not be a purely intellectual discovery but must have practical application (in US law this is referred to as utility whereas in European law the test is one of industrial applicability); finally, the inventor must provide adequate written instructions to enable the invention to be performed by the relevant skilled person. For various reasons of public policy some kinds of invention are specifically excluded from patentability.

Notes

1   Linzey, Andrew (1993), 'An end to Aristotle', *Chemistry & Industry*, 17 May, 376, and elsewhere.
2   *Report of the Committee on the Ethics of Genetic Modification and Food Use*, Ministry of Agriculture, Fisheries and Food, London, HMSO, 1993.
3   *Report of the Committee to Consider the Ethical Implications of Emerging Technologies in the Breeding of Farm Animals*, Ministry of Agriculture, Fisheries and Food, London, HMSO, December 1994.
4   Relaxin, Official Journal EPO 6/1995, 388.
5   EC Directive Joint Text approved by Conciliation Committee, 21 February 1995, PE-CONS 3606/1/95 REV 1.
6   Lord Adrian, President, Research Defence Society, at Royal Society Conference, London, 26 April 1991.
7   Plant Genetic Systems - EP 242,236, Opposition Division Decision.
8   Nuffield Council on Bioethics, Nuffield Foundation, 28 Bedford Square, London WC1B 3EG.
9   Council of Europe, Directorate of Legal Affairs (1994), *Draft Convention for the Protection of Human Rights and Dignity of Human Beings with regard to the application of Biology and Medicine: Bioethics Convention and Explanatory Report*, Strasbourg.
10 Vogel, F. and Grunwald, R. (eds) (1994), *Patenting of Genes and Living Organisms*, Springer Verlag: Berlin, Heidelberg p. 135.
11 Shiva, V. and Hollar-Bhar, R. (1993), 'Intellectual Piracy and the Neem Tree', *The Ecologist*, Vol. 23, No. 6, Nov/Dec.
12 UNESCO (1990), 'Conserving and Managing our Genetic Resources', *UNESCO Journal Impact*, No. 158, Vol. 40, No. 2.
13 Co-operative Exploitation of the Phytochemical Resources of Developing Countries, Final Reports, Biotics Ltd, School of Chemistry and Molecular Sciences, University of Sussex, Falmer, Brighton BN1 9QJ.
14 FAO Conference C 93/REP/5, December 1993.
15 Bioprospecting and Biopiracy, compiled by RAFI with assistance from J. Kloppenberg, GRAIN, Accion Ecologica and Darrell Posey.
16 Overseas Development Institute (1993), *Patenting Plants - The Implications for Developing Countries*, Briefing Paper November, London.
17 OECD (1992), *Biotechnology, Agriculture and Food*, Paris.
18 Ibid., pp. 42-3.

19 Duesing, J. (1993), 'The Convention on Biological Diversity. Its Impact on Biotechnology Research', *Agro-Food Industry High-Tech*, July/August, pp. 19-23.
20 FAO Conference 25th Session , Rome 11-29 November 1989, Resolution 4/89.
21 FAO Conference, op. cit., Resolution 5/89.
22 International Development Research Centre (The Crucible Group) (1994), *People, Plants, and Patents - The Impact of Intellectual Property on Trade, Plant Diversity, and Rural Society.*

References

Council of Europe, Directorate of Legal Affairs (1994), *Draft Convention for the Protection of Human Rights and Dignity of Human Beings with regard to the application of Biology and Medicine: Bioethics Convention and Explanatory Report,* Strasbourg.
Duesing, J. (1993), 'The Convention on Biological Diversity. Its Impact on Biotechnology Research', *Agro-Food Industry High-Tech,* July/August, pp. 19-23.
EC Directive Joint Text approved by Conciliation Committee, 21 February 1995, PE-CONS 3606/1/95 REV 1.
International Development Research Centre (The Crucible Group) (1994), *People, Plants, and Patents — The Impact of Intellectual Property on Trade, Plant Diversity, and Rural Society.*
Linzey, Andrew (1993), 'An end to Aristotle', *Chemistry & Industry,* 17 May, 376.
Ministry of Agriculture, Fisheries and Food, *Report of the Committee on the Ethics of Genetic Modification and Food Use* (Polkinghorne Committee), London HMSO, 1993.
..... *Report of the Committee to consider the Ethical Implications of Emerging Technologies in the Breeding of Farm Animals* (Banner Committee), London HMSO, 1994.
OECD (1992), *Biotechnology, Agriculture and Food,* Paris.
Overseas Development Institute (1993), *Patenting Plants — The Implications for Developing Countries,* Briefing Paper November, London.
Relaxin, Official Journal EPO 6/1995, 388.
Shiva, V. and Hollar-Bhar, R. (1993), 'Intellectual Piracy and the Neem Tree', *The Ecologist,* Vol. 23, No. 6, Nov/Dec.
UNESCO (1990), 'Conserving and Managing our Genetic Resources', *UNESCO Journal Impact,* No. 158, Vol. 40, No. 2.
Vogel, F. and Grunwald, R. (eds) (1994), *Patenting of Genes and Living Organisms,* Springer Verlag: Berlin, Heidelberg.

238

27 Summary of the questions

Sigrid Sterckx

Professor Désiré Collen asked a general question to Professor Marc Van Montagu, namely whether he thought that in his field (plant biotechnology), progress would be possible to any significant extent without the acceptance of patent protection, or whether he considered patents as an absolute necessity. Professor Van Montagu replied that in his view patents are necessary. He referred to the examples in Plant Genetic Systems on the rape seed and on the Bt stories. If there would be no patents, everybody could copy because it is not so complicated. It took years to discover that in bacteria there are sequences that are unstable in the plant, which you have to remove. This knowledge was the source of some patents. Once it is known what has to be removed from the Bt gene, everybody can remove this type of sequence and make his own construction. Once the value of this type of research is lost, nobody would go into this type of research for financial reasons, he said, so it would only be done at universities. But each plant made in a laboratory has accumulated many 'errors' in the regeneration which have to be crossed out. Therefore, if one has an engineered plant at that moment, one has to do hundreds of crosses and hundreds of field trials. This involves an enormous amount of repetitive work and therefore, according to Professor Van Montagu, it is typically an industry work that the university cannot undertake. He realises that some people will say that it has to be done by plant breeding stations and national programmes. However, he asserted, those who are familiar with the way national agricultural institutes work world-wide know that, as is always the case in science, you need a drive and a will to develop and you need incentives. And very often these incentives in our type of society are financial. We could try to identify some people who do it because of moral incentives and we could dream of a society where everybody would do something because he feels that it is important for humanity. But, said Professor Van Montagu, we have all seen how many of these dreams have failed and what came out of the power struggle of people who took over. He thinks that each time we look back at our type of economy, our type of society, if you a simple finding has to be made, people can make an enormous effort; on the other hand, if this has to be going on and on and there is a lot of routine work involved which gives no intellectual stimulation or reward, then there is nothing else but the financial reward to motivate people to make major efforts in carrying out this type of research, just like in scientific research. He said to fear that this is the only way out. So he would answer in the affirmative to Professor Collen's question whether it would be fair to say that, without these patents, Plant Genetic Systems would not survive.

Professor James Houghton (National Diagnostics Centre, BioResearch Ireland, University College, Galway, Ireland) put a question to Professor Van Montagu concerning his statement that only about 25 plants have been investigated, whereas there are thousands and thousands of plants which have never been investigated. Professor Houghton wondered how the two, the investigation of new species, new plant types, and the engineering of the existing two dozen, should be balanced. Professor Van Montagu explained that the situation he described is only the result of the situation with the classical tools, because of the repetitive work that was needed during hundreds of years. If one would know what the corn the Indians in the US were growing looked like in the past and what it looks like now, he said, one would not even recognise the plant. For many of the other plants it will have to be like that. Professor Van Montagu admitted that the road to go seems immense compared with the final result, but he noted that the molecular biologists know that. And since there is a limited number of genes and some of these genes can show spectacular changes, it is worthwhile to do this molecular biology and this engineering. Thousands and thousands of plants produce a lot of interesting compounds in minute amounts in the small structure of the plant. Possibly we will be able to make these interesting compounds in a much larger amount, in this plant or in another plant where we can have a much higher yield, a much higher harvest of this structure, or express it in others. So according to Professor Van Montagu there are many reasons to be optimistic.

Dr Ulrich Schatz reminded Professor Van Montagu of his statement that in all these cases the genetic engineering of plants is much less, or not at all, detrimental to the environment or to nature, compared to the classical methods like fertilisers or herbicides. So that would be the dangers of the 'old technology'. Dr Schatz wondered what would be the dangers of the 'new technology'. He gave an example from the field of animals, namely the breeding of salmon. One problem with the breeding of salmon, he said, is that it has an instinct to go up the rivers and reproduce itself near the sources. Through genetic engineering, however, one has taken off the gene at the root of that instinct, because it is supposed that the industrial breeding of salmon is much facilitated by this modification. If this transgenic salmon escapes, remarked Dr Schatz, it might have properties of reproduction which are far superior to the natural salmon. As a consequence, within a few years we might not find anymore natural salmon in the rivers of Scotland or Alaska. He thinks this is the type of danger people are very concerned about and he wanted to know from Professor Van Montagu whether this danger also exists in the field of plant engineering. Professor Van Montagu replied that biologists have to study, case by case, what the potential danger would be. On the other hand, he said, there is the reality of the toxicities of the used chemicals, for which we have to find a solution. If there is a potential danger, biologists always have to ask themselves whether it can really be stopped

240

and how it has to be handled. We consider this hypothetically, said Professor Van Montagu, because we do not know. In nature species taking over dramatically are exceptional cases in special habitats. For plants, he could imagine that at certain moments there can be one problem or another. That is why he feels that it is probably wise for us to have this kind of discussion, so that everybody can come up with rational facts. But, he added, if you say you want protection against things that you cannot even imagine, against things that you do not know, there cannot be a very fruitful discussion. According to Professor Van Montagu, we have to analyse whether there are rational arguments. We have to build a society that provides this type of control system. But at the moment, he said, for plants this kind of 'science fiction' dramatic scenario (cf the salmon) does not exist. Weeds could become resistant against one particular herbicide, but then we can always use another herbicide. That is the maximum of scenarios Professor Van Montagu could imagine.

Dr Ulrich Schatz remarked that he shares Professor Collen's view that one of the disadvantages of the patent system, particularly of the European system, is that if you publish something before you file a patent application it becomes a prior art against your application. This is one of the misgivings of the patent law in Europe, said Dr Schatz, and it is specifically so in biotechnology. In this field the link between basic research and its technological applications is very close. This is typical for biotechnology, but it is also more and more the case for other disciplines. Therefore, he explained, the EPO advocates the introduction of a so-called 'grace period', which means that if the inventor has published no more than within one year before he filing the application, that will not be state of the art which could be opposed to this application. Dr Schatz informed the audience that the EPO is working on this, but he added that there is opposition from the industry.

Professor Jozef Schell wondered whether the study conducted by Professor Evers-Kiebooms would allow her to say what would happen if the information she provides to the people who come to see her for genetic counselling would be of a less probabilistic nature. He asked this question in view of the fact that part of the focus that is presently on human genome analysis is to provide other ways of genetic counselling, which would give a higher degree of certainty. Professor Evers-Kiebooms is convinced that for a number of people, that would have a very important effect on their decision making. People are very different in the way they cope with uncertainty. One of the major achievements of human genetics in the last years, the Human Genome Project, has played an important part in giving certainty in some situations. She gave the example of the sister of a child with cystic fibrosis: the only thing one could say ten years ago was 'you have two chances in three to be a carrier; you also have one chance in 25 that you have children with a partner who is carrier; if you are both carriers you have one chance in four to have a CF child'. Today, much more can be done. First of all, Professor Evers-Kiebooms explained, we can test the sister and see whether she carries the gene or not. If the result is positive, we know for sure that she is a carrier. If the result is negative, due to the high number of mutations in CF, the result is not certain but there is already a high probability of not being a carrier. Thanks to the identification of the gene, every individual in the population

241

can be tested with the same limitations about negative test results. It is never sure because one does not detect all mutations. Nevertheless, in CF families, this is a very important achievement that makes it a lot easier for a number of people to make a decision. On the other hand, she observed, the emotional factors still play a part and growing up with a brother or a sister with CF can have such a tremendous impact on the child's development that even a small residual risk may be very important and carry a very high weight.

Professor James Houghton (National Diagnostics Centre, BioResearch Ireland, University College, Galway, Ireland) told Professor Evers-Kiebooms that her last conclusion highlighted a problem that he encounters. Professor Houghton is involved in genetic testing in Ireland, and there is still a great deal of public uncertainty about genetic testing in that country. Pregnancy termination is prohibited in Ireland. Whenever the topic is discussed, whenever genetic counselling or genetic testing is discussed, he reported, the media and the newspapers all concentrate on the termination of pregnancy and that creates aversion in the public. Professor Evers-Kiebooms agreed and stressed that this was one of the major findings in her study with adolescents.

Professor Jozef Schell wondered whether he would be correct in rephrasing Mr Rio Praaning's opinion as follows: it would be wrong to try to oppose biotechnology by opposing patents on biotechnological inventions. Mr Praaning confirmed that in his view opposing biotechnology patents seems entirely the wrong starting point, for a patent has already surpassed the moment of research. If one wants to have control, he said, one has to go to the earlier point. The patent stage is not the right moment to present objections. According to Mr Praaning it is an easier moment to do it, though, and he suspects that this is the big temptation for which the European Parliament has fallen. Mr Stephen Crespi agreed with Mr Praaning and noted that there may be a rather confused attitude on the part of many who are unwilling to condemn the research, but are quite strong to condemn the patenting, which he finds very difficult to understand. Unless society is going to make up its mind that the research is immoral, or the final commercialisation or the practical use of the invention is immoral, he wondered, how can you possibly select the middle stage of this whole process (the patenting stage) and say this is particularly immoral? Mr Praaning remarked that there is a very strong plurality in the US and in Europe saying that research by all means must continue. He illustrated this with an opinion poll held in the Netherlands in 1993 by the National Foundation for Statistics: 19.8 per cent of the Dutch population said that it was 'necessary' to change the hereditary features of animals for the production of medicines for incurable diseases, 44.2 per cent said that it was 'desirable', 14 per cent said 'rather not' and 11 per cent said 'absolutely not'. So, he calculated, you come to a 64 per cent majority which says that this work must be continued because otherwise we will lose out in the end. In Mr Praaning's view it is obvious that this has very much to do with the idea that human health is at stake and that for human health and its progress, it is necessary to have the research. Of course, he asserted, the groups which are against the research have well understood that and have decided not to attack the research and totally concentrate on the patenting issue. It is the combination of patenting and life which is doing the trick. Mr Praaning doubts whether this is a very responsible thing to do.

Dr Schatz thinks that these complex attitudes may have something to do with what he called a 'closeness' of biotechnology to our own being: the more we know that the basic structure of DNA is the same for all living entities, everything that has to do with biotechnology is very close to our body. On the other side, patents are perceived as an appropriation of those research results by somebody who is not good and that is the industry. Dr Schatz thinks there is a complex relationship between these elements. Mr Praaning completely agreed. He added that some of these things are perhaps approached better if we would go to the future and you look back from the year 2020. If we would hold an opinion poll and ask the public 'do you wish these and these diseases to be cured, particularly if they are genetic?', Mr Praaning predicted, everybody would say 'yes'. Next, we should ask the following question: 'how would we have gotten there, how much money would it have cost, where do you think the money came from, who should pay for that, will you, will taxes be raised?'. Finally, we would land at the patent issue. Brought up in that context the whole relevance of the patenting issue is at least different, Mr Praaning concluded.

Mrs Larissa Gruszow wanted to hear from Mr Lars Klüver how the Danish Board of Technology chooses people for the lay panel of the consensus conferences. Mr Klüver explained that his organisation advertises in regional newspapers, so that all of Denmark is covered. The people who react have to write one page about themselves: their age and education, what they do and how they are employed, of which sex they are and so on. Then a mixed panel is composed. The Board uses a planning group of stakeholders, usually a balanced group of different stakeholders in the field, to check that the panel that is proposed by the secretariat is able. This planning group usually changes one or two out of the fourteen panel members. So what is composed is not a representative panel, since that is not possible with fourteen members. A mixed panel is set up and the Board tries to have as many different values in that panel as possible. Mr Praaning observed that the system which is used in Denmark resembles somewhat the US legal system with a jury. He said that in the Simpson case, you could see how it *can* work out, essentially. In the opinion of Mr Praaning, who reported that he has tried to research carefully what has happened both in Denmark and in the Netherlands, the consensus conference is a perfect means to create bridges to a general public and to make them part of the dialogue. But at the same time, he warned, if you give instructions over the weekend through certain experts, and if you already create a consensus over the weekend over whatever the issue is, then you have the risk that the fourteen people rather come to a conclusion, if that is at all possible. If you have in that group of fourteen, for instance, someone with a genetic disease, the whole picture tends to tumble. Mr Praaning supported this view by referring to the study presented by Professor Evers-Kiebooms, which showed the influence in a certain population when one member of that population had a certain disease. Mr Praaning further said that he was interested in hearing how it works when the Danish Board of Technology calls on the European Parliament or on the European Commission to do something. He wondered whether the Board is actively engaged in the further announcing of what has been concluded by those fourteen people. Mr Praaning referred to a conference of the European Commission which was interesting to the extent that the Danish representative of the Foreign Ministry said that he was instructed to involve himself with the European Parliament to express the vision of the Danes, that animal patenting should be just as forbidden as human patenting. Apparently this was a result of the activity of the Danish Board, Mr Praaning stated. Mr Klüver clarified that the Danish Board does not try to sell the conclusions further in the political systems. If the Board would do this, he said, it would be dead tomorrow, simply because it is made by the Danish Parliament, and there are many different opinions inside the Danish Parliament. So if the Board took some of the lay people's conclusions, and for instance tried to make them end up as laws, made them into what could be called 'preparing laws' for the politicians, that would be suicide. What the Board is doing, Mr Klüver elucidated, is selling the whole package: 'we have organised this process, this is what came out of it, look at it and see if you can use it for anything'. And in most cases the politicians make use of it. As to Mr Praaning's first question — 'can you trust these results?', 'could they be manipulated?', 'what happens with the lay people during the process?' — Mr Klüver replied that the Board tries to make the process transparent and open. There is a process consultant who is chosen by the preparatory group, which is a balanced group of different actors and stakeholders; this person works on his own with the lay panel. The secretariat keeps an eye on this process consultant and he or she keeps an eye on the secretariat. There is an agreement that, if anything goes wrong, if the secretariat tries to manipulate, the process consultant will tell the preparatory group at once. This is an extra check that is built in. Mr Klüver does not think this is necessary, but the Board does it anyway.

Mr Gert Devries (Biotech consultant, The Netherlands) remarked that he does not question the integrity of the organisers, but that the very low number of fourteen (Danes) makes the panel a very selective group. Mr Klüver recognised that fourteen is a low number, but he added that fourteen is a number close to the number of people within commissions in Parliament which make the Danish laws. Mr Ronald Schapira made another observation on something which in his view might have resulted in bias. Among the experts who had to inform the panel, there was agreement that genes are not patentable and they could not explain why they might be, when in fact that is totally wrong, Mr Schapira said. Genes are patentable now in every country of the world, because they are not claimed in their natural state. It is clear that you cannot find genes, they do not just lie around on the ground to be picked up, they have to be identified by arduous procedures, he commented. Often the proteins that they encode were not known beforehand, and so the new proteins go hand in hand with the genes that encode the proteins. Therefore, according to Mr Schapira, there was a natural bias in the panel in that none of the experts knew that genes are patentable or could explain why, or had any real understanding about what the issue is all about. We have to pay attention that the experts we use are really experts, and do not have an axe to grind on their own, he concluded. Mr Klüver answered this remark by saying that the expert the Danish Board used at that particular consensus conference was the Danish representative in the European Patent Office.

Dr Ulrich Schatz stated that he admires Professor Vermeersch's analysis of the problem of the moral or ethical admissibility of certain technologies. He did wonder, however, to what extent this has an effect on the question of patentability. Professor Vermeersch answered that once you accept the 'morality clause' (Art. 9.1) of the proposed Directive on the Legal Protection of Biotechnological Inventions, you should take into account that several ethical considerations concerning biotechnology may be relevant to patenting (as special cases of the 'morality clause').

Mr Stephen Crespi recognised that Professor Vermeersch had given a very thought-provoking analysis from a philosopher's point of view, but he wondered whether it is fair to subject the patent law to criticism from an aspect of human thought which it was never intended to deal with. Patent law has evolved over centuries to be concerned with the advancement of new processes, new products, manufactures, which are technologically applicable, industrially applicable, he said. It has therefore been addressed over the centuries to the economic life of man, not the intellectual life, and that is why pure discoveries are not patentable. Mr Crespi admitted that he did not understand why, even from a philosophical point of view, there would be a problem with the 'blurring of the distinction between the natural and the artificial', as Professor Vermeersch called it. Man has been doing that ever since he came into existence. For over a century, synthetic chemists have been producing substances which did not previously exist in nature, new organic and inorganic chemical compounds which are biologically active. Mr Crespi does not see a problem with that kind of human activity, simply because these entities did not previously exist, and he would extend that to living organisms. What we do is not make a totally new organism but a modified version of what previously existed. But as the Supreme Court in the US held in 1980, what is being done here is not the work of nature, it is not a product of nature but a product of human ingenuity, and this they held to be patentable within the existing framework of patent law, he recalled. If some people want to change patent law, Mr Crespi stated, then that is a democratic possibility open to us, to take a new approach for the future. But, he added, we must have some good reasons for departing from a system that, on the whole, has been very successful in the past. As to Professor Vermeersch's statement that Watson and Crick may have anticipated Cohen and Boyer, Mr Crespi felt that this is untenable. Watson and Crick discovered the structure of DNA and how DNA replicates. That is far removed from Cohen and Boyer's building upon that knowledge to tailor micro-organisms to introduce foreign genetic material, he asserted. Professor Vermeersch maintained that he cannot accept such a clear-cut distinction between the work of Watson and Crick and that of Cohen and Boyer. As he explained in his lecture, both discoveries were made within a *theoretical framework* and the second one was not possible without the first one. He drew a parallel between the Watson and Crick / Cohen and Boyer example and the relationship between Maxwell, Herz and Marconi. From the equations of Maxwell follows the existence, for instance, of electromagnetic waves. These electromagnetic waves have been produced by Hertz. Then these waves have been technologically used, or have been modified in such a way that they had an immediate technological application, by Marconi. The point is, Professor Vermeersch said, that it is not so easy to say who is the one that should get the money. What Marconi has done is making the product. What Hertz has done is making the waves and what Maxwell has done is showing the existence of these waves, so that Hertz did not have to work out of the blue. So Professor Vermeersch's point is that even if you say there is a distinction between discovery and invention, you have to admit that this distinction is becoming more and more artificial — and that is where philosophy of science comes in. The distinction between invention and discovery, which is always used in patenting, is blurred. The distinction between the natural and the artificial exists of course, he admitted, but when an alien comes from Mars or somewhere else, he can see after a while that a stone is a stone, but when he sees something like a motorcycle he will think 'this is strange biology' or he will come to the conclusion 'this has been made'. Normally, after some time, he will see that it is made. Everybody on Earth can see that a motorcycle is artificial, but when there is a genetically modified organism, let us say a new type of mouse or cow or plant, and you ask even a specialist, after fifty years he will no longer be able to say whether it is a human product or a natural product. So the distinction is really blurred, Professor Vermeersch concluded.

Dr Ulrich Schatz reacted on what Mr Schapira said about the patentability of therapies. It is absolutely clear, also under European patent law, he said, that e.g. the treatment of blood taken out of the human body and reinjected in the human body or any other *tools* that are used in medical treatment are patentable under the existing European Law, so there is no difference with US law in that respect. 'Therapy' is only meant in the sense of *what a doctor does* on the body of his patient and that is not patentable, Dr Schatz elucidated. The reason for exclusion is of course not that therapy is immoral, which would be absurd. The reason for exclusion is that it is considered undesirable to bother the medical profession in the exercise of their job on the patient by any consideration of patent law.

Mr Gert Devries (Biotech consultant, The Netherlands) wondered how the 'Directive on the legal protection of biotechnological inventions' would be implemented if it were approved by the European Parliament. Mr Dominique Vandergheynst explained that the Directive will say that the delay for the Member States to implement the new rules normally, following all prognostics, would be until the beginning of 2000. After the first reading in the European Parliament, the Commission has the possibility to propose an amended proposal, a second reading in Parliament, a second reading by the Council, and maybe even a new Conciliation Committee. According to Mr Vandergheynst, the European Commission's prognostic is that the final decision by the Parliament and the Council could be taken at the end of 1997. Generally, in all the Directives to harmonise the law of the Member States, the Council and the Parliament give 18 months to all the national parliaments to change the different national laws. Mr Vandergheynst emphasised that the first goal of the Directive is to harmonise the national patent laws. Normally, there is no link between the Directive and the European Patent Office and there is of course no obligation for the EPO to improve the wording of the European Patent Convention after the final adoption by the Council and the Parliament of the new Directive. It will be the Contracting States which could decide at the moment. After a discussion inside the administrative council of the EPO, Mr Vandergheynst explained, the Contracting States will have to decide whether there is a necessity to start diplomatic negotiation to change the European Patent Convention.

Mr Stephen Crespi told Mr Vandergheynst that his first impression after reading the new proposal for Directive is that the Commission has stuck to its position on many of the points with which the Parliament had problems. He said to welcome this, but since so much is in common with the previous Directive text, he was wondering whether the Parliament has in the meantime been more educated to understand the issues that have been raised and whether the European Commission expects a more sympathetic audience next time around. Mr Vandergheynst explained that the Commission was going to send the new proposal to the Parliament and to the Council. The first step inside the Parliament would be to know which Committees of the Parliament will have to try to vote some amendments. Of course the main Committee of the Parliament will be the Legal Committee. However, other Committees will have to be consulted as well: the Research Committee, the Committee on the Environment, etc. The discussion to try to explain to the Parliament the real subject-matter of the new proposal takes place during all these discussions in the different Committees, Mr Vandergheynst elucidated, like in every national parliament. Of course, he added, all the interested parties have the possibility to try to explain to all the MEPs their concerns.

Mr Steve Emmott put two questions to Mr Vandergheynst. First, he wondered why, although the European Commission consulted widely with industry and parliamentarians before producing the new draft, the NGO's were not consulted. Mr Emmott found this surprising since it were the NGO's objections which in large part led to the defeat in Parliament of the previous proposal. Mr Vandergheynst replied that the responsibility to initiate a new proposal rests on the shoulders of the Commission, but that the public opinion will have all the possibilities to intervene. It is impossible for the Commission, he stated, to take fully into account all the concerns of the pubic opinion at a pre-embryonic stage of the process. According to Mr Vandergheynst, it should be kept in mind that a patent is a tool for an economic operator to try to compete against other economic operators. Therefore, he said, it is normal to try to consult the future patentees to know what their concerns are. As to the concerns of the so-called 'silent majority' of the public opinion, Mr Vandergheynst assured that the NGO's would have the possibility to express their concerns and that these concerns would be taken into account by the MEPs, the Council, and the Commission itself. Mr Emmott's second remark was that the new proposal does not include a statement that surgical, therapeutic and diagnostic methods are excluded from patentability, whereas the previous text did. He wondered whether this was accidental or deliberate. Mr Vandergheynst replied that this exclusion from patentability is already in Art. 52(4) of the European Patent Convention and that repeating this Convention is not harmonisation of the patent law of the Member States.

Ms Sigrid Sterckx observed that in paragraph 22 of the new proposal, which is the same as paragraph 10 in the previous version, it is said that 'the body of the Directive should also include a list of inventions excluded from patentability, so as to provide national courts and patent offices with a general guide to interpreting the reference to public policy or morality'. She felt that it would be very difficult to make up such a list. Mr Vandergheynst replied that, when the discussion started on the initial proposal in 1988, it appeared that the basic problems vis-à-vis the ethical dimension, beyond the problem of the patentability of elements coming from the human body, concerned the patentability of germ line gene therapy and the patentability of animals. The discussion that was going on at this workshop, he said, demonstrates that these are ethical problems. That is why the Commission tried to use the umbrella of public order and morality and to give two examples to try to indicate to the judges and to the patent offices two possibilities of implementation of this general umbrella. This is an attempt, Mr Vandergheynst added, and of course everybody would have the possibility to discuss it. Dr Ulrich Schatz also commented on Ms Sterckx's remark. He considers Art. 53(a) and corresponding national provisions as making a reference to underlying basic law, regulating certain acts. Since Mr Vandergheynst had explained that the example on the suffering of animals and proportionality with the objective is taken from the EC Directive on protection of animals used for test purposes, it was clear to Dr Schatz that he follows the same basic approach. The only difference, he said, is that the Commission gives an example of how Art. 53(a) is to be applied. From the point of view of a technique of law making, Dr Schatz feels that it is sufficient that we have laws on animal protection which say precisely the same on that issue, and if they do not say the same the Commission should harmonise the laws on animal protection, and then the exception under the patent law would have the same meaning in all the contracting states. As to the possibility of making a list of inventions that are excluded from patentability, according to Dr Schatz this is possible as long as we are referring to nothing else than to norms of public order that do exist in the law of the member countries. He does not regard it as a good example of legislation, but it is not basically wrong. However, Dr Schatz continued, when we make a list and add things which are perfectly allowed to do under public order in the different states, which means that we deviate in this list from what the law actually says, then the Commission becomes a lawmaker in a field which is not patent law, but 'ordre public'. He doubted whether the Commission has a competence in that field. But the current 'list', providing the two afore-mentioned examples, simply acknowledges the fact that public order in the member states does *not* permit germ line gene therapy and does *not* permit causing animal suffering without a reasonable proportion to the objective. So it is a reminder of the common law of the member states.

Mr Crespi asked Mr Jan Mertens to explain what exactly he understands the problem to be with the Neem tree, for Mr Crespi had been told by his contacts in

India that there was never any patent granted on the Neem tree. There was a patent on the Agracetus cotton, he said, but not on the Neem tree. According to Mr Crespi, the only patents concerning the Neem tree are US patents, which have nothing to do with the tree or the seed itself. Mr Mertens admitted that he did not know all the details of this matter because he is not the person dealing with it within the Green group. He only intended to clarify that you cannot separate the discussion about patents from the political and economical context in which they are used. In this case, Mr Mertens stated, there was a serious debate in India and there was strong opposition by thousands of Indian farmers against the efforts to change Indian patent rules under the heading of GATT/WTO agreements.

Dr Schatz emphasised that the patent system as such is certainly to be seen in the context of market economy. In his opinion there must be limits to market economy since the mere mechanism of the market does not solve all social problems. Otherwise, he said, we would be happy to do away with the entire political class: we would not need any politician or any lawmaker anymore. The market economy would do the whole thing. In view of the link between market economy and patents, Dr Schatz wondered what Mr Mertens would expect from excluding living matter, or transgenic plants and animals alone, from patentability and yet letting these things be made and sold on the market. He was curious about Mr Mertens' expectations concerning the benefits or disadvantages of that strategy in terms of welfare and employment in Europe, i.e. competitivity of European industry, and in terms of protection of the environment and maintaining the dignity of animals. Mr Mertens called this a very tricky question, but replied by saying that for the Greens the non-patentability of living organisms is a matter of principle and therefore they try to use all the instruments they have. He agreed with Dr Schatz that it is not only about patents. But, he added, we have to look for ways and measures and room for political debate to question the mechanism itself; we have to try to look for ways of dealing with the fundamental underlying debate.

Mrs Larissa Gruszow remarked that she finds Mr Daniel Alexander's idea to shift the examination on morality for granting patents in the EPO and in other European patent offices a very good proposal, for this would get us rid of a lot of problems. She wonders whether such a body — which would be placed after grant and on the moment that the exploitation would begin — would be to some extent similar to the authorities which give permission for medicaments to come on the market. She also pointed to a difficulty: a patentee will be in doubt after having received the patent and will not be very sure whether he has to continue the preparations for exploitation. Mr Alexander replied not to have thought out this proposal in sufficient detail to be able to give a clear and focused answer on this issue. He did make some general observations, first of all that it is very important that one should preserve, in any system, a possibility of open debate, public participation, and an opportunity to challenge the conduct of a patentee. Someone who is given valuable proprietary rights by the state in the public interest ought to be open to public challenge and ought to be held to public account, Mr Alexander said. So whatever institutional mechanism would be developed would have to have those characteristics. He went on to explain that what he had in mind was something a little bit different from Mrs Gruszow's suggestion, namely either an assessment additional to the assessment that goes on in the course of the process of deciding whether to grant a patent or in the course of an opposition, or alternatively to it, the courts who come to enforce a patent in due course are able to consider the extent to which the exploitation of the patent — 'exploitation' being understood in its widest sense: both actually technically putting the invention into effect and also the reaping of commercial gain from that activity — is or is not in the public interest and in particular in the environmental interest. To Mr Alexander this would seem to be a way of encouraging particular kinds of technology to be developed and discouraging the exploitation of other kinds of technology. So he is thinking of an analysis at a later stage. As to the difficulty mentioned by Mrs Gruszow, Mr Alexander agreed that there would inevitably be a degree of uncertainty, but, he said, that is the price one has to pay for any system which involves opening up people's proprietary rights to challenge. The patent proprietors will never know whether they are absolutely secure until that has been decided upon. But if one decides that it is in the public interest that there should be such a forum, then that is the private burden that patent proprietors will have to live with.

Mr Ronald Schapira admitted that he was a little surprised because at least two speakers mentioned the real possibility that the world would come to an end at some foreseeable future by virtue of overpopulation, namely Dr Schatz and Professor Van Montagu. In the view of many people, Mr Schapira said, biotechnology provides at least one hope for avoiding that result, at least short term — and possibly also long term — by increasing the availability of food products, especially under changing environmental conditions, which will require man to adapt much more quickly than he has been used to adapting to changing environmental conditions. Mr Schapira wondered whether Mr Jan Mertens or Mr Steve Emmott have taken into consideration that, by attacking the patent system and in particular the biotechnology of plants, they may be in effect going against what we ought to do. Mr Emmott replied that his first reaction to this remark would be that he has heard proponents of biotechnology saying for a long time that biotechnology will feed the world, clean up the environment and improve human health. Perhaps with the exception of the last, he said, there is no evidence at all that this is what is happening. Instead, what we get is tomatoes that do not rot and injections for cows to make them produce more milk that we do not need. Although Mr Emmott did not express the view that we must condemn this technology, he does not think that we have seen quite a lot of evidence so far in terms of the socially responsible use of biotechnology.

Dr Ulrich Schatz reacted to Dr Christoph Then's statement that we have a high degree of industrial concentration in the field of plant breeding. He admitted that this degree is remarkably high. However, Dr Schatz had a problem with Mr Then's assertion that, with patents, this situation will go even worse. He referred to the fact that in the field of genetic engineering only 29 per cent of all patents are owned by large companies, whereas small and medium size companies have as many as 27 per cent, and university research institutes and charities 34 per cent. Dr Then remarked that, if one makes such a statistical analysis, one should also add that nearly all these patents are in the hands of Northern based institutions, and not in the hands of institutions from regions where the biodiversity is very rich. As to Dr Schatz's question, he said that it is clear that small firms can use patents to survive on the market, but they are not the users of the patents. They have to give the patents away to firms which have the money to develop the products, for example plant varieties. So according to Dr Then Dr Schatz's analysis does not say very much about the concentration on the seed market. He also added that if we think of patents, it is necessary to think not only of the seed market but also of the food market which lies behind the seed market.

Dr Leo Fretz made an observation concerning Mr Stephen Crespi's attempt to construct a moral argument against the opponents of various forms of biotechnology. He noted that Mr Crespi does not respect the laws of logic and provided two examples. First, Mr Crespi says very clearly that patenting is an ethically neutral act, whereas in Dr Fretz's opinion you cannot say that any human act is ethically neutral. Secondly and more importantly, Mr Crespi's thesis that patenting is ethically neutral is inconsistent with other remarks he made, e.g. that patents are a way of encouraging people. Dr Fretz also referred to a syllogism built up by Mr Crespi: 'Life is an abstract notion', first premise; second premise: 'abstractions are not patentable'. Conclusion: 'living organisms are patentable'. This conclusion does not follow from the two premises, Dr Fretz said, so Mr Crespi is arguing in a sophistic way. Mr Crespi admitted that possibly he should have included 'life is an abstract term, abstractions are not patentable, therefore life cannot be patented'. But the point is, he argued, that living organisms *can* be patented. He explained that in fact he is trying to overcome the power of the slogan, which suggests that we are somehow committing an immoral act in trying to monopolise the very stuff of life. This is a very powerful slogan, according to Mr Crespi, which can be used to distort the truth in public opinion surveys, for the public can very easily get the impression that the researcher or the industrialist is monopolising the very basis of life, which is completely invalid. As to Dr Fretz's observation on 'ethical neutrality', Mr Crespi replied that he was looking at it from the point of view of the applicant of the patent or the inventor, who are seeking a right which can be neither morally praised nor morally condemned. It does not have any moral significance, he repeated. The encouragement feature is what society provides, not what the inventor does. Society has decided that it will encourage innovation, so the attitude of the inventor and the ethical status of the inventor is not the same thing as the ethical standpoint of society, which created patent laws.

Mr Daniel Alexander reminded Mr Crespi of his statement that he had difficulty in engaging with the arguments of the speaker that came before him (Mr Luc Vankrunkelsven). Perhaps, Mr Alexander said, that is a fundamental difficulty in this debate: unless one is predisposed for one point of view or the other in this debate, it is quite difficult to accept the positions on the other side. Mr Alexander also made an observation concerning the view, expressed by Mr Crespi, that the debate in the context of animal patenting is being conducted in the wrong way, because if one is for example challenging patents to something like the Oncomouse, one ought to be addressing arguments of the morality of the underlying research, rather than going round by the side door and challenging a patent in that context. According to Mr Alexander there is a rationale for challenging patents here, which is not merely instrumental. It is not merely that people are challenging a patent because they believe that it is likely to 'mess up' that particular aspect of research or make it less likely that it will be done. It may be, he said, that there is a respectable moral basis for challenging a patent that says that there are some areas where there are necessary evils in life. One has to do very unpleasant things because one believes that there are socially useful goals for them. But it does not follow from that that everything in connection with that necessary evil is thereby authorised. At the time of the Oncomouse case, Mr Alexander illustrated, there was a very interesting article in the *Independent* newspaper, which is a much respected newspaper in England. This article said that lots of people do not object to animal testing as such, but what they do object to is that large commercial organisations should be able to 'make a fast buck' out of the cruelty and suffering that was involved. Mr Alexander believes that there are many people, interested lay people, who may think that it is one thing to engage in this kind of research, but it is quite another thing for a public body to go about putting its imprimatur on it, unless it can be shown that the relevant research will not go on without a patent. People do have very strong feelings about this and there is a sense in which the arguments presented by Mr Crespi, which are consequentialist arguments, do not do those very strong feelings justice.

Mr Crespi explained that he had difficulty with Mr Vankrunkelsven's presentation because there were so many predictions of doom. How can you know what will happen, Mr Crespi wondered, unless there is some evidence or some logical connection? Where is the evidence that any transgenic plant is going to cause genetic erosion? Mr Crespi does not see any evidence and he finds it hard to grapple with assertions which are unsubstantiated. As to Mr Alexander's remark, he said that the idea of big firms making a fast buck is confusing many issues. If something is immoral, Mr Crespi argued, it is immoral whether you get money for it or not. Would the making of these oncomice be any more morally acceptable if it was done by a small firm? Mr Crespi would not think so. He did admit that some companies make life difficult for his argument, because the way they actually exploit their rights is questionable, but, he asserted, that has nothing to do with patenting.

Mr Simon Gentry (Manager External Scientific Affairs, SmithKline Beecham, UK) informed the audience that SmithKline Beecham is very involved in DNA research and in the patenting of DNA sequences. SmithKline's motivation, he said, is creating new medicines to relieve some of the diseases which afflict millions of people around the world today. He wanted to say something on the sloganeering and the use of the slogan 'No patents on life'. To his knowledge no company has ever claimed a patent on the quality of life, the spiritual essence which constitutes life. Industry is constantly accused of reductionism but it is the opponents of this technology, Mr Gentry said, who are being truly reductionist when they claim that life is actually constituted out of four nucleic acids or twenty amino acids. He called the slogan about patenting life fundamentally dishonest and misleading to the public, as well as deeply damaging to the debate about morality and ethics and to the prospects for improving health care and improving agriculture. Mr Crespi fully subscribed to this.

Dr Christoph Then said that in his opinion Mr Crespi is doing exactly what he says other people should not do: he permanently mixes up the reasons for doing gene technology and the reasons for patenting, whereas these are two different questions. If one wants to promote gene technology, Dr Then stated, that is fine, but then we have to discuss how to promote it. He added that the slogan 'No patents on life' is perfectly correct, because what is being done is not the patenting of a single animal or a concrete organism but the patenting of a principle, e.g. the principle to have a gene which can be introduced into any organism in which the gene can be incorporated.

Mr Crespi replied that he does not profess to tell society how we should or should not promote gene technology because that is not the function of a patent attorney. As to Dr Then's second critique, Mr Crespi admitted that some patents are indeed extremely and unjustifiably broad. But he feels this is a matter of patent law. Although some of these patents are pre-empting unreasonably wide areas of technology and are being in a sense anti-competitive, according to Mr Crespi this should be dealt with in disputes between rivals in research and industry.

Part Six
THE DEBATE

28 Biotechnology and the public debate: some philosophical reflections

Leo Fretz

In my country, The Netherlands, there have been two public debates on bio-ethical issues. In 1993, a lay panel of fifteen people discussed for two and a half days with an expert panel on the social and moral dilemmas around the genetic modification of animals. In 1995, the second public debate focused on predictive genetic research.

Although the Dutch public debate model has a lot in common with the Danish consensus conference, a crucial difference between the two is that, in The Netherlands, a substantial consensus to be expressed in the final report is not the most important and certainly not the only aim of the organisers. At least equally important for them is to see whether the final statements of the lay people really influence the opinion of experts, of policy makers and of the public at large, and whether lay people — after a short but intensive introduction — are capable of forming independent views on biotechnological and biomedical questions and of defending these views with cogent arguments. Although a number of the experts who participated in the debates, either as members of the expert panel or as part of the public, are rather critical about the lay panel statements, without any doubt many experts were also very impressed by the high level of the discussions with the lay people.

I do not intend to give an overall assessment of the two debates. I refer to the book by Simon Joss and John Durant, which was published a few weeks ago under the title '*Public participation in science. The role of consensus conferences in Europe*' and which contains an evaluation by Igor Mayer and others of the effects of participation in the Dutch 1995 Conference.[1]

I will confine myself to summing up a number of interesting points from the final document of the first debate, with the help of which I will categorise the ethical objections that are often put forward by the public against the genetic modification of animals. I will also quote a very interesting passage from the final document of the second debate, to which I will return in the second part of my lecture.

The final document on the genetic modification of animals contains a majority and a minority statement. A majority of nine of the fifteen members of the lay panel is very critical about genetic modification of animals and holds the view that it should be allowed only in the case of scientific applications about which there is a wide public consensus. These members of the lay panel consider genetic modification as an infringement of the intrinsic value of the animal and think that there is an inadequate scientific insight Into how the consequences for the animal's health and well-being can be measured. They wonder whether we, human beings, have the right to intervene in the process of evolution. They have doubts about the moral acceptability of patenting animal life and they believe that the gap between our society and the Third World will become even greater. They are persuaded that the disturbance of ecosystems by genetic modification is a real and difficult to control risk.[2]

From the final document of the 1995 debate on predictive genetic research, I will only quote one passage, which deals with delicate possible consequences of this research for some people:

> Within the current system of private insurance, there is a group of high risk people who are uninsurable. [They are speaking about life insurance] At this point, the group includes people who might be a carrier of Huntington's disease and muscular dystrophy. In the near future, this group will grow because diagnostic tests will become available for an increasing number of genetic disorders. It is the opinion of the panel that our insurance system is based on the solidarity principle. The principle could be expressed by covering the risks of the uninsurable by all those insured. The panel would like to see research being undertaken to integrate this in the current system. This could be done in a legal manner — compare the provisions for disasters — in the form of a surcharge for all remaining insured. Should this prove to be impossible, then even a collective provision insurance could be considered. If solidarity is the basis, the term "uninsurable" has no meaning. If solidarity cannot form the basis in this way, the increase of diagnostic capabilities will eventually lead to a substantial premium reduction. Finally, the insured group will consist primarily of low risk people.[3]

I will return to this insurance issue in connection with predictive genetic research in the context of what could be called 'the need for a European ethics'. First, however, let us consider again the variety of objections raised by the majority of the lay panel in 1993. I will try to translate them, as far as necessary, in more philosophical terms, and divide them into four categories by using two criteria: the degree of assessibility and the moral status of the possible negative consequences of genetic modification of animals.

With regard to the first criterion, I refer to David Collingridge's book 'The social control of technology', in which Collingridge states that, today, politicians take decisions about modern technology again and again, under conditions of uncertainty or even ignorance.[4] Concerning the second criterion, the moral status of the possible negative consequences of genetic modification of animals, we have to conclude that some of the risks of modern biotechnology can be described exhaustively or mainly in scientific terms, whereas others are describable primarily in philosophical, theological and ethical terms.

258

Using both criteria and distinguishing between, one the one hand, high or fair assessibility and low or non-assessibility, and, on the other hand, between so-called consequentialist and deontological risks, we can categorise the above mentioned negative consequences in the following diagram.[5]

Degree of assessability	(Moral) Risks Consequentialist	Deontological
High/fair assessability	Category (A) Impairment of the health/ well-being of animals/ human beings	Category (B) Infringement of the integrity and/or the intrinsic value of the animal. Instrumentalization of the animal.
Low/non-assessability	Category C Environmental damage. Health impairment of future generations. Genetic erosion	Category (D) Intervention in Creation/the process of evolution. Mechanisation of the image of the the animal/the human being. Metaphysical/moral erosion.

Before reading the diagram, I would like to make two remarks. Looking at this diagram, some of you may be surprised by my distinction between consequentialist and deontological risks. It might be argued that the term 'risk' refers by definition to possible consequences. Of course it should be clear that the phrase 'consequentialist risks' denotes risks which can be described in terms of the possibility of negative consequences for the life and health of the animal and the human being, and of environmental damage. However, these risks are already moral risks in the sense that the acceptability or non-acceptability of taking them is a moral issue, but risks which should be described primarily in terms of the possibility of negative consequences for the moral life are moral risks because, even if neither the impairment of human or animal life is at stake, taking these risks means that actions are performed which must not be done. I could refer to the work of the philosopher Bernard Williams, 'Ethics and the limits of philosophy', in which he says that the term 'deontological' is sometimes said to come from the ancient Greek word for 'duty', but there is no ancient Greek word for 'duty'. It comes from the Greek word for 'what one must do'.[6] The second remark I would like to make is that this diagram does not only order the objections of the lay panel of 1993. Some objections, that were raised by other opponents of the genetic modification of animals, have been added. Contrary to what the members of the lay panel suggested, the impairment of the well-being of animals is classed in category (a) and not in category (b), for reasons that can be found in the book 'Unanswered safety questions when employing GMO's'.[7]

At the left side of the diagram, you have the degree of assessibility and above you have risks and moral risks respectively. I make a distinction between consequentialist and deontological risks, as I already explained. I categorise the risks in four categories: (a) consequentialist moral risks with a high or fair assessibility; (b) deontological moral risks with a high or fair assessibility (the infringement of the integrity or the intrinsic value of the animal, the instrumentalisation of the animal); (c) consequentialist moral risks with a low or a non-assessibility (environmental damage, health impairment of future generations and so-called genetic erosion) and (d) deontological moral risks with a low or a non-assessibility (intervention in creation or in the process of evolution, mechanisation of the image of the animal or human being and metaphysical and moral erosion).

I have categorised the objections against the genetic modification of animals in this diagram because this makes it easier to see which risks should be evaluated by scientists alone and which ones should also be assessed by emancipated citizens without expertise in the biotechnological field. Although it is true that scientists should not worry only about consequentialist risks, but also about deontological ones, this does not imply that they are the only people who should care about what, from a moral point of view, must or must not be done. In an ideal democracy, all citizens have the right — and perhaps the duty — to reflect critically upon the norms and values in their and their children's society. Lay people without expertise in biotechnology cannot evaluate the risks of the categories (a) and (c). By contrast, interest groups, pressure groups, consumer organisations etc. may — and even should — engage experts to provide contra-expertise if they consider this necessary.

The division of the evaluation work between the different actors in the biotechnological field can be pictured in the following diagram.[8]

Moral risks	Assessment by			
	A (I) Scientists	A (II) Sociologists/ Theologians/ Philosophers	A (III) Trade and Industry, Interest Groups/ pressure Groups/ consumer organisations	A (IV) Unorganised citizens/ lay people
Category A	X	(X)	X	
Category B	X	X	X	X
Category C	X		X	
Category D	(X)	X		X

Category (a) are the risks — moral risks, if you want — with a high or fair assessibility and which are of a consequentialist character. They should be assessed by scientists and sometimes also by sociologists, theologians and philosophers. Of course they should also be assessed by industry and trade, interest groups, pressure groups and consumer organisations, because they can give the contra-expertise. In my opinion, unorganised citizens, lay people, should not assess these risks because they are not competent. Category (b) are deontological risks with a high or fair assessibility. They should be assessed by scientists, sociologists, theologians and of course philosophers, because these are 'moral' risks in a more precise sense of the word, but also by unorganised citizens and lay people. Category (c) are consequentialist risks with a low or a non-assessibility. They should of course be assessed by scientists and by trade and industry interest groups, who can give contra-expertise, but not by theologians or philosophers or unorganised people because they are not competent. Risks in category (d), 'moral' risks in a more precise sense of the word with a low or a non-assessibility, should be assessed by scientists, trade and industry, sociologists, theologians, philosophers and unorganised citizens.

I come to the second part of my lecture. In the foregoing, I have tried to argue why, on the national level, the social and moral implications of biotechnology should be discussed not only by experts but also by emancipated citizens. Now I would like to defend the thesis that, if an evaluation of the moral risks of biotechnology on the national level is very desirable, it is necessary *a fortiori* that such an evaluation takes place on the European level, and that also and mainly on this level, citizens should participate in the biotechnology debate as European citizens.[9]

One of the most serious complaints about European Union politics is that the distance between the European Parliament and the population in the different member states is considerable and that many people are not interested in the decisions of Straatsburg and Brussels and that a growing unity of Europe is incompatible with national interests. In my opinion, it is an illusion to believe that, in the long run, a united Europe will be realised and will hold, if there is no moral basis on which Gorbachov's 'the European house' or Delors' 'European village' can be built. Such a basis cannot be built by policy makers and industrialists. Of course, as some industrialists and politicians demonstrated earlier, they carry the material and the ideas, but the process of construction itself must be supported by the European citizens. Without a minimal moral consensus between them, the process of broadening and deepening the European Union will stagnate very soon. This moral consensus should not necessarily be a substantial consensus. More important is a procedural consensus about how to handle moral controversies. In the building of such a consensus, a European ethics can play an important role. The term 'European ethics', 'une éthique Européenne', is used by Jacques Delors in his book '*Le nouveau concert Européen*', published in 1992.[10] With this term, the author denotes more or less intuitively a basic moral attitude of the European citizen towards the environment, an attitude of respect towards the equilibrium between man and nature. In this context, I would like to use the term more precisely. A European ethics should be understood here as the systematic and critical reflection on the moral dilemmas facing the European population. As such, it is not the business of philosophers and theologians only, but essentially of all emancipated European citizens. This definition is not a descriptive but a normative one. Such a European ethics does not exist already. It is an ideal of a new forum of ethical communication, of discussing moral issues such as e.g. the moral risks of modern biotechnology. In Delors' book, many concrete problems of European politics are dealt with, but more important for our debate here are four

261

meta-questions, which emerge again and again in his discussion of concrete European problems that need to be solved. These meta-questions are the following. First, does a so-called 'European identity' exist, and if it does, how can it be defined? Secondly, what can be decided on the national or regional level and what on the central European level? How should the principle of subsidiarity be interpreted in practice? Thirdly, what has priority in European politics, goals or methods? In Delors' terms, the 'que faire' or the 'comment faire', the 'what to do' or the 'how to do it'? Fourthly, should the European Union be deepened or broadened at short notice? Or should both be done? I call these four questions meta-questions because they transcend the single problems of daily political life and because they have a regulative function in the process of solving these problems. These meta-questions are also fundamental questions of a European ethics, which can be illustrated in the following short remarks. As to identity: it is very difficult to give a sufficient definition of the European identity. Nevertheless, such a definition is necessary if we want to describe the essence of a European ethics. As to subsidiarity: the principle of subsidiarity counterbalances too strong a central European power. But even if we hope that never again a central ethical power will arise in Europe, we have to answer the question which ethical problems should be dealt with on the central European level and which ones on the national level. As to goals versus methods: very often in political practice, the problems are quite clear from the very beginning of the discussions, which cannot be said of the solutions. Frequently, the 'how' of the political relations determines the 'what' of the results. Also in ethical discourse, the complexity of the dilemmas is visible from the first moment, whereas the way out takes time and is dependent on the method used. As to deepening versus broadening: no matter which political constellation will ever be realised in Europe, one thing is clear: an ethical collaboration in Europe is very desirable, but also with regard to such a collaboration, the question has to be answered in which instrumental framework it should be effected.

Before trying to answer these four meta-questions and to show which consequences these answers could have for the biotechnology debate today, I would like to draw your attention to a phenomenon which could be characterised as ethical reductionism. For a good understanding of this term in the context of this debate, something must be said about two critics of so-called 'biological reductionism'. The first is André Pichot, who, in his book *'Histoire de la notion de vie'*[11] criticised the attempts of some biologists to reduce biology to biochemistry, because they eliminate the proper subject of their discipline, i.e. life itself. He characterises such a biology as a suicidal reductionism, which neglects the fact that time is inherent to the very phenomenon of life. Likewise, at the first world congress on medicine and philosophy, held in Paris in 1994, another French scientist, Henri Atlan, contested the view that the genes contain the essence of life. The genome, he said, insofar as it is considered a collection of DNA molecules, is a piece of organised, but nevertheless not living matter. It is interesting to observe that some moralists, warning against the moral risks of modern biotechnology, take biological reductionism as the universally accepted view of life, although this view is opposed by many biologists, not just by a few. After accepting this view, they fight this view by arguing in a way analogous to that of biological reductionists. Like the latter, they eliminate the temporal dimension from moral life and refuse to acknowledge that there are perhaps timeless values, but that concrete ethical insights are the products of a historical process. By preferring timeless to historical truth, they do not accept that we cannot know *a priori* what is good and what is bad in modern biotechnology, but that we have to find this out by trial and error. Ethical reductionism is not only a threat to the progress of science and technology, but also the death of morality and ethics themselves. I would like not to be misunderstood: I said that I consider *some* moralists as ethical reductionists, but this does not imply that I consider some biotechnologists as ethical reductionists as well.

But let us return to the four meta-questions: how should a European ethics be defined, how should the ethical work be divided between the national states and Europe as a whole, which methods should a European ethics apply and in what institutional framework should a European ethics do its work? The question of the identity of a European ethics and the question of the methods of such an ethics can be combined in a natural way. The essence of a European ethics consists in the roads it goes. In his book '*Penser l'Europe*', Edgar Morin tries to answer the question of the European identity.[12] He introduces the term 'dialogique'. The meaning of this word, in the context of his argumentation, is quite different from that of 'dialogue' and 'dialectics'. The distinction between 'dialogique' and the two other words consists in the fact that a 'dialogique' is not trying to conceal opposite positions or trying to suppress them, but to make them visible and let them impinge one against the other. Seeking an answer to the identity question, Morin says 'the unity of Europe consists in the vitality of its antagonisms, i.e. its dialogique'.[13] In my opinion, such a 'dialogique' should be considered the ideal of a European ethics. In such an ethics, European citizens are fighting for a new Europe in which there is room for opposite ethical viewpoints and in which they are seeking for a procedural consensus, i.e. an agreement on how to live with opposite viewpoints. By practising such a dialogique, people will acquire the insight that the European identity is not an unchangeable ghost in the machine, but that it is always in the making and never completed. At this point, it will be clear that such a dialogique is the opposite of ethical reductionism and that it provides an adequate structure for the biotechnology debate today.

It remains to be seen what the division of labour and the institutional framework of a European ethics would be, the questions two and four. In the motivation of the draft resolution of the European Parliament about the ethical aspects of new medical technology, in particular of prenatal diagnostics, rapporteur Pompidou observes that — unlike science and technology — philosophy, wisdom, morals and spirituality have made little progress in our society. I will give a short summary of his insights. Because new medical possibilities bring not only blessings but also dangers, a purely scientifical judgement of these possibilities is insufficient, he says. It is also necessary to reflect on the social and ethical aspects of the new technologies. This requires a democratic debate about the future of our society, in which experts, politicians and citizens should take part. Problems related to the control of procreation and death are the results of different cultural identities. The cultural differences should be respected at any rate, because they constitute the richness and variety of European society. How to think about the status of embryo's and foetuses is also determined by the various views in the member states of the European Union and in the other European states. It is too early to attempt harmonisation in this area. In particular the European Parliament and the Parliamentary Assembly of the Council of Europe should consult with one another about these matters. So far Pompidou. We can easily agree with these views. It would even be dangerous to attempt a European consensus in all areas of bio-ethics, but with regard to certain bio-ethical issues, we can no longer afford to postpone the formulation of a European point of view. This could be illustrated with various examples. I will confine myself to one example. Henri Atlan, whom I mentioned earlier, sharply criticises the opponents of genetic research that I characterised as ethical reductionists. As Atlan points out, the real and most difficult ethical problems facing clinical genetics consist in the fact that an early diagnosis of diseases that have not yet manifested themselves would widen the gap between diagnostic and therapeutic possibilities. The problem thus created involves the social and psychological state of individuals who enjoy good health and whose only suffering exists in anxiety and fear, caused by the diagnosis. Just how realistic these anxieties and fears are, can be seen from the threatening likelihood that individuals, who carry the risk of falling victim to a serious incurable hereditary disease, will be discriminated genetically in the future Europe. In The Netherlands, e.g., we are having serious discussions about the viewpoints adopted by the previous cabinet, and relativised again by the present cabinet, that people who carry the risk of Huntington's disease or muscular dystrophy might be refused a disability or life insurance. This danger of genetic discrimination threatens all European countries. In the opinion of many, such a discrimination is morally unacceptable and should be made legally impossible at the earliest occasion. Perhaps it should be included in Art. 14 of the European Convention of Human Rights that genetic discrimination is unacceptable. Although the need for this should perhaps be left to the judgement of European legal experts, my suggestion makes it clear that, with regard to certain ethical issues, a decision on the European level is unavoidable and cannot be left to the individual member states of the European Union.

But there is more at stake than merely a division of ethical labour between individual states and Europe as a whole. Hereditary diseases do not stop at the frontiers of the European Union. Therefore, ethical dilemmas of this kind should also be discussed extensively within the framework of the Council of Europe. But no matter how necessary the institutional framework of the European Union and the Council of Europe may be for a democratic discussion about ethical issues in present day Europe, they are not sufficient for a democratic decision process. Precisely because the ethical dilemmas in Europe concern every emancipated citizen, these citizens should themselves directly participate in these discussions. To this end, public debates should be organised in which lay people from various European countries discuss social and ethical dilemmas with experts. The results of such debates should be presented to the European Parliament and the Council.

We may summarise the answers to our four meta-questions in the following two provocative theses. First, the identity of a European ethics consists in its methods. Secondly, the ethical issues in today's Europe should not only be discussed in the institutional framework of the European Union and the Council of Europe. It is also necessary to organise public debates on these issues between citizens and experts from all European countries. The results of these debates should be presented to the European politicians.

Notes

1   Mayer, Igor S., de Vries, Jolanda and Geurts, Jac. (1995), 'An evaluation of the effects of participation in a consensus conference', in Joss, Simon and Durant, John, *Public participation in science. The role of consensus conferences in Europe*, Science Museum: London, pp. 109-24. See also Mayer, Igor S., de Vries, Jolanda and Geurts, Jac. (1996), *Effects of Participation. A Quasi experimental Evaluation of a Consensus Conference on Human Genetics Research*, Tilburg: Report Platform for Science and Ethics, Tilburg University, Department of Policy and Organisation Science.

2   NOTA/Rathenau Instituut (1993), *Report to Parliament. Public debate on the genetic modification of animals*, The Hague, pp. 7-14.

3   NOTA/Rathenau Instituut (1995), *Predictive genetic research, where are we going? Report to Parliament*, The Hague, pp. 2-3.

4   Collingridge, David (1980), *The Social Control of Technology*, Bristol.

5   This diagram has been developed in the Dutch report by Fretz, L. and Vorstenbosch, J. (1994), *Reëel, Rationeel en Redelijk. Een onderzoek naar normatieve vragen rond alternatieven voor transgenese bij dieren* (In opdracht van het Ministerie van Landbouw, Natuurbeheer en Visserij), Amersfoort/Utrecht.

6   Williams, Bernard (1985), *Ethics and the Limits of Philosophy*, Cambridge Massachusetts, p. 16.

7   Co-ordination Commission Risk Assessment Research (1995), *Unanswered Safety Questions when Employing GMO's*, Workshop Proceedings Leeuwenhorst Congress Centre, Noordwijkerhout (The Netherlands).

8   Fretz, L. and Vorstenbosch, J., op. cit.

9   A German version of the following pages has been published in Fretz, L. (1995), 'Möglichkeiten und Grenzen der Ethik im europäischen Integrationsprozess', *Wiener Blätter*, Vol. 84, pp. 43-52.

10 Delors, Jacques (1992), *Le nouveau concert européen*, Paris, p. 290.

11 Pichot, André (1993), *Histoire de la notion de vie*, Paris.

12 Morin, Edgard (1987), *Penser l'Europe*, Paris.

13 Ibid., p. 198.

29 Summary of the discussions

Sigrid Sterckx

Dr Fretz, the chairman of the discussions, explained that he had asked four people to prepare a short statement, to initiate the debate.

The first statement was made by Dr Huub Schellekens,[1] the Chairman of the Dutch Committee on Genetic Modification and a member of the Committee on Biotechnology and Animals. He said to be optimistic about recombinant DNA pharmaceuticals and vaccines and called these 'the real success story in medical biotechnology'. Thousands of patients are alive today who would not be alive without these products. Concerning other areas of medical biotechnology, however, Dr Schellekens is rather pessimistic. As to gene therapy, he explained, we have now about 150 studies world-wide, employing about 2,000 patients. Some of these trials are on-going, but a number of them have been completed and up to now we have not a single proof of the beneficial clinical effects of gene therapy in a single patient. According to Dr Schellekens this is not only caused by the lack of appropriate technology, but also because old concepts are being employed, which have not been successful in the past and will probably not be successful now. We have to improve on the concept of the diseases we would like to treat with gene therapy, he underlined. About germ line therapy Dr Schellekens was very short: the improvement of diagnostic methods such as pre-implantation diagnostics will make germ line gene therapy in the end unnecessary. As far as transgenic animals are concerned, they can be divided into two categories: using them as reactors to make pharmaceuticals and using them as scientific models, disease models. Regarding the first group of animals, Dr Schellekens is not yet convinced that they are really superior to other sources of these products like animal cells or yeasts or whatever host cell you use for the production of pharmaceuticals. In The Netherlands, animals are not used if there are alternatives. As to the second category, transgenic animals used as scientific models have become important scientific tools. However, Dr Schellekens warned, their commercial value should not be overestimated. The Oncomouse, for example, is not a commercial success. No commercial value means no reason to patent. In Dr Schellekens' view there are moral and technical problems with the patenting of genes and animals. He provided an example: it is difficult or even impossible to unambiguously describe your invention, not only if it relates to animals but also to

266

viruses. If new variants of a virus are discovered later, are they part of the patent or not? Companies and others deserve to get their investments back, Dr Schellekens stated, but there are alternative ways to protect inventions. He gave an example from the field of pharmaceuticals: the main investment in pharmaceuticals is in clinical trials, which are designed to find an application for the product. You could protect the companies on the level of registration of the products. If you found an original application of a product, you could be given the exclusive right, for a number of years, to sell it for that application. That is the system which is used now in the US to protect companies who invest money in orphan drugs. There are alternative ways to protect people and to see that they get their money back, Dr Schellekens concluded.

The second statement was made by Isabelle Meister, Greenpeace International's Genetic Engineering Co-ordinator. In her view, the presentations given at this workshop concerning the state of the art were very optimistic. Mrs Meister wondered whether it is possible to feed the world with genetic engineering and which risks are involved. This kind of information was lacking on the first day of the workshop, she stated. However, she observed, there were also some very interesting ideas, some new trends in the discussion. A very interesting idea was to consider the linking of patent rights with responsibilities for any damage or any suffering that arises. An argument that is becoming stronger and stronger, Mrs Meister noted, is that the patent system is not the right place to discuss morality. It is said that morality should be discussed when it comes to research or exploitation, but not in the context of the patent system. This worries her a lot because the reality is that it is discussed in this context and this discussion should not be denied. Mrs Meister emphasised that people should not try to shift this discussion to the level of research and marketing.

The third statement was made by Dr Tom Claes, post-doctoral researcher at the Department of Philosophy and Moral Science at the University of Ghent. He began by admitting that he is a very convinced proponent of biotechnology and patenting: he was a convinced proponent at the start of the workshop and he still was on the third and last day. However, in the course of the workshop, Dr Claes had somewhat adapted his triumphalistic view on science. His view had become somewhat milder. The only things we need to agree on according to Dr Claes are the following statements:

– some biotechnology is allowed, but not all biotechnology is allowed

– some patenting is allowed, but not all patenting is allowed

– some production is allowed, but not all production is allowed

He referred to the Oncomouse case and said that he considers oncomice as patentable subject-matter. Dr Claes does not know, however, whether *this* specific Oncomouse can be patented, because that is in part a practical problem. But oncomice in general are patentable in his view. Dr Claes labelled himself as a 'convinced antropocentrist'. He likes human beings. He likes animals too and would not want to see them suffer too much, but in cases where the suffering of animals and the suffering of people need to be balanced against each other, Dr Claes believes that the benefits have to be for people. Of course we have to be careful, he noted: we have to make sure that not too many irrelevant oncomice are produced, but this is a practical problem. It would be interesting to draw up a list with typical cases that are unpatentable, like the letter-bomb, Dr Claes observed. He asked those who are pro patenting and biotechnology which cases they think should be excluded, and he also asked those who are against biotechnology and biotechnology-patents whether they see any cases — and if so, why — in which the patenting of genetically engineered animals or plants is justified.

Dr Fretz, the Chairman, asked Dr Claes whether he felt that the public fear towards biotechnology is irrational and whether a public consensus should be reached before political decisions are taken. Dr Claes' answer was that the general public is not either rational or irrational. Sometimes, people behave rationally and sometimes irrationally. The concept of 'the general public' is a concept which is linked to democracy. Whenever you make a democratic choice, however rational or irrational, in a democratic society this choice should be respected. An important question, Dr Claes noted, is whether the general public should be educated up to the level of scientists, so that they will see that biotechnology and patenting are worthwhile. He could not answer this question, though. As to Dr Fretz's second question, Dr Claes replied that he does not think that a public consensus should be reached before taking political decisions in the field of biotechnology. It would be desirable, he added, but postponing political decisions until a consensus has been reached, would mean waiting for Godot.

The last statement was made by Dr Ulrich Schatz, Principal Director International Affairs at the European Patent Office. For him, all the parts of the workshop had been a learning experience. However, he remarked, consensus may not be within reach because each party, with the exception of those having discussed ethical aspects as such, has made winning speeches. Those involved in the biotechnological research depicted their wonderful products without saying what these products would be capable of if the inventors had less good intentions than they actually have. The patent people — and Dr Schatz included himself here — made winning speeches in the sense that they have not sufficiently indicated what they consider as the real problems in the field. Two specific problems are associated with biotechnology patents, he explained. The first is inventiveness: Dr Schatz predicted that this will become a critical problem within a few years, unless completely new ways of performing genetic engineering come up. The real solution of the debate, he said, may be that genetic engineering as a whole becomes trivial, or 'obvious', in patent law terms. According to Dr Schatz this might happen within a decade and it would be the end of patentability. The second problem with biotechnology patents has to do with the breadth of claims. A claim like in the Oncomouse case, which comprises all non-human mammalian onco-animals, is inconceivable to the general public. Is it really so that what is done with one mammal is technically under the same conditions as when it is done with another, Dr Schatz wondered. He explained that it must be possible to apply the invention without undue effort over the whole range claimed. Dr Schatz is not so sure whether that is always the case with biotechnology-patents. As to the presentations of the opponents of patenting, those were winning speeches as well. Dr Schatz admitted that he had some intellectual problems in this context, because in his view the opponents had not made sufficiently clear what their values are. He gave his personal view: the patent system as such has no value. It is a tool within market economy. We cannot solve all problems with market economy and we cannot and should not solve all problems with patent law. We should discuss the limits of our own proposals. Dr Schatz emphasised that, as a representative of a European institution, he fully agrees with Dr Fretz that Europe should not impose super-structures which are not supported by public consent. We should be prudent here, he warned. The European Patent Office should not determine what public order and morality are in the different Contracting States. Dr Schatz expects that we will see a process in which these concepts are 'Europeanised' progressively, but to the extent that they are interpreted differently they should be observed in their differences. Germ line gene therapy is forbidden in Switzerland and in Germany, he illustrated, so we should not grant patents for germ line gene therapy in these countries. In France, however, on the basis of a law, substantial amounts of public money are invested into research in that field, so the French public order is different. Public order is not in the hands of the European Patent Office, Dr Schatz underlined, but in the hands of the French, the Dutch, the Germans, and so forth, and this is true about morality as well. We should not, because of the fact that European patents are bundle patents for all countries, postulate that the interpretations of morality and public order are the same in all countries, he remarked. They are the same in a high number of cases, but not in principle.

After having listened to this fourth statement, the Chairman declared the discussions open.

Mr Ronald Schapira (US Patent & Trademark Counsel) said that in his view Dr Schatz made an excellent point: the morality in Belgium might be perceived as one thing, and the morality in France would be another, and in England a third thing. The concerns of the Green parties may be adopted in one or more countries or parts of countries, he asserted, and not in others. Mr Schapira expected that we could at least arrive at a consensus on that. Michel Vandenbosch (Global Action in the Interest of Animals, Belgium), however, did not agree. He expressed great concern about such a 'nationalist' morality, as he called it, because it results from a provincial or regional spirit. Mr Vandenbosch seriously doubts whether there are such things as a Belgian morality, a French morality, and so on. He also remarked that the fact that a government invests large amounts of money in a particular field of research does not prove that the general public agrees with this.

Dr Geertrui Van Overwalle (Centre for Intellectual Property Rights, Catholic University of Leuven) had drawn up a scheme of the life cycle of a product: starting from the research, over to the patent procedure, from the patent application until the patent grant, and then testing, exploitation, and in some cases patent infringement. She has the impression that the issue of morality can arise at each of these moments.

Different audiences put the question of morality at different moments of the life cycle of a product, Dr Van Overwalle noted. She feels that sometimes this makes the discussion very unclear. Many of the opponents of biotechnology patents who spoke at the workshop raise the question of morality at the beginning of the life cycle of a biotechnological product. Even before research has started, they want to ask if it is legitimate and what the risks are, Dr Van Overwalle asserted. Some people ask the question during the patent stage: they ask whether it is morally acceptable that a patent is granted on such or such technology. Still other people raise the issue at the time of exploitation. Others, like Mr Alexander, want to take up the problem at the moment of patent infringement. Dr Van Overwalle underlined that people have to make clear to one another at which level they want to raise the question. She wondered whether the question of morality needs to be asked at only one stage, or has to be repeated over and over again, at different stages? Dr Dani De Waele (researcher at the Department Philosophy and Moral Science, University of Ghent) remarked that the patenting issue is but a ventile, through which commotion on different levels *can* be vented. But of course, the commotion concerns research topics, production, consumption, and so on. If one wants to vent one's opinion on specific kinds of research, the only way to do this is through one's behaviour as a consumer, Dr De Waele pointed out. If one has the choice, one can decide not to buy certain products. But unfortunately citizens do not have many opportunities to show their opinions.

Dr Christoph Then (Keine Patente auf Leben, München) argued that there are different problems at different stages of the life cycle of a product. It is not possible to solve all these problems in the discussion about the technology as such, he said, because *specific* problems arise in the context of the patenting issue. So it is not possible to say we just have a discussion on the technical base and we can show there is no ecological risk or a risk for human health. Even if everybody would agree — which is not the case — that we all want biotechnology and we all want gene technology, we would still have to discuss the problem of patenting those things, Dr Then stated.

Mrs Larissa Gruszow (Principal Administrator International Legal Affairs, European Patent Office) referred to Dr Fretz's statement that we should not leave the duty to legislate on these problems to policy makers alone. In our democratic societies, we have always had movements of people who were organised in that or in another way and who influenced the politicians by what they did, she noted. The most apparent situation are strikes. However, she asserted, we should not minimise the importance of policy makers, because in our democratic system they are elected by the public. A second remark made by Mrs Gruszow was that she cannot completely agree with the opponents of biotechnology patents when they say that the patent system is the place to show their opinions. Of course that is a possibility, she said, but there is something not completely just in that because when they come to the patent offices they are alone, i.e. they are without a certain part of the public which is pro or has no problem with that. Mr Dirk Holemans (researcher at the Department Philosophy and Moral Science, University of Ghent) reacted to this remark. He stated that the fact that opponents of biotechnology patents are not backed by society is not really an argument, because the people from industry are also alone. They are not backed by society either. They do not have x per cent of the people saying: we want this patent, we want this product. Mr Holemans also made an observation concerning opinion polls. Advocates as well as opponents use the results of questionnaires in a biased way, to prove that they are right. The use of questionnaires in this context is problematic, he said, because it means that you go to somebody who is not informed — there has not been any debate — and then you ask ten questions. Moreover, the way you ask these questions influences the answer. The difference with a real democratic decision is that, in this case, people have time to inform themselves, there is a debate, and then a large proportion of the people votes, not just a thousand people out of ten million.

Mr Michael Linskens (Dutch Society for the Protection of Animals) took the point and admitted that questionnaires can be tricky. Nevertheless, he said, they can be a useful tool, but there should be a consensus on the questions that are asked. If simple questions are raised — e.g. 'Do you agree with patenting or not?' — you can do a simple check. Of course, when you inform people, you can do another check. But there are some tricks that can be played, Mr Linskens repeated. He also reacted on Mrs Gruszow's statement that the hundred opponents who were in Munich last November [Oncomouse case], were only a hundred. In fact, Mr Linskens explained, they were supported by signatures from over two hundred organisations all over Europe. So there *is* public concern about patenting and this should be recognised. For the European Patent Office, it was one of the first times that public order and morality had to be dealt with. This is a huge problem, Mr Linskens acknowledged, and the Oncomouse case has shown that the EPO is not able to cope with this problem. However, the morality provision is there in European patent law and it has to be dealt with. As to the statement of Dr Tom Claes that some biotechnology and some patenting is allowed but not all, Mr Linskens wanted to reply from an animal welfare point of view, but also from the point of view of a citizen. If you ask animal welfare organisations about genetic engineering of animals, he said, they will say no. If you ask them about patenting, they will also say no. The difference between these two no's, he explained, is that the no to patenting is a fundamental one. There cannot be 'a little bit patenting'. It is either patenting or no patenting. What we can see in society is that there is a lot of opposition to patenting as such, Mr Linskens said. Patenting of living organisms is considered to be immoral by a lot of organisations and a lot of people and politicians, including the European Parliament. So in the case of patenting we have a black and white situation. The situation regarding genetic engineering is different, he clarified. In the Netherlands a consensus has been reached about the policy. The Dutch Society for the Protection of Animals opposes genetic engineering but agrees with the law, which says that genetic engineering is not allowed, *unless* it can be proved that it is necessary and that there are no alternatives to reach the same goal. So on the level of legislation there is a certain form of consensus in the Netherlands concerning genetic engineering of animals. Of course, he added, patient organisations and industry may say that genetic engineering is important for society, and probably it is a good thing that animal welfare organisations

271

are not the only ones who make the rules in a country. Rules have to be made on the basis of a democratic process.

Mr Linskens further said that he is very worried about the view, expressed by Dr Tom Claes, that we should educate people to make clear to them what the opportunities of biotechnology and patenting are; why biotechnology and patenting are worthwhile. This, he argued, is a fundamental mistake of education. In the Netherlands, industry took the same position: they wanted to educate people with the goal of acceptance of biotechnology. NGO's, however, feel that a critical reflection on biotechnology is needed and that people should be educated from different points of view. At this moment, he added, the attitude of the Dutch industry is changing: they are willing to communicate on a level which is also accepted by NGO's. However, the majority of scientists has not reached that point yet. Scientists have to learn that science is not a holy cow. The people should be educated to be able to make a critical reflection on biotechnology. Mr Linskens emphasised that a lot of work remains to be done because the policy of the European Commission is only about competitiveness and promotion of biotechnology.

Dr Tom Claes elucidated that he was just arguing against those who say that the public is too stupid. The general public has the capacity of being rational *and* irrational, but of course they are still citizens in a democracy. So Dr Claes rather agreed with Mr Linskens.

The Chairman made a few summarising remarks. He said that the question raised by Mrs Van Overwalle had big consequences, because indeed each stage in the life cycle of a biotechnological product implies moral questions, but not the same moral questions. Another very important question, he noted, was the question about the complexity of the political decisions. These decisions have a big impact on public morality, whereas the public does not always know what the moral consequences of political decisions are.

Ms Sigrid Sterckx tried to combine some of the written questions, formulated by the participants in the course of the first and second day. Many of those questions were about scientific aspects and many people focused on the same kind of scientific questions. To put it generally: 'is there a fundamental difference between traditional biotechnology and modern biotechnology?' Somebody made the following remark in a written question: 'many committees, workshops, consensus conferences, etc... have come to the consensus view that the method of transgenesis does *not* by itself create a new situation, that produces an intrinsically new and therefore potentially dangerous type of living organism. What is said here, is that no new situation arises and, more importantly, no new ethical questions arise because of these methods of recombinant DNA technology'. The participants who had written down observations of this kind felt that we have been doing the same thing for centuries with traditional breeding methods. So in their view there is no fundamental difference.

Mr Ronald Schapira said that such remarks are easy to respond to: they are false. The traditional breeding methods involved the same species, he explained, but now we are putting human genes into animals, we are putting insect genes into plants. This is cross-species transfer that is virtually never possible in nature. So there is certainly a difference in kind. Dr Jan Van Rompaey (Manager Technology Protection, Plant Genetic Systems NV, Ghent) did not agree with the statement that cross-species barriers are not crossed in nature. When you look at plant biotechnology, he said, the major instrument is Agrobacterium-mediated transformation, and that is an example of a situation in which genes of bacteria are introduced into a plant genome. Dr Dani De Waele replied that this is not true. When biotechnologists say that there is horizontal gene transfer in nature, in evolution, between an Agrobacterium and a plant, that is true, she pointed out. But when Agrobacterium colonises a plant in nature, the plant dies from it. A plant containing Agrobacterium genes cannot be reproduced out of it. The point in biotechnology, Dr De Waele underlined, is that when a gene transfer between two species is made, the descendants should carry the recombinant gene and these descendants should be able to reproduce. So there is a fundamental difference between what happens in nature and what happens in biotechnology. There is also a fundamental difference between classical breeding and biotechnology, she noted, because in classical breeding one could only work with phenotypic characteristics. One could not directly choose the genetic component. Since modern biotechnology, recombinant DNA and restriction enzymes came in, one can — if one knows what the genetic component of the characteristic is — choose the gene, pick it out of the chromosome and put it in whatever other organism, since the genetic code is universal.

According to Dr Huub Schellekens, the scientific truth is somewhere in the middle: natural transfer of genes exists, but there are examples of gene transfer which we do in a laboratory which are unnatural. What we have to take in mind, he argued, is a matter of scale, because scientists now come up with at least ten new transgenic animals per week. That is much more than what we see through the natural evolution or the natural occurrence of mutants. In the other fields of biotechnology, Dr Schellekens estimated, 50,000 people throughout the world are working on genetic modification. Of course this creates mutants in a much more directed way than nature was ever able to do.

Dr Ulrich Schatz remarked that he fully agrees with Mr Schapira's view that there is a difference in kind between classical breeding and modern biotechnology. It is not just a difference in scale, he said. The best proof of this is that the results of breeding are not patentable, but the results of genetic engineering are. We would be completely wrong in denying that difference in kind.

Mr Stephen Crespi (Patent Consultant, London) referred to Professor Schell's talk on the first day, in which he said that it is wrong to assume that genetic modification must necessarily be dangerous. He said it could be more safe than traditional methods. Here is an expert, said Mr Crespi, and I have to listen to an expert. Professor Schell said gene technology was not designed to make monsters. Because something happens in nature, this does not mean that it is safe. Professor Schell gave examples of quantification, showing that in traditional breeding you might be introducing as much as two per cent foreign DNA, whereas in the transgenic approach you are getting small percentages of changes of genes: less than 0.6 per cent, he quoted. Mr Crespi repeated that we have to listen to the scientific experts.

Dr Fretz, the Chairman, remarked that the problem is of course that the experts are not unanimous.

Ms Sigrid Sterckx noted that many people had written down questions about the economic aspects of patenting. She summarised these questions by quoting somebody who wondered 'how broad a patent can morally be' and added 'the aim of a patent is to regulate competition. That is OK, but what about the ruling out of healthy competition? This is necessary for bringing out the best of science to the public.' In other words: can we still talk about healthy competition in the case of broad patents?

Dr Jan Van Rompaey noted that the problem of broad patents, in biotechnology in particular, is a question that is being put on the table in the patenting world as a whole. It is indeed sometimes true, he admitted, that the scope of a claim is not fair. Any patentee tries to get the broadest possible protection, and it is the duty of the patent office to regulate that. However, the patent offices do not always apply a fair view on this. But this problem, Dr Van Rompaey repeated, is being put on the table in industry as a whole and in the patenting world as a whole. Dr Christoph Then argued that broad claims in the context of biotechnology create a new problem because organisms can reproduce themselves. If you have a claim on a gene, you can have a claim to all organisms incorporating this gene and to all the following generations of these organisms. This is a totally new situation which arises in the patent system because of gene technology, he noted.

Dr Ulrich Schatz added that the problem is that applicants tend to cover by the claim — which describes the limits of the exclusive right conferred to them — things they have not invented. So they try to seek protection for things they have not invented. A claim may be broader than an example on which the applicant has actually worked, Dr Schatz said. That is fine as long as the disclosure makes it clear to the man skilled in the art that once you have done it with this organism, you obviously can do it with the other organism. Then, he argued, the invention must be protected and is fairly protected as covering the range to which the invention can be applied by any man skilled in the art in that profession. That is fair but if it goes beyond it becomes unfair. Dr Schatz observed that this problem also exists in chemistry, and that it specifically exists in all basically new technologies. He gave the example of the car industry, which is an old technology. Here, what you can add by inventions is a small improvement on a huge background of old technology. In a young technology, however, you see many pioneer inventions, which condition a whole technology, so you get broader claims which have a broader applicability because of the basic inventions which are made there, Dr Schatz explained. The problem of broad patents is recognised as a problem which exists in biotechnology and must be properly addressed by the patent offices, he concluded.

Dr Christoph Then argued that in the field of living organisms every claim which is directed to the organism, which derives from a claim which is directed to a gene or a method incorporating that gene, is a very broad claim because you did not invent the organism. This is a totally new situation, he said, because if you have a chemical compound and you add a new compound, the resulting compound is made by man. This is not the case for an organism in which you incorporate a gene. Dr Jan

Van Rompaey replied that what is being patented is not the organism in which the gene is put, but the organism that contains the gene. That is a *new* organism, he asserted, that did not exist in nature before. The patent rights do not extend to the organisms that do not have the gene.

The Chairman reminded the participants that Mr Crespi put a very important thesis, namely that the act of patenting is ethically neutral. If this is true, Dr Fretz said, the patenting of gene constructs could be ethically neutral as well. However, if it is not true that patenting as such is a neutral act, does this mean that then the question arises whether this has to do with the fact that things are being patented now that were not patented before? Dr Fretz was interested in hearing the opinion of the experts in the field of patenting.

Dr Ulrich Schatz began by stating that he cannot think of any human act which is ethically neutral. All human acts have an ethical quality, he said, because an act is something that you do with a certain purpose and something that has consequences. As to the 'ethical statute' of patenting, Dr Schatz emphasised that the main ethical aspect of patenting is that it ensures competition which is fair and right. To allow competition at equal conditions, timely limited protection is given. This is the moral aspect, Dr Schatz noted. The second aspect — which also has a moral quality but which is not a moral motive — he called the 'utility aspect'. If protection of inventions is available, then industry and research institutions become motivated for investing into research and development. In this context he quoted Lincoln: 'the patent system adds the fuel of interest to the fire of the genius'; the patent system makes inventing commercially viable. Dr Schatz argued that a patent does not confer a right on a thing. Concerning the arguments saying that patenting animals turns them into a commodity, he expressed the view that animals *are* a commodity. He referred to the stock markets, where animals are traded as commodities. The ownership of tradable goods makes them a commodity. Animals are an object of trade in our society and a patent does not change this status according to Dr Schatz. It only says who is allowed to trade this commodity: the inventor and his successor in title, be it a licensee or whatever. The patent system is nothing but a regulation of competition. In Dr Schatz's view the questions to be answered are 'does patenting, as an act, have other consequences, and which ones, in the field of biotechnology, than in any other field?' and 'if not, are the consequences a patent ordinarily has in any other field, acceptable in the field of biotechnology?' His personal assumption is that, unless there is evidence to the contrary, the functioning and the social effects of patenting in the field of biotechnology are the same as in any other field. However, he concedes to those who are against patenting that, as in other fields, patenting is giving a strong incentive to research and development. So, he addressed the opponents, if your objective is to slow down research and technical progress in biotechnology, you are right: you should stop patenting such inventions. But, Dr Schatz added, we should not answer the afore-mentioned questions in an abstract world. The real world is that we live in Europe and we can exclude in Europe, but what will be the consequences, he wondered, since all around Europe these things are patented.

Mrs Janne Kuil (Anti Vivisection Organisation, the Netherlands) remarked that the first reason why she is against the patenting of animals is that she is opposed to the research. However, she explained, there is another reason to be against the patenting of animals. Mrs Kuil described the situation in The Netherlands: fortunately, she said, many researchers first try to use other methods of research and a lot of alternatives are being developed. Only if that is not possible, animals are used. If patenting of animals becomes possible, she fears people will no longer look for alternatives, or they will do so to a lesser extent. Mrs Kuil also noted that Dr Schatz's point that patenting makes development and research possible is the view of *some* researchers. A lot of researchers have made their inventions known to the public not by patenting, but by putting them on the Internet, so that everybody can use the information. So, she concluded, among researchers not everybody is in favour of patenting. Mr Stephen Crespi replied that he cannot understand the logic of the argument that if you allow the patenting of animals people will stop seeking alternatives. Rather the reverse is true, he asserted. If you patent animals, you do not force anyone to adopt that particular technology. In fact people who want to avoid having to pay you for the right to use that technology *will* seek alternatives. According to Mr Crespi this is one of the objectives of the patent system: to stimulate alternatives to what is patented. There may be alternatives to patenting, he said, referring to Dr Huub Schellekens, but we are living in a free society so we must have the freedom to choose whichever way we think is appropriate to our interests.

The Chairman explained to Mr Crespi that if public policy stimulates certain technologies, whereas others are not stimulated, there is no freedom to choose. In The Netherlands e.g., he illustrated, biotechnology is stimulated very strongly by the politicians, but there is no political support for stimulating the alternatives. One can say that we are living in a free society, but if people do not have the possibility to choose, this freedom is a curious freedom. It is a freedom without force.

Mr Stephen Crespi came back to his point about ethical neutrality because he had noted that Dr Schatz does not agree with him. As to Dr Schatz's suggestion that all human acts are ethical, Mr Crespi observed that he can think of many human acts that are not ethical, one way or the other. He gave an example: deciding to let one's beard grow long or deciding to shave. This is not an act of ethical significance, Mr Crespi argued. It is an act of ethical indifference, neither right nor wrong. The same holds for patents: seeking a patent is simply seeking the right to exclude others, which is neither right nor wrong.

Ms Sigrid Sterckx asked Mr Crespi whether she would be correct in summarising his argument as follows. There are three options when there is a patent infringement: either the patent holder can close his eyes and pretend nothing has happened (when it is only a minor infringement); a second option is to try to arrive at some arrangement with the infringer and the third option is to go to court. So there are three options and you cannot know in advance which option the patent holder will take, so you cannot say 'this will be the *inevitable* consequence of the grant of a patent' and therefore patenting is ethically neutral. Mr Crespi agreed. This made Ms Sterckx ask Mr Crespi how many acts he could name which lead to an *inevitable* consequence?

Few if any such acts exist, she said, so according to Mr Crespi's argumentation only a few acts would be ethically relevant. Dr Schatz took up the point and repeated that in his mind, there exists no human act which would be ethically completely neutral. He gave an example, saying that if Ms Sterckx would cut her wonderful hair, that would do harm to him because he enjoys to see her with her wonderful hair. So in his view this has an ethical aspect.

Dr Schatz also reacted to Mrs Kuil's argument. He provided an example outside the field of biotechnology: IBM has a journal in which they publish things they have invented, but will not patent, because in their view there is no commercial interest in it which would justify the expenses for patenting. However, they publish it in order to avoid that somebody else says he has invented it and goes to the patent office — which would mean that IBM would be excluded from an art which IBM has developed. This is one more way of using the patent system, Dr Schatz explained. In many cases, patent applications are dropped just after publication because the only purpose of the applicant is to publish it and to abandon it.

The Chairman formulated a short conclusion about the discussion concerning the 'moral status' of the act of patenting. It is interesting, he remarked, to see that the experts in the field of patenting do not agree about the answer to the question whether patenting is an ethically neutral act. Dr Schatz says it is not, whereas Mr Crespi says it is. So not only the scientists disagree on scientific issues, but the lawyers disagree as well about the interpretation of the moral status of the act of patenting. Dr Fretz's second conclusion was that it is interesting to see that there is openness for discussion. He asked the patent experts whether they *could* imagine that the moral status of the act of patenting — which is, as such, acceptable to them from a moral point of view — *could* change by unforeseen consequences of patenting, or whether they would exclude that possibility. Dr Schatz replied that he does not exclude this at all. There are cases, he clarified, where precisely ethical considerations are at the root of exclusions from patentability; this is the case for medical treatment. The patent legislators say that they consider it not appropriate that the patent system, as a regulation of competition, becomes the basis for the exercise of medical art. That is why Art. 52(4) EPC exists. Dr Schatz provided a second example: the socio-economic milieu of industry and the milieu of traditional farming (a family living on its ground) are milieus which do not abide by the same economic conditions. They are not at equal footing under economic terms. Therefore, plant breeders' rights give the farmers a privilege, so that the farmer can run his farm and renew his seeds for the purposes of his own farming business, and he cannot be bothered by the owner of the exclusive right on the plant variety. The patent people say there is a serious argument in that and the same should be done when we patent an invention in the field of plant genetic engineering and animal genetic engineering. Of course the farmer's privilege must be limited to the effect that, *if* the farmer starts entering in commercial competition, namely when he starts producing seeds in big masses not for the purpose to use them on his ground, but to become a producer and a businessman in that field, *then* he enters the domain where we want to apply the competition rule set by the patent system. So according to Dr Schatz the patent legislator takes into account

277

substantiated socio-economic, moral or whatever other reasons to exempt an activity from the field where the competition rule applies.

Mr Ronald Schapira said that for him, the question whether the patent system is ethically neutral or not is analogous to the question whether producing gun powder is an ethically neutral act. Gun powder can be used for beneficial purposes or it can be used for horrible purposes. In Mr Schapira's view a patent is like the production of gun powder or the production of a lot of other things that can be used for beneficial or for harmful purposes. He does not exclude that a certain abuse could occur. The risk involved in biotechnology, he said, is that it is such a new technology. Patents in the domain of biotechnology may be an instrument for abuse as well as good.

Dr Huub Schellekens made a remark about the cultural differences that were discussed earlier. It was said, he recalled, that we have differences in opinion about morality in Europe. The way Dr Schellekens interpreted this is that some of the speakers want a system in which the technical and legal decisions on a patent are taken in Munich, whereas later on the individual countries will decide whether they will adopt the patent. That could be difficult, he warned, because in Europe we do not have a system in which the market has been declared holy in free trade. In the Netherlands, for instance, there is a system for the regulation of research concerning transgenic animals. The government has said that if somebody wants to make a transgenic animal, he has to go to a commission and seek a permission. However, if the transgenic animal comes from another country, the opinion of the Dutch government now is that the same system cannot be applied because Brussels will consider this an unfair trade inhibition.

Dr Christoph Then noted that certain economic aspects of patenting should also be evaluated under the heading of public order. Dr Tom Claes called this a very important observation, because the patent system is embedded in a specific social, political and economical system. It is a system some of us like and others would like to have it changed, Dr Claes said. A lot of arguments against patenting are embedded in a much larger societal programme. According to Dr Claes many of the arguments of the opponents are not so much arguments against patenting as an activity, but against patenting as a symptom of a certain economic and political system they do not like. Dr Then argued that as a result of patenting in the field of gene technology, very often economic situations arise which are ethically not tolerable.

Ms Sigrid Sterckx noted that it was getting late and that the discussions should have been closed already. She said there was room for one more question: if there is a fundamental difference between obtaining a patent on the one hand and obtaining the authorisation to exploit an invention on the other hand — and this is what is always said: it is not because you have a patent that you can exploit it and Art. 53(a), the 'morality provision', is not about the act of patenting but about the act of *exploitation* of the invention — well, then why does a European *Patent* Convention prescribe that the morality of the exploitation of inventions for which patent protection is sought should be assessed in order to decide on whether or not to grant a patent?

Mr Siegfried Van Duffel (philosophy student) said that two arguments are constantly being mixed up in this discussion. The first is the argument Dr Schatz called the 'utility argument'. The second argument is that when somebody claims a patent on an invention, what comes in is a property right to this invention and that has little to do with utility arguments. So, seen in this light patenting is certainly not ethically neutral, he argued, because the consequence of a patent is that everybody else is forbidden to exploit the invention, even if the owner of the patent is not allowed to exploit the invention himself. A patent is a property claim, so it is certainly not ethically neutral.

Dr Schatz reacted on the Chairman's question whether he could imagine that there should be exceptions to patenting because of moral reasons or economic consequences. He referred to the talk of Mr Luc Vankrunkelsven, who had given a list, a kind of doom scenario, of what would happen if we patent living matter. Mr Crespi had called these arguments totally arbitrary and without any basis in reality, but Dr Schatz said that this was not his point. He emphasised that if a substantiated case could be made that this list of points is correct, we should immediately abolish patenting in biotechnology.

The concluding remarks were made by Ms Sigrid Sterckx. She said that first of all she regretted that the discussions had to be closed already, because only about one third of the important issues she had written down during the preparation of the discussion had been handled. Ms Sterckx expressed her gratitude towards everybody who attended the workshop and especially the people who had accepted her invitation to give a lecture. She was happy to have been able to bring together people with such differing backgrounds because this made the discussions so much more interesting. Ms Sterckx felt that there had been a very stimulating exchange of ideas at the workshop.

Note

1   Dr Schellekens is a M.D. and his speciality is the use of recombinant DNA-products in medicine.

Part Seven
COMMENTS AND CONCLUSIONS

30 Comments on the proceedings of the conference on biotechnology, patents and morality

Harriet M. Strimpel

According to Art. 1, 8 of the US Constitution, signed into law in 1790, 'Congress shall have the power ... to promote the progress of science and useful arts by securing for limited times to authors and inventors, the exclusive right to their respective writings and discoveries'. The inclusion of this clause in the American Constitution is an acknowledgment that even 200 years ago patents were considered of great importance to the citizens of the United States. The patent system has survived hundreds of years of scrutiny and has been adopted in a relatively uniform manner by a large number of countries.

The principle of patenting is to provide the public with access to new inventions. The reward to the inventor for making public his invention is a promise to protect the inventor, for a limited period of time, from the exploitation of the fruits of his labors and ingenuity by others. Public access to descriptions of inventions that meet a market need has the effect of encouraging inventive activity around what had already been disclosed such that the subsequent competition fuels creativity and brings benefits to the public by access to new and improved products. The exclusionary right, awarded with the granting of the patent, may have a significant impact on the progress of science, business and public well-being. This impact may be deemed positive or negative depending on the bias of the observer.

Biotechnology is one area of commerce where patents have proved particularly important. The consequences of applying patent law to this industry, which encompasses health, the environment, agriculture, food and the chemical industry, has raised new and important patent issues that test the patent statutes in many countries. These issues relate to patentability of subject matter, scope of claimed subject matter, and sufficiency of disclosure. Some of these issues have been debated at the Conference on Biotechnology, Patents and Morality: Towards a Consensus.

Several broad areas of discussion emerged from the conference. These include:

– to what extent do European patent examiners have a responsibility to establish the morality of an invention according to Art. 53(a) of the European Patent Convention where, according to speakers at the conference, patents accord credibility to an invention and implied approval of the field of commerce;

- whether multicellular organisms should be excluded from patentability based on arguments concerning animal rights, the consequences of viewing animals as commodities, the encouragement of future developments of animal based commodities, and because inventions that are self reproducing might be potentially dangerous;

- whether the patenting of genetically engineered plants provides an unfair commercial advantage over protection of plant varieties under UPOV and whether such patenting would negatively impact the business of agriculture; and

- whether the scope of patentable inventions in biotechnology, particularly with respect to body parts, has invaded the area of fundamental discoveries which should not be patentable at all.

The above issues should be divided into two categories. The first category involves issues related to the impact of the inventions on society, while the second involves patent related issues such as what constitutes patentable subject matter and the scope of claims.

Many participants want abstract ethical issues to drive patent policy. Below, I shall discuss the approach taken by the US towards these issues which is more practical in nature. The United States patent policy is founded primarily on considerations relating to the encouragement of innovation and the advancement of the economy. Ethical principles are expressed by laws outside the patent system. The wording of the patent statutes is deliberately broad and the refinements and interpretations of this statute is left to the courts.

The US Supreme Court recognised that while there might be ethical issues associated with commercializing multicellular organisms, the patent system was not the appropriate forum [Diamond v. Chakrabarty, 447 US 303, 309 (1980)]. The morality issues that have been raised in oppositions to patents in Europe are important in the appropriate context and powerful if they are used to introduce legislation to protect animals and the environment. There are three possible outcomes to undermining the patent system by seeking to block the issuance of patents on the basis of morality. Firstly, inventors may be deterred from filing patents in those areas, with negative repercussions for research funding. This, in turn, may have a negative effect on the competitive environment that may otherwise yield improved alternative solutions to the technical solutions that are opposed on the grounds of morality. Secondly, inventors may continue to conduct research in the areas that are deemed to be harmful to the public, but will maintain this research in secret so that it will not be accessible to public scrutiny. Thirdly, inventors who have carried out research of the type deemed to be unpatentable, will publish work such that it will be freely available for exploitation without limit. Ultimately, the process of patenting encourages a variety of novel technological solutions to technical problems and in turn provides an opportunity for the public to shape the market through regulation rather than diminishing it through restricting both research and public access.

To some extent, the tension between public access to information and the withholding of patent protection can be observed with regard to the massive sequencing efforts now being pursued by genetics companies such as Sequana and Human Genome Sciences in the United States. These companies have filed mega-sequence applications with the United States Patent and Trademark Office (USPTO) containing hundreds of thousands of DNA sequences. The US PTO has reported that the cost of examining these applications far exceeds the budget of the PTO and so they languish unpublished and unexamined in group 1800 of the PTO. It is unclear that these applications would be allowed even if the claimed subject matter were novel and non-obvious, if the individual fragments lacked utility. In the meantime, unlike the European system which requires publication of applications at 18 months after the filing date, the United States patent process does not publish the invention until the date of issuance of the patent. Consequently, the public and the research community at large are deprived of access to these proprietary databases filed with the USPTO and work is being duplicated and resources wasted in industry and by university research laboratories as a consequence. Furthermore, since the US patent system depends on the first to invent and not the first to file as is the case in Europe, it is possible that an inventor who isolates a DNA sequence coding for a particular gene product with demonstrated utility, might be foreclosed from claiming the sequence because of the presence of a majority of such sequence in the proprietary databases on file at the PTO.

The second category identified above includes issues relating to the patent statutes. Selected sections from title 35 of the United States Code (35 U.S.C.) are provided here that are directed to patentability, scope of claims and sufficiency of disclosure.

35 U.S.C.§ 100 defines an 'invention' as an invention or discovery. Although the statute does not differentiate between the terms 'invention' and 'discovery' as has been done in Europe, the US patent laws require that an invention as defined above have utility and be disclosed in such manner that one skilled in the art could reproduce the invention. Furthermore, although the term 'discovery' is not a form of discrimination against patentability, nonetheless the fundamental nature of an invention is an important consideration with regard to scope of claims. An important decision that is valid today regarding patenting fundamental research can be found in O'Reilly v. Morse [56 US (15 How.) 62, 14 L Ed. 601 (1853)] in which Morse sought to patent electromagnetism. The scope of such claims were deemed too broad and he was limited to claims directed to message transfer by telegraphy. To have done otherwise could have resulted in far reaching control of a field that in turn would have restricted inventive activity in areas distantly related to the products described by Morse.

35 U.S.C.§ 101 identifies patentable inventions:

> Whoever manufactures or discovers any new or useful process, machine or composition of matter, or any new and useful improvement thereof, may obtain a patent thereof, subject to the conditions and requirements of this title.

In considering the patentable nature of a multicellular organism, the Supreme Court determined that 'anything under the sun that is made by man' is patentable (Diamond v. Chakrabarty). In considering the utility requirement of § 101, it is generally considered that the standard of industrial applicability as used by the European Patent Office (EPO) is a much lower hurdle than the standard of utility.

35 U.S.C.§§ 102 and 103 require a patentable invention to be novel and non-obvious.

35 U.S.C.§ 112 is directed to the specification:

> The specification shall contain a written description of the invention and of the manner and process of making and using it, in such full, clear, concise and exact terms as to enable any person skilled in the art to which it pertains, or with which it is most nearly connected, to make and use the same, and shall set forth the best mode contemplated by the inventor of carrying out his invention. The specification shall conclude with one or more claims particularly pointing out and distinctly claiming the subject matter which the applicant regards as his invention.

Because biotechnology is considered to be an unpredictable field, U.S. patent examiners have imposed relatively strict requirements concerning enablement and operability according to § 112 for claimed inventions (in Europe, 'enablement' is equivalent to 'sufficiency of disclosure'). Unlike the European rules which require a method for accomplishing the invention, the US patent rules demand provision of the best method known at the time, to achieve the invention that must teach those skilled in the art how to make and use the full scope of the invention as defined by the claims, without 'undue experimentation' [In re Wright, 999 F.2d 1557, 1561 (Fed. Cir. 1993); In re Vaeck, 947 F.2d 488, 495 (Fed. Cir. 1991)]. Furthermore, whereas in Europe a single example is required to exemplify the claimed subject matter, the biotechnology division of the US Patent and Trademark Office may require large numbers of examples and frequently attempts to limit the scope of the claims to the subject matter described in the examples. Consequently, it is generally perceived by patent practitioners in the US that patents may be obtained in Europe having broader claims for a particular invention than those obtained for an equivalent patent in the US.

Concerning operability, the US Patent and Trademark Office has requested and in some cases demanded clinical data supporting efficacy of the claimed invention. The imposition of this standard has given rise to the assertion by the public that the US is improperly applying legal standards governing proof of operability in particular for the treatment of claims directed to human disorders.

The application by the USPTO of selectively applied extraordinarily stringent criteria to patentability of biotechnology inventions led to a meeting between the USPTO and the biotechnology industry in October 1994 at which the frustrations of the industry were aired. The presentations made at that meeting were published by the Biotechnology Industry Organization (BIO). Part of the frustration expressed by

286

the biotechnology industry in 1994 was that the PTO was in effect threatening the survival of the industry which depended in significant part on patents to raise investment capital. This sentiment was expressed by Carl Feldman, President of BIO, as follows:

> Intellectual property protection is critical to the competitiveness of our nation in general and biotechnology in particular. Our investors will not risk their capital to create innovative state-of-the-art approaches to unique problems if meaningful patent protection cannot be secured. The USPTO therefore plays a critical role in our industry's ability to fund its research into life saving and life enhancing products.

The USPTO listened to the industry and has moderated its policies such that clinical trials are no longer demanded to demonstrate utility. There has also followed improved communication between the PTO and the applicants so that disagreements concerning the application of legal principles can be discussed and resolved so as to reduce the incidence of costly and lengthy appeals.

The reliance of the biotechnology industry on patents results from the need for some competitive benefit to offset the enormous cost of developing products and the extended product development time. For example, a novel genetically engineered drug may take 10 years to develop at a cost of about $400 million. A significant portion of that time involves clinical trials leading to regulatory approval. The approval process is directed to safety and efficacy and is relatively time insensitive. It is therefore not unexpected that by the time a drug is on the market, a mere 5 years may remain of the inventor's monopoly to provide the market protection so necessary to recover investment.

Despite the increasing number of hurdles to patentability, the biotechnology industry has successfully obtained many thousands of US patents for genetically engineered products since 1980. The industry now directly employs 60,000 people with many more being involved in related businesses that provide contract services and supplies. The investment of enormous amounts of resources into biotechnology has accelerated the rate of research and development to provide products that are deemed to have a substantial market — these markets reflecting the public desire for health products. According to Ernst and Young, biotechnology companies spent $7 billion on research in 1994, with an average expenditure of $59,000 per employee.[1] Furthermore, $15,690 million has been raised in 484 total public offerings greater than $10 million in the US since 1985. In 1996, $4 billion was raised in 110 deals. This is about four times the annual R&D budget of Merck ($1,331million).[2]

One particular segment of the biotechnology industry that has seen enormous growth in the last three years in the US is that of the genetics companies. The patent issues associated with this industry have demonstrated that a strategy has been adopted that focuses on patenting potential products of commercial significance rather than relying on patents for isolated sequences. The debate regarding patenting of DNA sequences was initiated by the government sponsored National Institute for Research when it filed patent applications in 1991 on the work of one of their scientists (Craig Venter). Intense international opposition as well as intense opposition from the pharmaceutical and biotechnology industry in the US followed this action. The patent application was directed to thousands of short DNA sequences identified as expressed sequence tags (EST). Whereas these fragments may have been novel and possibly non-obvious, no single utility for this invention proved persuasive. The patent applications were ultimately abandoned and in 1993, Dr Venter left the NIH to set up a non-profit institute funded by a new company called Human Genome Sciences (HGS). Two years later, in 1995, Human Genome Sciences reported that they owned 900,000 cDNA sequences in their proprietary databases for which patents had been filed but none obtained. In the meantime, the company is selling access to its database for approximately $70 million per customer, providing the company with a 1995 valuation of $385 million. The commercial carrot being offered to customers is the competitive advantage provided by the extensive proprietary database for obtaining full length genes of known function and therapeutic potential. HGS and its licensees expect to obtain patents on genes with a defined function coding for proteins having commercial importance. Ultimately, the commercial significance of patents claiming isolated genetic sequences may be less important than those protecting potentially commercial products. The issues raised by the Venter patent applications and by the mega-sequence applications have not been completely resolved in the United States and are under review by a committee appointed by President Clinton to study the commercial, legal, and public welfare issues associated with genetics. The outcome of this committee's decisions with regard to patenting isolated genetic sequences is not yet certain.

In the meantime, between January 1994 and September 1995, 42 per cent of 183 agreements in biotechnology concerned genetics. In February 1995, Amgen Inc. announced it would pay Rockefeller University $20 million as up-front payment for access to a gene that codes for a secreted protein associated with obesity. With development milestones, Amgen could pay Rockefeller more than $100 million. Analysts observed that the market valuation for Amgen rose $2 billion on news of the license. With deals such as these, there is a great interest in patenting genetic sequences. However, it should be noted that 70 per cent of obesity is thought to be due to genetic factors and there are more than 100 million medically obese people worldwide. It is no wonder that drug companies are interested in providing effective products for this market — but so, I imagine, are the obese public. Potentially everyone could benefit. What would have happened if the genetic information that Amgen licensed had remained in the public domain and was not patentable? The result would likely have been a much reduced investment in product development and consequently a considerably longer development time. However, the Amgen license has not provided the company with an exclusionary right to the entire field of obesity. For example, Millenium Pharmaceuticals has obtained a patent to the ob gene receptor and is developing competitive products in this field.

Another measure of the value of patents to the biotechnology industry has been evident as a consequence of patent infringement disputes in the United States. An example of such a case occurred between Amgen and Genetics Institute over the exclusionary right to erythropoietin [Amgen. Inc. v. Chugai Pharmaceutical Co. Ltd., 927 F.2d 1200 (Fed. Cir.)]. The consequences of the suit which Amgen won, was riches in the form of a multibillion dollar market in the US for Amgen and financial hardship for Genetics Institute (the licensor for Chugai Pharmaceutical) who was foreclosed from that market. In fact, the decision was reported on March 5, 1991 and the consequences were immediately felt in the market capitalisation of Amgen and Genetics Institute. Amgen gained about 25 per cent in value (about $1 billion) while Genetics Institute lost 50 per cent of its value (about $500 million). Twelve months later, Genetics Institute had still not regained any of its lost value while Amgen proceeded to go through many stock splits including two 5:1 splits.

A third area in which the value of patents contributes to the economy in the United States, has been in the area of industrial sponsorship of university research. In 1980, the Bayh-Dole Act was passed that allowed private parties to obtain patent rights in federally funded research. This act was later modified to permit exclusive licenses for companies in federally sponsored technology. This has resulted in substantial financial benefits for universities. For example, University of California received $45 million in royalties in 1993 and Stanford University received $31million for the same year. Some academic institutions, notably teaching hospitals such as the Dana Farber Cancer Institute and Massachusetts General Hospital, have entered into multimillion dollar research sponsorship arrangements with large pharmaceutical companies where first options on patent rights are an important aspect of the sponsored research program. Others, typified by Massachusetts Institute of Technology, have successfully licensed patents that have formed the basis of numerous start-up biotechnology companies bringing prosperity to the environment in which they are located.

Without extensive patent protection of innovations resulting from research, the majority of the funding from private, commercial, and public sources would quickly disappear. Patenting biotechnology has provided many public benefits since 1980.

Most significantly, with patents as a prerequisite, financing has poured into the new industry and resulted in the birth and growth of many new companies. Much needed employment opportunities have arisen for young people coming out of college with technical knowledge and lots of enthusiasm. The needs of the public for cures for the most pervasive conditions including allergies, heart disease, obesity, and infectious diseases are being addressed. The various states in the United States are competing to have production plants built locally by providing tax incentives and various financial packages. With these plants, and the prospects for employment, come prosperity for the communities. Whereas patents are by no means the only requirement of a start-up, there would be no biotechnology industry without patents, no health care products to regulate, and finally, no opportunity to contemplate the moral cost of our desire for immortality.

Ultimately, the public debate will be decided with regard to the level of risk that society is prepared to take for the perceived benefit. Unfortunately, the large majority of the public understands little of the details and challenges of the scientific discoveries that have occurred. As we have developed new and more efficient tools for moving forward, a cry is being raised louder and louder, 'we do not understand what is happening, everything is happening too fast, we do not trust the scientists to have our best interests at heart, commerce is bad and we are being manipulated'. Are things actually out of the control of the public? To some extent — yes. Education of the general public in science is inadequate, so that a reasoned public debate cannot be maintained. People have been manipulated by the media and do not trust the information that reaches them. On the other hand, the public wants jobs. The public wants prosperity. The public wants cures for their diseases. The AIDS activists demand that scientists provide immediate treatment for their disease. The adult who has had cystic fibrosis since childhood, who struggled stoically through youth, choking on mucus and knowing the end was round the corner, now occupies a useful job and lives a happy life because of a genetically engineered drug produced by a biotechnology company. The legislature and the judiciary is being called upon to balance conflicting interests that arise out of advances in health care so that the fortune of some does not give rise to an unacceptable cost to others. This balancing act is being conducted on far-ranging issues including the fair distribution of health care benefits, the impact of prognostic tests on individual employment and insurance opportunities, reproductive technologies, voluntary euthanasia and so on. These issues are complex, and require considerable input from experts in many fields including ethics, law, politics, sociology, anthropology, science and history. I propose that the discussions should not be directed to banning the technologies that allow these issues to arise or condemning man's curiosity by barring research subjects that may be investigated. But rather, the issue should be on how to distribute resources and to ensure that everyone in the public domain has a fair and reasonable access to the market.

The Conference 'Biotechnology, Patents and Morality: Towards a Consensus' has raised important issues regarding awarding certain exclusionary rights on the basis of morality. The impact of denying patentability to biotechnology inventions, will not prevent their use. The economic significance of denying patentability should be carefully considered in the context of public interest and welfare.

Notes

1   Ernst and Young (1994), *Biotech, Ninth Annual report on the Biotechnology Industry IX.*
2   Unpublished report by Robertson, Coleman and Stephens.

290

31 Patents and morality: a philosophical commentary on the conference 'Biotechnology, patents and morality'

Michele Svatos

What moral and jurisprudential issues are raised by patenting? And how may we reach or at least work towards a consensus? As Sigrid Sterckx admirably suggests in her welcoming speech, I will take it that we are all concerned with animal welfare and environmental safety, as well as scientific progress. This conference has shown the possibility for meaningful dialogue between opponents on issues in patenting and morality. What further consensus can be reached, and what are the essential philosophical issues and principles? I shall address these questions in what follows.

It has sometimes been said that patents have little or nothing to do with morality. This claim might actually mean a few different things. Does the patent system have nothing to do with morality? Or does granting a patent on a particular invention have nothing to do with morality? This raises the point that morality can come into play (or fail to do so) at various levels in the legal practice of patenting. One might argue that moral issues can arise:

– in the justification of the patent system as a whole

– in the justification for extending the patent system to a previously unanticipated new technological discipline, e.g. computer science or biotechnology

– in the justification for extending the patent system to a particular type of invention within a new technology, e.g. computer programs, medical therapies, human genes, genetically engineered animals

– in the granting of a particular patent to a particular invention, e.g. a letter bomb or germline genetic engineering

– in applying patent protection for a particular invention, such as germline gene therapy, to a particular context, e.g. germline genetic engineering *of humans*, or with a certain degree of breadth, e.g. granting the patent for all mammals

- in allowing for certain types of exceptions to patent protection, e.g. a derogation for farmers or for pure research

- in enforcing patents, particularly in an international context

The list could probably go on, but these are the main levels where patents and morality may intersect (although a case to the contrary can obviously be made for any one of these areas). Any one of these is worthy of an entire paper (or rather, an entire library shelf), and indeed the speakers at this conference have addressed controversies at each of these levels. It is encouraging to note that there is substantial agreement over many of the issues, for example the derogation for farmers. I would like to briefly examine each of these areas, to see what more agreement can be found, and to clarify what disagreement remains and what the fundamental principles are. In the first two sections, I will be borrowing heavily from my article entitled, 'Biotechnology and the Utilitarian Argument for Patents'.[1]

The justification of the patent system

What is the justification of the patent system as a whole? One might suggest that it is raw economic need — Europe needs patents to compete in our increasingly technological world. While Europe may well need patents, surely moral reasons underlie the economic justification. It is not a matter of moral indifference whether inventors are able to protect their inventions, whether others are excluded from competing for 20 years, whether society enjoys few inventions or many, or whether some countries suffer while others prosper. As Mr Mertens said, 'Philosophy has everything to do with economy. Economy is not a value in itself.' Mr Van Duffel made a similar point in the concluding debate. I hope that we can most of us agree that morality *is* relevant to patents, in the sense of justifying (or failing to justify) the patent system as a whole, although there may be some people such as Mr Crespi who remain unconvinced.

A utilitarian argument is most often given in attempting to justify the patent system or its extension to a new industry. Patents provide a necessary incentive; without the promise of being able to control an invention and thereby potentially recoup costs, a company or individual would have little reason to invest in research and development (R&D). Since society benefits from the disclosure and use of inventions as well as from the resulting increased economic activity, patents are justified. According to this argument, the rights granted are a matter of social convention, not natural right or distributive justice. They are justified by their contribution to social utility, says the utilitarian proponent, despite their history in monopoly privilege, and despite any welfare losses due to the restrictions in disseminating an invention. As Mr Crespi points out, patents are not a reward, and they do not give a right to exploit the invention. However, they are intended as an encouragement or incentive to do research, make the research public, and commercialise it, and this incentive, as Mr Alexander points out, is supposed to be in the public interest.

Although the utilitarian or incentive argument is the most common justification for the patent system, when challenged opponents of this view often retreat to a distributive justice view: 'It's only fair that the inventor profits from her invention, and that others are excluded from doing so' or a natural rights view: 'The inventor has a right to the fruits of her intellectual labors, just as we all have a right to the fruits of our physical labors'.

Given these attempted justifications for the patent system as a whole, what about the justification of extending patent protection to some new technological discipline? Our discussion of this next topic will also bring up points relevant to the justification of the patent system as a whole.

The justification for extending the patent system to modern biotechnology

Normally the extension of the patent system to a new technological discipline is automatic and may be assumed. However, there are sometimes certain features of the new technology which were unanticipated and indeed undreamt of when patent law was originally created or most recently changed. Automatically extending patent law to these new technologies results in an awkward fit, and as Professor Vermeersch argued, the responses to these unique features are often rather piecemeal. I will speak about computer technology first because I think that we can learn something from looking at a discipline which is in many ways similar, but which many of us do not have such strong feelings about.

Computer programs and algorithms fit awkwardly within intellectual property law; algorithms have some features of discoveries or mathematical formula, but also some features of inventions which would normally be patentable. (Indeed, why do we call the discipline 'computer science' rather than 'computer technology'?) Likewise, computer programs seem to fit between copyright and patent law. For both, there may be bad consequences of allowing patent protection. The birth of a new technological discipline is an especially sensitive time for encouraging or stifling competition, and software or algorithm patents may do both. This is in part because patents over the first discoveries in the field may be very broad and may limit further growth in the discipline. Imagine if basic innovations within computer science, such as the blinking cursor, had been patented! Many computer scientists, at least within the U.S., feel that patents are slowing development within the field and giving too much control to software giants such as Microsoft.

How does this apply to biotechnology? As Professor Vermeersch pointed out in his talk, the distinction between discovery and invention is becoming blurred in biotechnology, as it is with computer algorithms. Also, he points out that there are questions of the breadth of one's claim; 'Did Cohen and Boyer put their flag on the whole continent of rDNA? Or only on a part? Or perhaps were Watson and Crick there already before.' Modern biotechnological inventions are unique in further ways; not only are some of them alive, but also they are able to reproduce on their own, unlike other patentable inventions. Along with this comes the fact that the products are not standardised, easily described, and so on. They are released into the environment and interact with it (changing it and being changed by it) in ways which we cannot even begin to track or predict accurately. Many biotechnological innovations do bring patent law into previously undreamt-of areas. As Mr Emmott said, 'It is true, there is not much morality if you are patenting a photocopy machine, but if you are looking at the issue of living material and the extent to which patent law should apply to them, it is inescapable.' These are good reasons to pause, rather than extending patent protection automatically to these new areas. Our old definitions and criteria need to be reconsidered and sometimes redrawn.

If we do pause to consider whether patent law should extend to modern bio-technology — and of course it already has been, for decades — we should consider whether the general philosophical justifications of the patent system are naturally extended to this new technological discipline. Are there any special reasons to believe that patents are more or less justified in biotechnology, putting aside controversial cases such as the patenting of genes, animals, and human germline genetic engineering? As Dr Schatz puts it, 'The patent system is nothing but a regulation of competition. The question is: does patenting, as an act, have other consequences, and which ones, in the field of biotechnology, than in any other field? And if not, are the consequences a patent has in any other field, acceptable in the field of biotechnology?' I will argue that there are good reasons to be concerned about the application of patenting to the field of biotechnology in particular.

Several participants of this conference, such as Professor Van Montagu, have stated that patents are absolutely needed in biotechnology. Without them, research will not be done. I think that both sides are sometimes guilty of overstating their claims (though perhaps not intentionally), and that this is one of those cases. One wonders how research into medical therapies is done, if patents are not available for them. Or how scientific discoveries are made, since discoveries are not patentable. Mr Linskens wondered how the claims about the absolute necessity of patents for biotechnology could be justified, in view of the fact that 'already since the beginning of the 1980s, probably over a thousand transgenic mice lines have been produced by scientists without patents at all.' As Dr Then points out, there is a tendency for industry lobbyists 'to expand their claimed monopolies and look for a new distribution of markets ... Society is called upon to resist this development.' Industry claims regarding the absolute necessity of patent protection, particularly patent protection for this or that new technology, tend to be overstated and we must guard against them.

Lest I too overstate my claim, let me be explicit: Although it may be true that *less* research will occur without patents (in general, or in a particular discipline, or for a particular type of invention), it is not true that research will come to a grinding halt. R&D often continues even when it's uncertain whether an industry's products will receive patent protection. Uncertainty is the hallmark of biotechnology patenting. There are often profits to be made, even without a limited monopoly, from being first to market or being able to make investment decisions based on the new invention before others know about it. It's questionable just how necessary patents are to the majority of inventions once the considerable costs of patenting and patent litigation are factored in.

The question of whether or not patents encourage innovation may depend on an industry's stage of development, as it does in computer science. Patents in an early stage of development may slow the industry down and make it difficult for late-comers to enter. In 1992, a former U.S. National Institutes of Health (NIH) scientist working on the Human Genome Project tried to patent 6,122 gene fragments of unknown function, useful only in tagging a gene. As usual, the patent application was written as broadly as possible, claiming not only the gene fragment but also the unknown gene it would point to, and whatever protein the gene codes for (although the last part of the claim was eventually dropped). The entire human genome of some 100,000 genes could have been patented this way, in bits and pieces, with a small number of corporations, universities, and governments owning humanity's genetic code — even though scientists understand the function of less than 1,500 human genes.

NIH's thousands of patent applications were eventually rejected. However, had they been upheld on appeal, it would have slowed research, discouraged cooperation among scientists and nations, and further slowed the development of therapies once the genes' functions were discovered. A company which wished to work on developing a diagnostic kit or therapy using a gene which had been tagged and thereby patented would have to obtain a license to do so, if the patent owner even offered such a license. Some companies were already working on genes which NIH had filed patents for. Even *applying* for the patents slowed research and discouraged cooperation, as other companies and nations scrambled to file for patents on the sequences they had found, in case the NIH bid were to be successful. Britain postponed release of their findings of more than 1,000 gene fragments for a year while battling the U.S. and European patent systems.[2] Relatedly, Mr Linskens argued that the broadness of claims in the Oncomouse case might actually frustrate scientific research.

It is a false dilemma to say that either we have patents (e.g., biotechnology patents) or we do not encourage invention and/or protect inventors at all. There are certainly other alternatives for doing both, as Dr Schellekens points out. For example, we might levy a tax, much like a sales tax, on all products of technology. We could then redistribute this tax to fund technologies which society (perhaps through lay person technology consensus groups, or public referendums) considers most valuable, or which scientists (perhaps through peer review) or politicians consider most valuable. Although the tax would increase the total cost of technologies to consumers, increased competition from the lack of patent protection might compensate for that. On the other hand, or in addition, we could protect inventors through laws against unfair competition. If a competitor simply takes the end product of another company's long and expensive research process, and she is able to reproduce and sell the invention cheaply (as with computer software), we could accuse her of unfair competition.

These are not meant as serious, well-considered proposals! The point is, that there are alternatives. On a more serious and less creative note, corporate tax policy is one of the most influential factors in the funding of corporate R&D; perhaps intelligent tax policies would have the desired result of stimulating innovation. If one compares the utility of having the patent system with the utility of simply abolishing it, perhaps it would be justified. However, this presents a false dichotomy. Patent policy is hardly the only incentive affecting biotechnology. A 1984 U.S. Office of Technology Assessment report identifies 9 other important factors: financing and tax climate, health and environmental regulation, university-industry relations, government funding for research, availability of trained personnel, antitrust law, international market and trade arrangements, government policies relating directly to biotechnology, and public opinion.[3]

If we wish to encourage useful inventions, let us do so. The patent system encourages inventions, but not necessarily useful ones — particularly if we acknowledge that there may be considerable a difference between the market value of an invention, and its social utility. For example, large chemical companies which have acquired seed companies often try to genetically engineer their seeds to be specially compatible with or dependent upon the particular insecticides, pesticides, and so on which they market. Research may be geared more toward increasing demand for the company's chemical products than producing the sort of seeds farmers would most like to see.[4] Dr Then quotes a 1994 OECD report on 'Biotechnology, Agriculture and Food' which says that 'the crucial focus in this sector is on the reorganization of the seed market resulting in stronger integration into the agrochemical sector'. As Dr Then says, gene technology is being used as a tool to reorganise the agriculture and food industries and to increase monopolistic control, which is not the goal of modern patent law and which indeed may be contrary to the public interest:

> Biotechnology offers new opportunities to tailor crops to specific needs, a reduction in the use of purchased inputs, an emphasis on nutritional quality, and fewer environmental problems in agriculture. Alternatively, biotechnology could induce further concentration in farm structure and further industrialization of agriculture, with the highly monopolized input and output sectors of the agribusiness community capturing the bulk of the benefit.[5]

Perhaps Professor Van Montagu and others here have given examples of new biotechnologies which will benefit society and even less developed countries. However, in an age when seed companies have mostly been bought out by large chemical corporations, it's unlikely that biotechnology will be allowed to achieve its full potential for reducing chemical inputs and providing other benefits:[6]

> Much of the evidence ... suggests that just the reverse is happening. An associate dean of a U.S. college of agriculture illustrated this conflict when he related the following, "In speaking to a group of state agribusiness leaders recently, I observed that in the distant future I could foresee a perennial corn crop which fixes nitrogen, performs photosynthesis more efficiently and is weed and pest resistant." With that one statement he alienated nearly every sector of the agribusiness community.[7]

Moreover, blind encouragement of research in agricultural biotechnology is likely not justified, because over-production is already a major problem for European agriculture, as Professor Van Montagu notes (although he dismisses the problem). Genetically engineered crops designed for higher yields, or designed to withstand more fertilizers or pesticides and thereby also increase yields, are hardly the sorts of research we need to encourage while overproduction remains such a huge problem. On the other hand, research which would reduce chemical inputs and thereby cut costs for farmers and benefit the environment should be encouraged. However, seed companies owned by chemical companies have little incentive to conduct research to reduce the use of their stock-in-trade. The patent system does not provide a vehicle whereby decisions can be made regarding which areas of research might be most socially beneficial, and in fact it skews the sorts of research which are funded. It is the blind promotion of technological innovation in general, with the mere hope that the outcome will maximise utility. As Professor Van Montagu argued, there are drawbacks to taking financial incentives out of research entirely, but again, the patent system is not the only possible way of providing financial incentives.

Another set of issues arises when we turn to the international arena. When patents are assessed on utilitarian grounds, the version of utilitarianism used is usually (a) limited to the economic realm, and (b) limited to Europe or the U.S., or at least to technologically advanced nations. Given the vast number of people living in less developed countries (LDCs), it's doubtful that the patent system can be justified on truly universalist utilitarian grounds if it fails to work to their advantage. Patents often work to the disadvantage of LDCs, as their resistance to recognising western patents or adopting intellectual property laws within their own countries demonstrates. Under-utilisation of inventions can become a very dramatic problem in a country with great need and little money to spend on technology. Although Western agricultural technology and medicine often do 'trickle down' to LDCs and thereby benefit them, the Western patent system certainly was not designed (and does not act) to protect their interests and maximise utility on a global scale. As Mr Vankrunkelsven of the Task Force for a Just and Responsible Agriculture argues, patenting living beings encourages the dominant industrial system of agriculture at the expense of sustainable family farming. Even though biotechnologists may have the explicit goal of helping sustainable agriculture, family farms, and LDCs, the results may sometimes be about as effective as the Green Revolution, and indeed may be the opposite of what was intended — which is not to say that biotechnology can never be genuinely helpful on these counts. The possible bad economic effects of biotechnology which Mr Vankrunkelsven cites are relevant to the justification of the patent system and in particular its extension to plants and animals.

At any rate, it's misleading to assert that the patent system, or its extension to biotechnology, is justified on utilitarian grounds if one does not carefully consider its effects on a large portion of the world's population. Europeans and Americans are often concerned that they must have a strong patent system, so that their countries do not lag behind in the biotechnology race. It seems that one is always concerned with the finish line, and those who are ahead, rather than the vast majority of people who are behind. If Europe provides strong patent protection for biotechnology, where will that leave other countries, besides America and Japan? Many LDCs consider genetic information to be the common heritage of mankind, and do not consider patents appropriate. How will they survive — and they hold much of the genetic diversity of the world in their hands — if the U.S., Japan, and Europe unite against their view? Biotechnology patents are not justified as a form of international self-defense.

Furthermore, patents are not necessarily a particularly effective vehicle for promoting useful inventions, although people on both sides of the table often assume that they are. In fact, economists themselves disagree about the economic effect of patents (in biotechnology, or in general). Indeed, I find it interesting that economists are often not invited when patents and biotechnology are being debated, probably because the economic facts are generally assumed to be pretty clear. As one economist argues:

> The justification of the patent system is that by slowing down the diffusion of technical progress it ensures that there will be more progress to diffuse ... Since it is rooted in a contradiction, there can be no such thing as an ideally beneficial patent system, and it is bound to produce negative results in particular instances, impeding progress unnecessarily even if its general effect is favorable on balance.[8]

Economists have been unable to determine empirically whether the economic benefits of the patent system outweigh its costs. Indeed, one must ask whether patents are meant to (a) simply allow inventors to make a normal profit by internalising positive externalities — that is, seeing that those who benefit from the invention pay for it — which might not happen (due to the special nature of intellectual property) without patent protection, or rather (b) allow inventors to make a super-normal profit on their investment, as a special inducement to innovation? Helping to capture normal profits versus economic profits may be the difference between preventing under-investment in R&D versus promoting over-investment. There is little agreement amongst economists as to which the patent system does, or is designed to do, and they are far from decisive about the actual influence of the patent system on innovation.

One should not simply assume that patents are pro- or anti-competitive, or whether they control the market system or allow it to function properly. The question of whether patents prevent under-investment or promote over-investment presumes that economics can tell us what the optimal level of investment in technology R&D is — another problematic assumption.

There are numerous drawbacks to using patents as incentives to invent, as I argue in 'Biotechnology and the Utilitarian Argument for Patents'. The drawbacks include the inefficient allocation of resources, encouragement of 'work-arounds' or copycat inventions, changes in the allocation of research funds, increased secrecy, substantial legal and administrative costs, an arbitrary incentive to focus on the sorts of research which are patentable, and differential effects depending on the stage of industry development.

298

One commentator argues that the patent system has become a patent 'arms race,' where 'everyone must run harder and incur significant additional costs just to stay even. The effect has been to increase quite dramatically the cost of innovation,' and researchers must now spend considerable time assisting patent attorneys.[9] From 1980 to 1990, patent litigation increased by 52 percent, and in biotechnology 'legal briefs outweigh scientific papers by orders of magnitude, and lawyers are as eagerly sought as PhDs.'[10] One speaker at the conference mentioned a 15 million dollar biotechnology patent lawsuit.

Not only can small companies, which are sometimes simply intimidated out of business or out of the field, ill afford high patent costs, it's wasteful of potential research funds. If we want to stimulate invention, diverting millions from biotechnology R&D into the legal system is hardly productive. Choosing in favor of the patent system, rather than other ways of encouraging research (or other, non-research allocations of resources), constitutes a choice to allocate very considerable resources to the legal system. Good news for lawyers, perhaps, but certainly a utilitarian cost of the patent system which cannot be ignored.

The increasing presence of patents in scientific research has made many scientists — even within universities, much less industry — more wary of sharing information, reagents, tissue samples, and so on with other researchers. The lesson many scientists learned from the dispute over the interferon cell line was to keep basic research materials to themselves. Ownership and priority disputes have become more common since the early 1980s. 'Scientists who once shared prepublication information freely and exchanged cell lines without hesitation are now much more reluctant to do so ... The fragile network of informal communication ... is likely to rupture.'[11]

The patent system has considerable disadvantages. When taken together, it's far from obvious they are offset by the benefits provided by the stimulation of innovation, even in simple economic terms. The benefit of the patent system is not the sum total of inventions patented; rather, it is only that portion of inventions which would not have been made, or made so quickly, in the absence of the system, and it is difficult to tell what that portion is. Although industry often claims that patents are utterly essential, this is not necessarily so. For more than a century, economists have been unable to agree whether the system is a net social benefit, whether it prevents under-investment or promotes over-investment in R&D, or even whether the patent system or its absence is most consistent with a capitalist economic system. This must give someone hoping to justify the patent system considerable pause.

A utilitarian argument for the (further) extension of the patent system to bio-technology, would have to show three things: (a) the patent system makes a significant contribution to stimulating invention, (b) the patent system is the best way among the alternatives to stimulate invention, and (c) at a more basic level, stimulating technological invention in general — or in our case, stimulating biotechnological invention — is itself justified on utilitarian grounds. There are strong reasons to doubt each of these three necessary points, despite the fact that they are often simply assumed to be true, and the last point is seldom even mentioned. Although economists don't have the facts to completely support or reject (a) and (b), and aren't particularly qualified to assess the truth of (c), there are many strong reasons to be dubious of the patent system's efficacy in maximizing utility.

Finally, we return to the specific practical problems in patent law. To what extent should patent law be applied to biotechnology? Definitions of terms in patent law are often quite ambiguous or arbitrary. What constitutes non-obviousness or usefulness? How minor may an improvement on, or deviation from, an existing product or process be in order to be patentable? Should the mere fact of isolating a gene or purifying a substance found in nature make it fully appropriable? Should a company which finds one way to genetically engineer cotton, be given a patent for all genetically engineered cotton, even if it is engineered by a completely different method and for a completely different trait? Should a company which works on a gene in the human immune system be given a patent over all possible medical uses of that gene? Should a company be allowed to own a patent on all germ line genetic engineering, which has the potential to permanently alter the genetic makeup of one's descendants in perpetuity?

When patent law is extended to a new area, or interpreted in a particular way, decisions are made about what research is to be encouraged, and to what extent. If the putative justification of the patent system is utilitarian, we should see the encouragement of inventive activity as a problem of allocation of resources — a policy issue, like any other, as Mr Alexander says. As he also suggests, we should explore alternatives to the patent system. Beyond Mr Alexander's suggestions, we should perhaps encourage innovation only in needed and underdeveloped areas, rather than across the board; and recognise that the patent system is probably not a good vehicle for promoting utility, and that the alternatives need to be considered more seriously.

Extending the patent system to particular types of biotechnological inventions

Since the landmark 1980 Supreme Court decision in Diamond v. Chakrabarty, micro-organisms and higher forms of life — 'anything under the sun made by man'[12] — can be patented in the U.S. Explains one commentator, 'There was no legal ... reason to differentiate animate and inanimate products of human manufacture. Biotechnology was recognised as a field of endeavor in which inventions could secure monopoly rewards.'[13] Observers have claimed that Chakrabarty 'opened the floodgates' of investment in biotechnology R&D, calling it 'the driving force behind the commercial development of biotechnology.'[14]

As several speakers at the conference have noted, there is a tendency for patent applications to claim not just anything but rather everything under the sun. I hope we can all agree that overly broad patents are a problem in biotechnology and that they are inappropriate, and Mr Van Rompaey gives us some hope that this is recognised within the patent world as a whole. Patents have been awarded for all genetically engineered cotton and soybeans, regardless of the method of genetic engineering or the traits engineered. Other crops will surely follow. Allowing such broad patents can hardly be in the best interests of developing nations, some of which already depend upon genetically engineered rice or other crops to feed their growing population and perhaps produce a small amount for export. Even within Europe, very broad patents may benefit one firm, but only at the expense of other firms and often also the public interest.

What about patenting genes, animals, and plants? The point which Mr Crespi and others have made, that genes are not patented in their natural state in the human body (or animal, or plant) is well taken, as is the point that we are not patenting 'life itself', whatever that is. Let me address the latter point first.

Both sides of the debate are prone to excesses of rhetoric, usually in order to make an important point in a perhaps overstated way. Of course one cannot patent life itself, although one can patent living things and that alone is bad enough, according to some groups. But it's the scientists themselves who have encouraged the idea that we are 'patenting life'. The human genome has been referred to as 'the book of life', and James Watson, co-discoverer of the double helix structure of DNA, says that this research is aimed at the understanding of human beings and of life itself. Another prominent scientist, I cannot remember who, said something to the effect of this: 'Once we have mapped the human genome, we will understand what it is to be human.' Such overblown rhetoric on the part of the scientists is not without its effect on the public. It is no wonder that the opponents of patenting genes (or animals and plants), some of whom include scientists, say that they are against 'patenting life'.

If a gene, its function, and its structure are scientific discoveries, how can the purified and isolated gene become an invention? The public is perplexed and accuses patent lawyers of playing word games, as Mr Klüver tells us. Mr Schapira responds, 'Among your experts, they all agree that genes are not patentable and they could not explain why they might be, when in fact that is totally wrong. Genes are patentable now in every country of the world, because they are not claimed in their natural state, because you cannot find genes, they do not just lie around on the ground to be picked up, they have to be identified by arduous procedures. Often the proteins that they encode were not known beforehand, and so the new proteins go hand in hand with the genes that encode the proteins.' Genes which have been isolated and purified are a technical solution to a technical problem, and they do not naturally occur in this state.

I think it's a mistake to completely blame the public for misunderstanding the practical, patent law definition of the difference between discoveries (unpatentable) and inventions (patentable). Indeed, some molecular biologists agree with the public on this point, and I do not think that it is simply because of their ignorance of patent law. Science and technology, discovery and invention, are indeed as Professor Vermeersch argues becoming increasingly intertwined in biotechnology.

I do not think that we should prohibit the patenting of genes, particularly human genes, on the grounds that it is contrary to the public order. As Dr Schatz and others point out, Article 53(a) refers to the exploitation of the invention, not the mere fact of patenting it. Patenting human or other genes may be immoral, but not in a way that is best addressed by Art. 53(a) or the examiners at the patent office. Furthermore, human genes, like other compounds which have been isolated and purified, probably are technically patentable. However, this does not end the debate. If patent law technically considers human genes to be inventions because they are a technical solution to a technical problem, then the public may respond that patent law has focused too much on technical definitions at the expense of the common sense 'fact' that human genes are discoveries, not inventions. The proper call is for a directive to the patent office, not action within the patent office or — beyond a certain point — interpretations and explanations by patent lawyers. As Mr Emmott, himself a barrister, says, 'If we insist on leaving ethical and moral judgements concerning patents on life to the patent lawyers and patent administrators, we cannot get a humane, compassionate set of decisions, but rather reductionistic and mechanistic ones.' This problem is not unique to patent law or patent lawyers. Let us examine this view against the patenting of 'life' in more detail.

Dr Schatz wonders whether the complex attitudes toward patenting and morality have 'something to do with what I would call a "closeness" of biotechnology to our own being? The more we know that the basic structure of DNA is the same for all living entities, everything that has to do with biotechnology is very close to my body. On the other side, patents are perceived as an appropriation of those research results by somebody who is not good and that is industry. I think there is a complex relationship between these elements.' I think he is correct in this, and I shall elaborate below.

Mr Crespi argues that opponents of biotechnology patents should attack the research itself, or perhaps the exploitation, not the middle step of patenting. Professor Vermeersch shows that there is indeed an important point to be made here. His view is that, 'The methods of germ line engineering should not be patentable. That does not imply that I am opposed to all types of germ line engineering. For instance, I am not necessarily opposed to the correction type of germ line engineering, but it should never be introduced in the commercial domain.' One may well not be opposed to the research itself, but rather its commercialisation.

Likewise, most people support genetic research which would lead to new forms of human gene therapy. They might even be in favor of commercialising gene therapy, to the extent of making it a private for-profit venture and allowing many aspects of the research to be patented. However, many proponents of gene therapy are opposed to patenting human genes, some of them scientists who certainly understand the difference between a gene in its natural state, and a gene which has been isolated through a laborious technical process, and which represents a technical solution to a technical problem. The idea here is that human genes are the common heritage of mankind (and many people would extend this to plant and animal genes as well), and the fact that the genes have been isolated from the body, and that they represent a solution to a problem, is quite beside the point. They are still human genes, and that is precisely why they are such a good solution to a medical problem. The gene for human growth hormone or Factor VIII should not be yours or mine, it should and in some sense does belong to everyone and no one. Every researcher should be free to work on them and try to use them to benefit mankind, without having to get a license and pay a royalty. Of course, if researchers are not able to patent a gene but only the process which they use to make it into a successful therapy, that will be difficult for them, but to stake a claim on human genes, within the human body or not, is to go too far.

Obviously it would take a great deal of work to fully develop this argument, as well as to apply it to plant and animal genes (or to plants and animals), but this is the view which Mr Emmott and the Patent Concern Coalition and Greenpeace and GAIA and others hold, and it should not be confused with a criticism of research into a particular area. It is rather an argument for limiting the subjects of product patents, regardless of whether we wish to limit the research itself, and regardless of the fact that patenting genes probably does meet the criteria of current patent law. Mr Emmott provides one example of such a proposal, on behalf of the Patent Concern Coalition.

I will discuss the patenting of animals, and of plants only very briefly; what I say about animals and human genes may also be largely applied to plants and plant and animal genes. On the question of patenting plants and animals, I think that Mr Alexander's and Mr Linskens' points about Article 53(b) and new plant and animal varieties is right on target. If the plant or animal has been genetically engineered, then it constitutes a new variety — but plant and animal varieties are not patentable. I find the argument that genetically engineered plants and animals constitute new species, or any taxonomic slot other than variety just so long as it's patentable, to be a bit of opportunistic and unconvincing reasoning, both biologically and philosophically.

Certainly animals have been owned and treated and traded as commodities for centuries, as several speakers have rightly pointed out. However, there may be grounds for making a distinction between the ownership of particular animals, and the ownership of, say, all mammals with a certain gene. There is, one might argue, a difference in attitude (which is simply a fact of our society) between the owner of a cow or dog, and the owner of a technology/invention. This takes the mechanistic view of nature even farther than it has previously been extended. As the participants in Mr Klüver's Danish consensus conference said, the patenting of animals is problematic; 'It tends to make us consider animals as things, objects.' We may think that we have finally refuted and rejected Descartes' view that animals are simply machines — but wait! They are after all but our own technologies, our own inventions. To say that we own them, is different than saying that we invented them or that they are a technical solution to a technical problem.

Similarly, there is a distinction between objecting to the suffering of research animals, and objecting to the patenting of animals. This parallels a difference in the animal welfare community, between those who merely object to the suffering of animals, versus those who argue that animals actually have rights. Those who argue that animals have rights, of course consider the suffering of animals to be very important and unfortunate, but they see the primary problem as a violation of rights; even if the animal did not suffer, its rights would be violated. Surely even those who do not believe in animal rights, can see how this claim could be made of humans.

Likewise, the opponents of patents on animals may view the primary problem as patenting itself — Mr Vandenbosch from GAIA articulated this view — rather than, or above and beyond, the suffering which occurs. Thus, GAIA would presumably object to the patenting of animals even in cases where the animal does not suffer, or actually benefits, from a genetic modification. The No to patenting is a fundamental one, explains Mr Linskens; you can have a little bit of animal research or a little bit of animal suffering, in certain circumstances, but not a little bit of patenting. As Mr Then says, 'Even if we would agree ... that we all want biotechnology and we all want gene technology, we would still have to discuss the problem of patenting those things.' It is not a criticism of the research itself necessarily, but rather of patenting living beings. And that is a question of the appropriate scope of patent law, as opposed to its actual scope.

Granting a patent for a particular invention, or for an invention applied in a particular context: letter bombs versus germline genetic engineering

In the European patent system, there is an explicit recognition that although society wishes to encourage invention, it does not wish to encourage all inventions; there are limits. The letter bomb is the usual example of an invention which is beyond the limit, because its exploitation would be contrary to the public order or morality. Of course, the patent office cannot be in the position of having to scrutinise every invention to determine whether it is socially useful or not, and this is a matter of judgement in which patent examiners are not particularly trained or competent. Most people will agree, I think, that we do not want the patent office deciding whether a new contraceptive is contrary to the public order or morality (this is one of the historical cases which Dr Schatz mentions). Although we have set a limit to the things we will encourage by patents, that limit is and must remain a very loose one, within the patent system itself. Moreover, as Dr Schatz points out, the inventor of something which would clearly be contrary to the public order would probably have good reasons not to file for a patent.

How does this, or how should this, apply to biotechnology? The first and least controversial limit, is that society will obviously not permit the patenting of a human being or human body parts, including a human embryo or fetus. There is very little disagreement about this, although there is some disagreement over the implications of, or rather the extension to, this for patenting human genes.

Secondly, society does not wish to encourage human germline genetic engineering, because the exploitation of such a patent (which of course may well be prohibited by law) would be contrary to the public order. The main controversy here, judging from the proceedings of the European Parliament on 1 March 1995, is whether this limitation should be made to appear more temporary and tentative or rather more permanent. I see little reason for the former approach; of course Parliament may change its mind later on (although many people strongly desire that it will not), and there is no reason to underline that fact in this case. At any rate, there seems to be a consensus on the need, at least for the present, to exclude human germline genetic engineering. Let us focus here on the consensus, and not on the 'for the present' clause, for the debate about the future acceptability of some forms of human germline genetic engineering is a complex topic in its own right.

When it comes to the idea of requiring a balancing exercise for inventions involving the suffering of animals, as in the Harvard Oncomouse case, this is obviously a decision that we as a society do not want to encourage every invention whatsoever. In the unique case that the invention itself (an animal) may suffer, the patent system will only encourage those inventions where the benefits would clearly outweigh the risks. In the Netherlands, Mr Linskens reports, all political parties wanted a ban on patenting animals; failing this, they embraced the principle that animal patents were to be granted only if the research is necessary, there is a lack of alternatives, and there is no unnecessary suffering.

Allowing for certain types of exceptions to patent protection

There seems to be a great deal of consensus that exceptions to patent protection should be allowed, such as the derogation for farmers or for pure research, or for medical treatments, as Mr Schatz explains in the final debate.

Conclusion

Today we are seeing greater public and interdisciplinary involvement in technology policy, and this should be welcomed and encouraged. In this conference, we have heard encouraging stories of programs such as the Danish technology consensus conferences, and indeed this conference itself is an encouraging example of greater interdisciplinary cooperation in technology policy issues. Dr Van Rompaey argues that 'society already provides sufficient instruments to cope with potential problems associated with the exploitation of inventions.' However, I think that Greenpeace, GAIA, and other organizations and individuals would disagree with this assessment of the current situation. Strong public concern about these and related issues suggests that further social mechanisms, or a refinement of those already in place, may be called for, and meaningful dialogue and even consensus or consensus-building are possible. This is true also in the realm of patenting, where we need to acknowledge the numerous philosophical issues raised by patenting in biotechnology, and work for a coherent and well-justified response to the unique problems facing the patent system. I would like to thank Sigrid Sterckx for inviting me to make my own modest contribution to this most interesting and valuable discussion.

Michele Svatos (svatos@iastate.edu)

Notes

1   Svatos, Michele (1996), 'Biotechnology and the Utilitarian Argument for Patents', *Social Philosophy and Policy,* Vol. 13, No. 2. Also published in *Scientific Innovation, Philosophy, and Public Policy,* Cambridge University Press, 1996.

2   Wuethrich, Bernice, 'All Rights Reserved: How the Gene-Patenting Race is Affecting Science', *Science News,* Vol. 144, pp. 154-7. Also see the continued discussion of the issue over the past several years in *Science* and *Nature.*

3   U.S. Congress Office of Technology Assessment (1984), *Commercial Biotechnology: An International Analysis* (USGPO).

4   See, for example, Kloppenburg, Jack (1988), *First the Seed: The Political Economy of Plant Biotechnology, 1492-2000,* Cambridge University Press: New York; and Lacy, William B., Bush, Lawrence and Cole, William D. (1990), 'Biotechnology and Agricultural Cooperatives: Opportunities, Challenges, and Strategies for the Future', in Webber, David J. (ed.), *Biotechnology: Assessing Social Impacts and Policy Implications,* Greenwood Press: New York, p. 77.

5   Lacy, Bush, and Cole, op. cit., p. 82.

6   See Kloppenburg, op. cit.

7   Lacy, Bush, and Cole, op. cit., p. 80. The authors cite an anonymous personal interview.

8   Robinson, Joan (1956), *The Accumulation of Capital,* Macmillan: London, p. 87, quoted in Nelkin, Dorothy (1984), *Science as Intellectual Property,* Macmillan: New York, p. 15.

9   Quillen, Cecil D., Jr., 'Innovation and the United States Patent System Today' (paper presented to the Antitrust and Patent Sections of the American Bar Association meeting, 19 October 1992), quoted in Warshofsky, Fred (1994), *The Patent Wars: The Battle to Own the World's Technology,* John Wiley & Sons: New York, p. 246.

10 Warshofsky, op. cit., p. 247.

11 Kennedy, Donald, 'Health Research: Can Utility and Quality Co-exist?' (Lecture given at the University of Pennsylvania, 6 December 1980), quoted in Nelkin, op. cit., p. 12.

12 *Diamond v. Chakrabarty,* 447 U.S. 303, 308-9 (1980).

13 Kenney, Martin (1986), *Biotechnology: The University-Industrial Complex,* Yale University Press: New Haven, CT, p. 257.

14 Plein, L. Christopher, 'Biotechnology: Issue Development and Evolution', in Webber, David J. (ed.), op. cit., *supra* note 4, p. 158. Plein is himself describing the views of several observers.

32 Conclusions

Sigrid Sterckx

Several important observations have been made at the workshop. We will only recapitulate a few.

Tensions between the academic community and the industry

The patenting-related tensions between the academic environment and the industry were clearly demonstrated by Professor Collen's outline of the tPA patent proceedings. He rightfully stated that the European patent system seriously disadvantages academic research. Whereas in the industry environment know-how is usually kept as a trade secret within the company, academic research is about communicating and publishing results. Professor Collen was very clear about this:

> When we discover something in our laboratories ... it is in our interest and actually it is our duty to publish these [findings] as soon as possible.

If one decides to file a patent application, one's own publications can become prior art against this application, so the academic community is injured in that communication and publication become unattractive options. The introduction of a so-called 'grace period', following the example of the US, could be (part of) a remedy, but as Dr Schatz noted, the industry is opposing this. Moreover, even in the US patent system, with its one year grace period, academic scientists in the field of biology are complaining about the effects of the patent system on the way research is being conducted:

> In recent years several authors have addressed the increasing "fuzziness" of the interface between biological research and its commercial exploitation. The growing importance of intellectual property has been singled out — often with concern — as particularly representative of the new climate in which biologists pursue their research.[1]

The 'appropriate stages for attack'

We believe that the view, typically formulated by Mr Crespi and Mr Praaning, that opponents of biotechnology patents — to be more precise: those opponents who are no members of the scientific and commercial community, e.g. NGO's — choose the 'patenting route' because this is the easiest option and that they condemn the patenting because they are unwilling to condemn the research since that would harm their 'public relations', is unwarranted. The appropriate 'stages for attack', it is said, are the research into and the marketing of the inventions in question. We cannot imagine anyone who has ever attended an oral hearing at the European Patent Office in the context of opposition or appeal proceedings in the domain of biotechnology — in cases where (some of) the Opponents or Appellants are NGO's or similar groups or groups of 'concerned individuals' — leaving the conference room thinking 'well, they sure did it the easiest way'. Although many of the arguments presented by NGO's and the like are rather unconvincing, some of their statements are highly relevant and yet no biotechnology patent has ever been revoked on 'ordre public' or morality grounds. Moreover, taking into account the way Art. 53(a) is being interpreted by the EPO, it is not very likely that this will happen in the (near) future. Unless the Opposition Division in the Oncomouse case would come up with a surprising decision ..., which we consider doubtful. As to the assertion that the afore-mentioned opponents solely focus on the 'patenting avenue' and do not 'attack' the technologies and products in question in their stages of research and marketing, this is clearly inconsistent with the variety of actions against the genetic modification of animals and foodstuffs. So 'attacks' *are* being directed also to research and marketing. The success of Greenpeace's campaign against Monsanto's genetically modified (glyphosate tolerant) soybean may turn out a lot 'easier' than the opposition they recently filed to Monsanto's 'glyphosate patent', although the EPO's most recent jurisprudence on transgenic plants[2] offers some prospects of success in this context as well.

Another problem with the statement that attacking the patenting of an invention is an entirely wrong starting point and that it is the research which should be the 'target' was very clearly pointed out by Mr Alexander. He reacted to Mr Crespi by saying that there are many people who think that it is one thing to engage in certain kinds of research ('necessary evils', e.g. animal testing), but that it is quite another thing for a public body (a patent office) to put its imprimatur on it. The wording 'necessary evil' is not merely used as a label by NGO's and the like: even Mr Bizley, Harvard's legal representative in the Oncomouse case, calls the experiments carried out on the oncomice 'necessary evils'. As Mr Alexander rightfully argued, 'one has to do very unpleasant things because one believes that there are socially useful goals for them, but it does not follow from that that everything in connection with that necessary evil is thereby authorised'. So in some cases there may be very good reasons to focus one's attacks on the patenting stage. According to another commentator, who discusses the ethical admissibility of patent applications directed to human embryo's or their use for research purposes:

In view of the degradation of human embryo's to technical objects, patent applications in this field would ... transcend the boundaries of what can be considered tolerable within the overall European *ordre public*. Reference to the fact that research on embryo's is allowed subject to strict conditions in certain Member States of the EPC does not provide a sufficient counter-argument. *The mere intention of commercialization, constituting the background of the patent application, is a violation of morality. ... a broad interpretation of the term "exploitation" in Art. 53(a) EPC must be applied: Even if the concrete use of the invention may possibly be justified from an ethical point of view, the sole circumstance that an exclusive right will be granted for it for the purpose of commercialization may violate fundamental ethical principles.*[3]

The assertion that attacking the biotechnology research would be bad for the NGO's 'public relations' was expounded by Mr Praaning: the public wants the research to continue, he said, because if you ask the public 'do you want these and these diseases to be cured, particularly if they are genetic?', everybody will say yes. Of course we would also say yes, but surely biotechnology is not only about curing diseases. Therefore information of the kind provided by Mr Praaning cannot serve as 'evidence' of a positive perception of biotechnology by the public.

Consensus conferences

Mr Klüver's presentation on the consensus conferences organised by the Danish Board of Technology felicitously struck us: we feel that this is a very encouraging 'story'. We were (almost) equally impressed after hearing the account[4] of the National Consensus Conference on Plant Biotechnology which took place in the UK in November 1994 (funded by the UK Biotechnology and Biological Sciences Research Council and organised by England's famous Science Museum).[5] Such consensus conferences, such 'negotiations between values' as Mr Klüver so marvellously called them, are indeed very valuable experiments in democracy. As John Durant, Head of Science Communication at the Science Museum, remarks in the preface to the Final Report of the UK consensus conference:

All too often, the public understanding of science is seen as a one-way process in which scientists communicate their knowledge and expertise to the public. In the consensus conference, by contrast, the public understanding of science is a two-way process in which scientists and lay people enter into a dialogue, the outcome of which is intended to be better mutual understanding.

Of course, just like the Danish panel, the UK panel pointed to certain risks and possible undesirable effects. Some people will say that this is due to the fact that 'they have still not completely understood what it is all about', just as some people maintain that the sole reason why the MEPs rejected the proposed Directive on the

Legal Protection of Biotechnological Inventions is that they are not sufficiently 'educated' to understand what this is all about. Although the reading of certain passages of the transcription of the discussions that took place in the European Parliament on 1 March 1995 (the day the proposal for Directive was rejected) makes one lose heart, various relevant critiques were voiced as well. We agree with Mr Jan Mertens when he says that the great amount of disrespect for the European Parliament which was shown after the 'no' vote is inappropriate.

Patent offices and the assessment of risks and benefits of inventions

Dr Schatz made some very important observations on the decision of the Board of Appeal in the PGS/Greenpeace case (T 356/93). In the context of its remarks concerning 'ordre public' (Art. 53a), the Board draws a dividing line between 'possible hazards' on the one hand and 'conclusively documented hazards' on the other. In the Board's view hazards which are only 'possible' escape the ambit of Art. 53(a) whereas hazards which are 'conclusively documented' fall under it. Of course, as we noted earlier,[6] an assessment of the hazards of an invention at the stage of the patent procedure is often problematic for in many cases only little information is available at this stage about the benefits and disadvantages that would result from the exploitation of the invention. This is all the more the case for environmental risks. Indeed, as the Board puts it:

> In most cases, potential risks in relation to the exploitation of a given invention for which a patent has been granted cannot be anticipated merely on the basis of the disclosure of the invention in the patent specification. ... [7]

According to the Board, patents should not be refused on grounds of *possible, not yet conclusively-documented hazards.*[8] As we noted in Chapter 1, this argument — which has already been used repeatedly by the EPO — implies that decisions on 'hazard issues' taken by Opposition Divisions and Boards of Appeal of the EPO will in most, not to say all cases, be in the interest of the patent proprietor for the onus of proof lies with the Opponents/Appellants.

The reason why the Board of Appeal in the PGS/Greenpeace case feels it would be unjustified to deny a patent under Article 53(a) EPC merely on the basis of possible, not yet conclusively-documented hazards, is given in § 18.7 of the Decision and was commented on by Dr Schatz. Should the 'competent authorities' (regulatory bodies), after having assessed the risks involved, prohibit the exploitation of the invention, says the Board, the patented subject matter could not be exploited anyhow. But if regulatory approval is given, then patent protection should be available. Dr Schatz rightly noted that this means that 'whatever we do, grant or reject, it makes no difference'. There is no clearer way of saying 'it is not our business to assess technological hazards', he correctly remarked. Perhaps even more importantly, Dr Schatz emphasised that:

If this means that Art. 53(a) EPC *only* comes into play when, on the basis of convincing documentary evidence, it is clear from the outset that regulatory approval for whatever form of use of the invention *will never be obtained*, then the Board would have made an important step back behind its starting point, namely that laws and regulatory provisions in all or some of the Contracting States are irrelevant, and would have come closer to the traditional way of applying the public order and morality bar on patenting, namely as a reference to a body of existing, generally accepted ethical and legal norms of fundamental importance for the cohesion of society.[9]

Dr Schatz made a very important point when he referred to this contradiction in the Board's reasoning: on the one hand the Board says (*cf* § 18.3 of its Decision) that, although the function of a patent office is to grant patents, patent offices 'are placed at the crossroads between science and public policy' and *'find themselves side-by-side* with an increasing number of other authorities and bodies, in particular regulatory authorities and bodies'. 'The assessment of the hazards stemming from the exploitation of a given technology is one of the important duties of such regulatory authorities and bodies.', the Board adds. On the other hand (*cf* the Board's remarks on the 'required extent of documentation' of the hazards in question), the way this assessment is interpreted by the Board, which in this context refers to the EPO's 'house rule' that the exceptions to patentability under Art. 53(a) EPC have to be narrowly construed, comes down to saying 'it is not our business to assess technological hazards'.

Dr Schatz's personal view is that patent offices should not play any role in the prevention of technological hazards of biotechnology or any other field of technology. Many commentators argue — as Dr Van Overwalle does — that the patent examiners working at patent offices are technical experts who are not competent to decide on 'ordre public' and morality issues in general and environmental hazard issues in particular. Personally, we consider Mr Alexander's position to be more pertinent. After all, he says, in granting or refusing patents, patent examiners are making decisions on public policy issues every day, for the granting of patents is a very important public policy issue. The weighing up of risks and benefits of particular technological processes and products is not an easy job, but a patent office is a body which has a very special qualification precisely to engage in this kind of activity, Mr Alexander notes. We would subscribe to his argument that:

... it is only in that kind of institution [a patent office] that these issues may be addressed on a sufficiently case-by-case basis, because a lot of these questions that one has to address are very particular questions, involving very particular facts and, for instance if one is to go into environmental risks, very detailed evidence as to what those risks and benefits are. That is one reason why one might say that an institution such as the EPO is in fact not unsuited, but might be particularly suited to making decisions of this kind and ought to do so.

311

Mr Alexander made another very important remark on how patent offices could 'design' such a case-by-case assessment of risks and benefits. He suggested that this assessment could (either additionally or perhaps better) be performed at the stage when a patent comes to be enforced. This would mean that the courts who come to enforce a patent take into consideration the extent to which the exploitation of the patent is or is not in the public interest (e.g. account would be taken of the extent to which the invention is beneficial to the environment). Of course, he added, the participation of the public in decision making of this kind would have to be ensured. We think Mr Alexander's suggestions are very valuable. This would indeed be a way of encouraging particular kinds of technology to be developed and discouraging other kinds of technology. 'Encouraging technological development' is often said to be a major goal of the patent system. But at present, as Professor Michele Svatos notes:

> The patent system does not provide a vehicle whereby decisions can be made regarding which areas of research might be most socially beneficial, and in fact it skews the sorts of research which are funded. It is the blind promotion of technological innovation in general, with the mere hope that the outcome will maximise utility.

Mrs Larissa Gruszow had a point when she remarked that in the kind of system proposed by Mr Alexander patentees would be in doubt after having received their patent and would not be sure whether they have to continue the preparations for the exploitation of the invention. Surely the present patent system is very sympathetic to patent applicants' 'need for certainty': the afore-mentioned 'house rule' of the EPO that exceptions to patentability should be narrowly construed is an often repeated statement. This principle appears to have been inferred from the historical documentation relating to the EPC, which shows that the view according to which 'the concept of patentability in the European patent law *must be as wide as possible*' predominated in the 'travaux préparatoires' of the EPC (*cf* references provided by the Board in § 8 of T 356/93). The question is, of course, *how wide* the concept of patentability *possibly ought to be*. All too often, people belonging to patent circles — members of the staff of patent offices as well as patent attorneys — forget that 'the whole foundation of the patent system is that patents are granted in the public interest' and that patents are not granted for the sake of providing 'certainty' or giving a particular private benefit to an inventor, as Mr Alexander, himself a patent attorney, reminds us. The system he proposes would inevitably imply a degree of uncertainty for patent proprietors, but we can only agree with Mr Alexander that 'someone who is given valuable property rights by the state in the public interest ought to be held to public account'.

The problem of broad patents

As Professor Svatos mentions, patents have been granted — to one company: Agracetus, owned by the US chemical enterprise W.R. Grace & Co. — for all genetically engineered cotton and soybeans, regardless of the method of genetic engineering or the traits engineered. Given the fact that the patent application for the cotton described only one method for making transgenic cotton, one realises how extremely broad such patents are. Commentators have noted that:

> Such "species" patents have been likened to Ford being given a patent on the automobile: other companies may go on to develop a different kind of automobile, but they would have to pay Ford a royalty for doing so. [10]

Even patent applicants themselves, like Professor Collen, admit that:

> ... if you have a good patent agent, the first claim of most patents will always be that you claim half the world, and then you try to trim down

This seems to be exactly what Genentech has done in the case of tPA. As one observer, himself an attorney, puts it:

> The court [the British court of appeal] found itself confronted with the wide claim: "Human tissue plasminogen activator (commonly abbreviated t-PA) as produced by recombinant DNA technology". As to the generality of the patent claim the Lords came to the unanimous conclusion which was most clearly phrased by Lord Justice Mustill: "I now consider it clear that this is a patent which should never have been granted, at least in its present form." Nevertheless, the British Patent Act 1977 ... did not allow undue and unsupported scope of a claim as a ground for invalidity.[11] [neither does the EPC: lack of clarity of a claim (Art. 84 EPC, *cf* Chapter 1) is not a ground for opposition (*cf* Art. 100 EPC), nor can it be a ground for revocation under the law of the Contracting States of the EPC (*cf* Art. 138 EPC)]

According to Dr Schatz, the problem is that patent applicants tend to seek protection for things they have not invented. Our analysis of a few concrete patent applications for biotechnological inventions left us with the same impression. Like Dr Schatz, we doubt whether it is really so that what is done with one plant or animal is technically under the same conditions as when it is done with another. Biotechnology patents do not always seem to fulfil the requirement that it must be possible to carry out the invention over the whole range claimed 'without undue effort'. As the EPO Board of Appeal in the Fuel Oils/Exxon case[12] decided:

> Thus, a claim may well be supported by the description in the sense that it corresponds to it, but still encompass subject-matter which is not sufficiently

313

disclosed within the meaning of Art. 83 EPC, as it cannot be performed without undue burden or vice versa. ...

In the Board's judgement, the disclosure of one way of performing the invention is only sufficient within the meaning of Art. 83 EPC if it allows the person skilled in the art to perform the invention in the whole range that is claimed. We would subscribe to the view that 'whether this is the case or not must be answered, not on the basis of a presumption in favour of the applicant, but by weighing the balance of probabilities in each individual case.'[13]

Mr Praaning made the following rather cynical statement:

> ... patenting drives industries and also the economic world to openness. After all, you must deliver your documentation, your papers, to a forum where everybody can see, read and know what you have done and what you plan to do; more or less, because of course you cannot be too explicit over what you do with a certain invention.

In an environment where patent applicants 'of course' cannot be too explicit about their inventions, of course patent offices cannot be expected to be too generous in their granting of monopolies.

We strongly disagree with Mr Crespi when he says that the problem that some patents are indeed extremely and unjustifiably broad should be dealt with in disputes between rivals in research and industry. As we argued in Chapter 1, the problem of broad patents is not only a legal-technical but also an ethical problem, so it certainly does not only concern competitors in research and industry.

The patenting of genes

> The very patentability of genetically engineered life forms is increasing patent activity, even amongst those opposed to the principle of patenting. Many scientists do not agree with the patenting of human genes, but because others are patenting them, feel obliged to do the same. If they do not, others could patent their work ... [14]

Indeed, in a joint statement by the (British) Clinical Genetics Society, the Clinical Molecular Genetics Society, the Association of Clinical Cytogeneticists and the Genetic Nurses and Social Workers Association, addressed to the Government of the United Kingdom and the European Patent Office[15], these notable organisations declare:

> ... we note that attempts at defensive patenting of natural gene sequences are already involving academic and commercial organisations in considerable work

and expense, to the detriment of the wider interests of society. There is an urgent need to send a clear signal that such patents will not be allowed.

As Professor Vermeersch and other speakers at the workshop pointed out, there is no agreement among scientists on issues of patentability of certain categories of subject-matter. The discussion on whether genes are patentable or not is a clear example. A remarkable general observation on scientists' attitudes towards patenting issues was made by Canadian researchers who have interviewed several scientists:[16]

... we noticed a certain amount of disagreement concerning what exactly is patentable and what is not. In general, scientists appeared to be very cautious, answering questions concerning patents more in the tentative than in the affirmative mode; their answers often referred to the patent attorney, who "knew better."

Inventiveness, a patentability requirement of major importance, has to be decided from the point of view of 'the person skilled in the art' (taking into account the state of the art at the time of the patent application, would the invention be obvious to a person skilled in the art?). As we mentioned in the context of Chapter 1 ('European patent law and biotechnological inventions'), this 'person skilled in the art' is described in the *Guidelines for Examination in the European Patent Office*. In part C, Chapter IV, § 9.6, it is said that what is meant is:

... an ordinary practitioner aware of what was common general knowledge in the art at the relevant date. He should also be presumed to have had access to everything in the "state of the art" ... and have had at his disposal the normal means and capacity for routine work and experimentation.

One might wonder where the patent attorney fits into this picture. Many people might come to a conclusion similar to Professor Collen's third one-liner about the patent system:

... the patent system, as it has evolved ... is a system that is made by lawyers and for lawyers.

As we explained in Chapter 1, the arguments that are provided to justify the patenting of genes are not convincing. They only show that the processes used to isolate, identify and purify genes are artificial processes in which 'the hand of Man' clearly intervenes, but this does not justify the conclusion that the genes themselves are 'artificial elements' and therefore constitute patentable subject-matter.

We would like to emphasise once again that we recognise that elements isolated from the human body have indeed enabled the development of many useful drugs. As Dr Rainer Moufang from the *Max-Planck-Institut für ausländisches und internationales Patent-, Urheber- und Wettbewerbsrecht* notes:

Prominent examples include the use of human DNA sequences in genetic engineering for the production of insulin, interferon or erythropoietin, as well as the production of monoclonal antibodies by means of new hybridomas.[17]

Yet, he adds:

> ... regardless of such examples the mere thought of patents on human genes and cells gives rise to a feeling of uneasiness not only in the general public. ...
> It is true that bio-ethical norms and postulates do not appear to offer a conclusive justification for a general prohibition on patenting parts of the human body. Nevertheless, in accordance with the problem-oriented character of ethics, they do reveal a number of important aspects, the cumulation of which can serve to justify serious reservations. The increasing commercialization of parts of the human body is, from a legal policy point of view, a very dubious development. ... Industrial property rights granted in respect of parts of the human body encourage such commercialization tendencies. ...
> Even if general civil law does recognize tangible property rights enabling control over severed parts of the human body, by no means does this dictate a parallel assessment under patent law.

As far as the 'legal-technical' arguments are concerned, we do not consider the argument that the mere fact of isolating a substance from its natural environment or purifying it by means of technical processes turns that substance from a discovery into an invention as a 'logical line of reasoning'. We would also like to underline that in our view this line of reasoning does not become any more logical when genes of micro-organisms, plants or animals are concerned. So we do not share the view that 'special' objections should be raised to the patenting of human genes because — for some dark reason — human genes are 'different' from other genes or because the patenting of human genes would be an attack on human rights or any other similar arguments.

A problem that was dealt with only summarily: The impact of (biotechnology) patents on developing countries

As Professor Svatos makes clear in her commentary, there are many levels where patents and morality intersect and any one of these is worthy of an entire library shelf. Hence inevitably — and unfortunately, as we said in our concluding remarks — some important aspects of the 'broader' questions surrounding biotechnology patents have only been treated superficially at the workshop or could not be discussed at all. One example concerns the consequences of the 'Northern' patenting practices for the 'Southern' countries. In view of the great significance of this aspect of the debate on biotechnology patents, we will briefly go into it here.

Unfortunately, at the time the workshop took place, about a year ago, we had only a very vague idea of the content of the so-called TRIPs Agreement, the agreement on trade-related aspects of intellectual property protection contained in Annex 1C of the WTO Agreement [the Marrakesh Agreement Establishing the World Trade Organization, ratified on 15 April 1994, outcome of the Uruguay Round of multilateral trade negotiations under the General Agreement on Tariffs and Trade (GATT)].[18] In the mean time we have made a start reading the patent-related provisions in TRIPs[19] and collecting literature reflecting the flood of reactions this agreement has elicited, especially from developing countries. However, the following remarks are by no means intended as thought-out and well-founded statements. In the light of what we already 'know', though, we find it rather surprising that the TRIPs agreement has been mentioned only a few times at the workshop. It isn't as if this is 'just another agreement'; indeed, 'the importance of TRIPS cannot be easily overemphasized'.[20] Many commentators suggest that the enforcement provisions, laid down in Part III ('Enforcement of Intellectual Property Rights', Articles 41-61) constitute the most 'revolutionary' aspect of TRIPs.[21] In this context, we will only make a few brief remarks on the 'Patents' section (section 5, Articles 27-34, of Part II, 'Standards Concerning the Availability, Scope and Use of Intellectual Property Rights').

Biotechnological inventions are not given any 'special treatment'. The first paragraph of Art. 27 ('Patentable Subject Matter') requires that:

... patents shall be available for any inventions whether products or processes, in all fields of technology, provided that they are new, involve an inventive step and are capable of industrial application.[22] ... patents shall be available and patent rights enjoyable without discrimination as to the place of invention, the field of technology and whether products are imported or locally produced.

Members who want to exclude certain inventions from patentability on grounds of 'ordre public' or morality are allowed to do so [cf Art. 27(2)]. This provision is comparable to Art. 53(a) EPC, accompanied by a few examples. The criterion seems to be the necessity of the prevention of the commercial exploitation of the invention in question:

Members may exclude from patentability inventions, the prevention within their territory of the commercial exploitation of which is necessary to protect *ordre public* or morality, including to protect human, animal or plant life or health or to avoid serious prejudice to the environment, provided that such exclusion is not made merely because the exploitation is prohibited by domestic law.

The third paragraph of Art. 27 allows the members to exclude from patentability '(a) diagnostic, therapeutic and surgical methods for the treatment of humans or animals'. In its original draft, TRIPs would have demanded the signatory countries to provide patent protection for plants and animals. As a result of protestation from developing countries, it was accepted that members may exclude from patentability

317

... plants and animals other than microorganisms, and essentially biological processes for the production of plants or animals other than non-biological and microbiological processes. (Art. 27, § 3b)

As far as *plant varieties* are concerned, however, all signatory countries must provide for the protection thereof 'either by patents or by an effective *sui generis* system or by any combination thereof' (ibid.). In view of TRIPs' emphasis on the harmonisation of intellectual property rules, it is likely that 'an effective sui generis system' is an allusion to the International Convention on the Protection of New Varieties of Plants (UPOV Convention, 1961). It is also interesting to note that Art. 27, § 3b, allowing the exclusion from patentability of plants and animals and essentially biological processes for their production, 'shall be reviewed four years after the entry into force of the Agreement Establishing the WTO' (*cf* last sentence of § 3b).

As to the term of patent protection, the European example is followed: a period of twenty years counted from the *filing* date (*cf* Art. 33).

The last provision of TRIPs' 'Patents' section is Art. 34, entitled 'Process Patents: Burden of Proof':

For the purposes of civil proceedings in respect of the infringement of the rights of the owner ... if the subject matter of a patent is a process for obtaining a product, the judicial authorities shall have the authority to order the *defendant* to prove that the process to obtain an identical product is different from the patented process. Therefore, Members shall provide, *in at least one of the following circumstances*, that any identical product when produced without the consent of the patent owner shall, in the absence of proof to the contrary, be deemed to have been obtained by the patented process:
(a) if the product obtained by the patented process is new;
(b) if there is a substantial likelihood that the identical product was made by the process and the owner of the patent has been unable through reasonable efforts to determine the process actually used. (our emphasis)

So the burden of proof is reversed (the onus of proof is said to be on the defendant). It is noteworthy that the rights of a process patent proprietor seem to apply also to 'identical products'. This may discourage the development of substitute products. In this light, Mr Crespi's statement that one of the objectives of the patent system is to stimulate alternatives to what is patented (*cf* Discussions) seems to become problematic. According to some commentators:

What this [Art. 34 TRIPs] implies for countries such as India, which have seen the generation of novel processes for the production of patented chemicals, including drugs, is that all the producers using such new processes would have to prove that they are not infringing any patent rights. The proposed change does not bode very well for the future of the local enterprise in these countries ...[23]

318

Concerning the implementation of the TRIPs Agreement, transitional arrangements have been decided upon, which are set out in Part VI (Articles 65-67). Members are obliged to apply the provisions of the Agreement upon the expiry of a general period of one year following the date of entry into force of the Agreement Establishing the WTO [*cf* Art. 65(1)]. Developing countries and countries which are 'in the process of transformation from a centrally-planned into a market, free-enterprise economy' and which are 'undertaking structural reform' of their intellectual property system and 'facing special problems in the preparation and implementation of intellectual property laws' are entitled to delay for four more years the date of application [*cf* Art. 65, §§ (2) and (3)]. To the extent that developing country members are obliged by the TRIPs Agreement to provide *product* patent protection in areas of technology not so protectable in their territory, the application date of the provisions concerning product patents may be delayed for another five years [*cf* Art. 65(4)]. Art. 66(1) allows 'Least-Developed Country' members to delay the date of application for a period of ten years and 'upon duly motivated request' this period can be extended. The second paragraph of this provision requires 'Developed Country' members to 'provide incentives to enterprises and institutions in their territories for the purpose of promoting and encouraging technology transfer to least-developed country Members', in order to enable the latter to 'create a sound and viable technological base'.

However, patent protection for pharmaceutical and agricultural chemical products must be provided immediately. The above-mentioned transitional arrangements do not apply for these categories of inventions. Art. 70, 8 of TRIPs requires that all members 'make available as of the date of entry into force of the Agreement Establishing the WTO patent protection for pharmaceutical and agricultural chemical products commensurate with its obligations under Article 27'. '[N]otwithstanding the provisions of Part VI' (transitional arrangements), the members that do not fulfil this requirement have to (i) provide a means by which such patent applications can be filed, (ii) apply to these patent applications the patentability criteria laid down in TRIPs and (iii) provide patent protection for these categories of inventions in accordance with Art. 33 (20 years counted from the filing date). Apparently, patent protection for pharmaceutical and agricultural chemical products is considered of utmost importance, whereas many developing countries exclude precisely these categories of inventions from patentability (*cf* infra).

As we already mentioned, the TRIPs Agreement has provoked a flood of reactions, which contribute to it that:

> ... disinterested observers still worry about the very different attitudes of the member states toward intellectual property protection and about the lack of consensus that continues to plague international intellectual property relations notwithstanding the TRIPS Agreement.[24]

Many commentators from developing countries are very unhappy with this agreement. As one of them summarises the objections to the 'Patents' section of TRIPs: instead of giving developing countries freedom and flexibility to exclude from patentability sectors of strategic importance for national development (e.g. agriculture, food, medicines, chemical products, and atomic energy), TRIPs forces them to patent everything, including agricultural and biogenetic innovations; instead of including only processes in patent grants as at present, TRIPs obliges the developing countries to extend patent protection also to products; instead of reducing the duration of patents, TRIPs extends it to 20 years; instead of strengthening the system of compulsory licensing, TRIPs practically abolishes it. 'In short', it is said, 'they [the TRIPs provisions] universalize the US system of intellectual property rights.'[25]

As to the exclusion from patentability of 'sectors of strategic importance for national development', it is true that many countries exclude pharmaceuticals (especially pharmaceutical *products*), food products, agricultural and chemical products.[26] Such exclusions are indeed motivated by the national interests of the countries in question. Motives of this kind may not be reconcilable with the 'philosophy' underlying the TRIPs Agreement, but the attitudes of developing countries towards intellectual property rights may be very different from those of developed countries. Apparently one of the — fundamental — differences which surfaced during the TRIPs negotiations was that 'Whereas the developed countries' texts all supported negotiation of a comprehensive agreement [on protection of pharmaceuticals], the developing countries viewed intellectual property not as a property right, but rather as an instrument of public policy.'[27] Yet the TRIPs provisions:

> ... provide universal subject matter protection for substantially all inventions and provide a uniform term of protection, irrespective of subject matter. This protection must be accorded by a member of GATT, regardless of its impact on social welfare in that country. ... these rights are [deemed] so important that individual member welfare should not stand in the way of their being protected as an entitlement of the creators. This invokes a counter-instrumentalist policy that members, regardless of their state of industrialization, should sacrifice their national interests in favor of the posited higher order of international trade.[28]

Concerning TRIPs' 'practically abolishing' the system of compulsory licensing, it is important to realise that compulsory licensing allows countries to ensure the local working of inventions ('local' meaning in the country that granted the patent). Art. 5(A)(2) of the Paris Convention (Paris Convention for the Protection of Industrial Property, 1883) provides that:

> Each country of the Union shall have the right to take legislative measures providing for the grant of compulsory licenses to prevent the abuses which might result from the exclusive rights conferred by the patent, for example, failure to work.

The TRIPs Agreement seriously diminishes the ability of the signatory countries to grant compulsory licenses to ensure the exploitation of the patent (*cf* Art. 31, 'Other

Use Without Authorization of the Right Holder'). Yet in the views of many, compulsory licensing is considered necessary to establish a balance between rights and obligations of patent holders. Moreover, it has been noted that the weakening of the compulsory licensing system, introduced by the TRIPs Agreement, marks 'a complete turnaround from the objectives of the *Revision of the Paris Convention* which the WIPO [World Intellectual Property Organisation] ... has been discussing for more than a decade and a half.':[29]

> These objectives [reference is made to WIPO (1985), *First Consultative Meeting on the Revision of the Paris Convention* (PR/CM/I/3), Geneva] provided that (1) actual working of inventions was to be promoted in each country itself; (2) importation of a patented article was not to be regarded as working; (3) each country should have the right to adopt legislative measures to prevent abuses arising from the exercise of rights afforded by a patent; (4) the domestic legislation of each country should be able to provide for working of patented inventions at any time, by the government or a third party authorized by the government subject to remuneration of the patentee when such working was dictated by public interest; and (5) a collection of sanctions and remedies, including nonvoluntary licenses and revocation, was to be available to countries granting the privilige.[30]

The abandoning of these objectives, these commentators argue, is a testimony to the fundamental changes brought about by the TRIPs Agreement. The future will show whether the effect this will have on the transfer of technology from developed countries to developing countries will indeed be a technology transfer 'to the mutual advantage of producers and users of technological knowledge and in a manner conducive to social and economic welfare' and whether TRIPs will contribute to 'a balance of rights and obligations' (*cf* TRIPs Art. 7, 'Objectives').

How realistic is the aim for consensus?

As Dr Schatz observed, 'consensus may not be within reach because each party ... has made winning speeches'. Indeed, this is to a great extent the case for the workshop and it is also a tendency which can be generally observed in the debate on biotechnology and biotechnology patents.

Opponents of biotechnology patents are sometimes guilty of making mere assertions, without substantiating them. We also have the impression that some of the people/ organisations (outside the research or commercial community) who oppose biotechnology patents are not very familiar with what the patent law actually says and with the case law of the EPO. This sometimes leads to statements like 'should patents on living matter be allowed...', whereas of course such patents are being granted for a long time. As Mr Alexander correctly noted, they also often do not realise the relevance of Art. 53(b) to their case and tend to focus mainly on the 'morality' provision.

Biotechnologists too are sometimes guilty of overstating their claims and making promises about a better world that will come about thanks to biotechnology, while remaining silent about potential risks or underestimating or denying them. It is true that it is very hard to predict such risks. As we mentioned in Chapter 1, the EPO's Oppisition Division in the Relaxin case noted that:

> ... experts all over the world have for at least the past fifteen years been intensively addressing themselves to the question of possible risks associated with genetic engineering and in particular with the release of genetically engineered organisms into the wild. Despite all this effort, there is still no agreement concerning the extent of these risks ...

But this very fact should inspire to some modesty on the part of the scientists.

Dr Schatz did not spare the 'patent people' either, and rightly so. He said that they have 'not sufficiently indicated what they consider as the real problems in the field' and named two specific problems: the breadth of patent claims and the fact that, unless entirely new methods of performing genetic engineering come into existence, it is likely that within a few years genetic engineering will become 'obvious'. As to the problem of broad patents (cf supra), Dr Van Rompaey noted that this is being discussed within patent circles. We hope a switch in policy will come out of these discussions very soon for, as we noted earlier, this is not a merely 'technical' problem. Allowing lack of clarity of patent claims (Art. 84 EPC) as a ground for opposition and a ground for revocation under the national patent laws of the EPC Contracting States would probably be a good start. Concerning the 'inventiveness problem', some of the processes in the field of genetic engineering are already automated, which makes it is hard to speak of processes involving inventivity. Another aspect of the 'winning speeches' of patent people is that the 'implantation' of the patent system in a broader socio-economical context is sometimes 'overlooked', except as regards the 'raw' economic statistics.

The inevitable 'selectivity' of each of the involved parties' stories in some respects creates yawning gaps and we fear that these will prove difficult to narrow. As things currently stand, it would be unrealistic to expect any consensus to come about. Nevertheless, we feel the way the workshop has proceeded is very encouraging for many of the participants have shown a great, often even surprising, degree of openness and — thanks to the 'variety of the gathering' — this has enabled a truly exciting exchange of ideas.

Notes

1   Mackenzie, M., Keating, P. and Cambrosio, A. (1990), 'Patents and Free Scientific Information in Biotechnology: Making Monoclonal Antibodies

Proprietary', *Science, Technology & Human Values*, Vol. 15, No. 1, p. 65. For similar observations by several other authors, see Svatos, M. (1996), 'Biotechnology and the Utilitarian Argument for Patents', *Social Philosophy & Policy*, Vol. 13, No. 2, pp. 113-44.

2   The decision in the PGS/Greenpeace case - *cf* Chapter 1 and several speakers' texts.

3   Moufang, R. (1994), 'Patenting of Human Genes, Cells and Parts of the Body? — The Ethical Dimensions of Patent Law', *IIC (International Review of Industrial Property and Copyright Law)*, Vol. 25, No. 4, p. 507, our emphasis.

4   From internationally-known science writer Dr Bernard Dixon, who served on the steering committee for this consensus conference and gave an account of it in September 1996 at Madingley Hall, Cambridge, in the context of a conference on 'The Ethical and Economic Aspects of Genetic Science' organised by the 21st Century Trust.

5   The Science Museum also published the Final Report of this conference.

6   *Cf* Chapter 1 - comments on Art. 9 of the European Commission's new proposal for Directive.

7   See § 18.4 of Decision T 356/93.

8   See § 18.7.

9   In § 7 of its Decision, the Board observed that the second half-sentence of Art. 53(a) EPC — 'the exploitation shall not be deemed to be so contrary merely because it is prohibited by law or regulation in some or all of the Contracting States' — makes clear that the assessment of whether or not a particular subject-matter is to be considered contrary to either 'ordre public' or morality is not dependent upon any national laws or regulations. Of course, if one postulates this 'independence', as the drafters of the EPC have done, the concept of 'ordre public' cannot be understood as referring to a body of certain laws, for this would mean that the exploitation of an invention could *only* be regarded as contrary to 'ordre public' if it is prohibited by those laws. However, according to Dr Schatz, this is how 'ordre public' is traditionally interpreted. The Board of Appeal in the PGS/Greenpeace case follows the 'independence interpretation' laid down in the EPC, thus no reference is made to any body of law. The Board says it is 'generally accepted' that the concept of 'ordre public' covers the protection of public security and the physical integrity of individuals as part of society and that this concept also encompasses the protection of the environment (T 356/93, § 5).

10   McNally, Ruth and Wheale, Peter (1996), 'Biopatenting and Biodiversity. Comparative Advantages in the New Global Order', *The Ecologist*, Vol. 26, No. 5, September/October, p. 224.

11   Brandi-Dohrn, M. (1994), 'The Unduly Broad Claim', *IIC (International Review of Industrial Property and Copyright Law)*, Vol. 25, No. 5, p. 650.

12   Decision T 409/91, *cf* Official Journal of the EPO, 1994, at 653; this decision is discussed in Brandi-Dohrn, op. cit.

13   Brandi-Dohrn, op. cit., p. 654.

14 McNally, R. and Wheale, P., op. cit., p. 224.

15 See also Chapter 1, p. 25.

16 See Mackenzie, M., Cambrosio, A. and Keating, P. (1988), 'The commercial application of a scientific discovery: The case of the hybridoma/monoclonal technique', *Research Policy*, Vol. 17, pp. 155-70.

17 Moufang, R., op. cit., p. 513.

18 GATT Secretariat (Ed.) (1994), Marrakesh Agreement Establishing the World Trade Organization, Annex 1C: Agreement on Trade-Related Aspects of Intellectual Property Rights, 15 april, in *The Results of the Uruguay Round of Multilateral Trade Negotiations — The Legal Texts 2-3* 6-19, 365-403.

19 TRIPs also contains provisions regarding copyright and related rights, trademarks, geographical indications, industrial designs, layout-designs (topographies) of integrated circuits, protection of undisclosed information and control of anti-competitive practices in contractual licenses.

20 Adelman, M.J. and Baldia, S. (1996), 'Prospects and Limits of the Patent Provision in the TRIPS Agreement: The Case of India', *Vanderbilt Journal of Transnational Law*, Vol. 29, No. 3, p. 511.

21 Art. 41 requires that 'Members shall ensure that enforcement procedures as specified in this Part are available under their national laws so as to permit effective action against any act of infringement of intellectual property rights covered by this Agreement, including expeditious remedies to prevent infringements and remedies which constitute a deterrent to further infringements. ...'.

22 The requirement of 'sufficiency of disclosure' is reflected in Art. 29(1).

23 Dhar, Biswajit and Rao, C. Niranjan (1996), 'Trade Relatedness of Intellectual Property Rights. Finding the Real Connections', *Science Communication*, Vol. 17, No. 3 (March), p. 317.

24 Reichman, J.H. (1996), 'Compliance with the TRIPS Agreement: Introduction to a Scholarly Debate', *Vanderbilt Journal of Transnational Law*, Vol. 29, No. 3, p. 369.

25 Patel, Surendra J. (1996), 'Can the intellectual property rights system serve the interests of indigenous knowledge?', in Brush, Stephen B. and Stabinsky, Doreen (eds), *Valuing Local Knowledge. Indigenous People and Intellectual Property Rights*, Island Press: Washington D.C., p. 316.

26 *Cf* World Intellectual Property Organisation (WIPO) (1988), *Existence, Scope and Form of Generally Internationally Accepted and Applied Standards / Norms for the Protection of Intellectual Property*, WO/INF/29 Sept., GATT Document MTN.GNG/NG11/W/24/Rev. 1.

27 *Cf* The GATT Uruguay Round: A negotiating history (1995), Chapter 2, quoted in Oddi, A. Samuel (1996), 'TRIPS — Natural Rights and a "Polite Form of Economic Imperialism"', *Vanderbilt Journal of Transnational Law*, Vol. 29, No. 3, p. 435.

28 Oddi, A. Samuel, op. cit., p. 440.

29 Dhar, Biswajit and Rao, C. Niranjan, op. cit., p. 322.

30 Ibid.